AF577136

Fluid Mixing IV

H. Benkreira (Editor)

Institution of Chemical Engineers,
Rugby, UK.

Hemisphere Publishing Corporation
A Member of the Taylor & Francis Group
New York Washington Philadelphia London

Fluid Mixing IV

Members of the Institution of Chemical Engineers should order as follows:

Worldwide	Institution of Chemical Engineers, Davis Building, 165–171 Railway Terrace, RUGBY, Warwickshire CV21 3HQ, U.K.
Australia only	R. M. Wood, School of Chemical Engineering and Industrial Chemistry, University of New South Wales, PO Box 1, Kensington, NSW, Australia 2033.

Non members' orders should be directed as follows:

UK, Eire and Australia	Institution of Chemical Engineers, Davis Building, 165–171 Railway Terrace, RUGBY, Warwickshire CV21 3HQ, U.K.
or	Taylor & Francis Ltd., Rankine Road, BASINGSTOKE, Hampshire RG24 0PR, U.K.
Rest of the World	Taylor & Francis Inc., 1900 Frost Road, Suite 101, Bristol, PA 19007, U.S.A.

Fluid Mixing IV

A three-day symposium organised by the Institution of Chemical Engineers (Yorkshire Branch) in association with the IChemE's Fluid Mixing Subject Group, the Royal Society of Chemistry Process Technology Group and the University of Bradford, and held at the University of Bradford, 11–13 September 1990.

Organising Committee

H. Benkreira (Chairman)	University of Bradford
K. J. Carpenter	ICI plc
R. Mann	UMIST

INSTITUTION OF CHEMICAL ENGINEERS

SYMPOSIUM SERIES No. 121
EFCE Event No. 432
EFCE Publication No. 86
ISBN 0 85295 257 0

Preface

Fluid Mixing IV continues the series of conferences which began at Bradford in March 1981. It is hoped to hold a fifth in the series in 1993. The conferences are intended to cover all aspects of mixing including the assessment of mixture quality, experimental and theoretical studies of mixing, chemical reaction and mass transfer, heat transfer, novel experimental techniques, scale-up and optimisation.

Preface

Fluid Mixing IV continues the series of conferences which began at Bradford in March 1981. It is hoped to hold a fifth in the series in 1993. The conferences are intended to cover all aspects of mixing including the assessment of mixture quality, experimental and theoretical studies of mixing, chemical reaction and mass-transfer, heat transfer, novel experimental techniques, scale-up and optimisation.

Contents

Evaluation of Mixing Performance

Powder Mixing

Non-Newtonian Fluids

Liquid-Liquid and Gas-Liquid

Mixing in Reactors

Solid-Liquid Mixing

Poster Papers

THE INFLUENCE OF VISCOSITY AND DENSITY DIFFERENCES ON MIXING TIMES IN STIRRED VESSELS

I. Bouwmans* and H.E.A. van den Akker*

Mixing time measurements have been carried out in a 0.29 m diameter vessel mounted with an inclined blade impeller. The viscosity ratio between the added liquid and the bulk liquid has been varied between 0.009 and 11. The density difference has been varied between -100 and +50 kg/m^3. The data show that for certain flow conditions the viscosity ratio and the density difference have a significant effect on the mixing time. For high stirrer speeds the mixing time can be predicted using the correlation $N \cdot t_m$=constant. For low stirrer speeds the mixing can be described by the theory of surface renewal.

INTRODUCTION

Blending miscible liquids of different viscosities or densities is a common operation in the process industries. This operation may involve difficulties in predicting the mixing time. The prediction of mixing time is needed for sampling for the purpose of quality control. Because a more viscous liquid is generally also denser, it is often not clear whether mixing problems arising in these cases are caused by viscosity differences or by density differences. Moreover, in many investigations in the past the effects of differences in viscosity or density have not been taken into account. The aim of the project of which the present work forms a part is to develop a model that includes these effects.

MODELLING MIXING

Model of the vessel

The history of a drop of liquid added to a liquid in a stirred tank is determined by a combination of factors. Depending on the conditions in the vessel, the properties of the liquids and the position of the addition point, the drop may pass through several of four main zones: the stirrer region, the free surface, the bottom of the vessel and the bulk. In each zone there is a chance that the drop will be divided into smaller drops or mixed into the bulk liquid. In addition, in each zone the drop has a chance to move to another zone. If all these chances are known, a prediction of the mixing time and of its expected standard deviation can be made.

* Kramers Laboratorium voor Fysische Technologie, Delft University of Technology, Prins Bernhardlaan 6, 2628 BW Delft, The Netherlands

The impeller's energy input is divided between large-scale flow and turbulence, depending on the type of impeller. The distribution of these two flow types over the regions in the vessel is also determined by the stirrer. Thus, each region makes its own characteristic contribution to the process of mixing, depending on the type of impeller and the flow conditions in the vessel.

- In the stirrer region the turbulence intensities as well as the shear forces are high, so that drops have a good chance of breaking up. Especially when the viscosity of the added liquid is high compared to that of the bulk, these shear forces are very important for the mixing process, because in other parts of the vessel the shear forces may be too weak to cause break-up. Drops may also be divided by physical contact with the stirrer. The circulation time t_c, i.e. the time between two passages of a liquid volume through the impeller, is determined by the large-scale flow. If a drop is not broken up during a passage through the impeller the mixing time will be increased by t_c before the drop will pass the stirrer again. For a Rushton turbine t_c has a log-normal distribution (Gavlak & Calabrese (1)). This results in an unavoidable uncertainty in the predicted mixing time.
- In the bulk the shear stresses and the level of turbulence are lower, but deformed drops may be deformed further or be mixed completely.
- The large-scale flow, combined with the net force resulting from density differences, determines whether a drop reaches the free surface or the bottom of the tank. If this happens, the added liquid will generally remain there provided the net force is in the appropriate direction. The liquid can then re-enter the bulk by turbulent eddies or disappear by means of diffusion. Both mechanisms yield relatively long mixing times. For a drop reaching the bottom of the vessel this has been described by Smith & Schoenmakers (2). For the surface this mechanism is called surface renewal. Danckwerts (3) assumed that the chance for a liquid element at the surface to be replaced is independent of the time for which it has already remained at the surface. According to this theory the fractional area of the free surface that is renewed within a period of time t is equal to $1-\exp(-f_s \cdot t)$, where f_s is the rate of surface renewal (Davies (4)).

Mixing time correlations

Hoogendoorn & Den Hartog (5) used several ways to represent their experimental mixing time data in graphs. When comparing similar types of impellers they plot the dimensionless mixing time, $N \cdot t_m$, against the impeller Reynolds number. The value of $N \cdot t_m$ can be interpreted as the number of stirrer revolutions needed for homogenisation. This is linked with the average number of times a fluid element passes the stirrer region during the mixing process. For the comparison of different stirrers $N \cdot t_m$ does not suffice, because it does not include the energy needed for homogenisation.

For this reason two dimensionless groups were introduced in (5):

$$\frac{P t_m^2}{\mu T^3} = f\left(\frac{\rho T^2}{\mu t_m}\right), \qquad (1)$$

where P is the power input, μ and ρ the viscosity and the density of the bulk liquid respectively, and T the tank diameter. The right-hand side of the equation is a modified Reynolds number: D and N have been substituted by T and $1/t_m$. The term on the left-hand side can be interpreted as:

$$\frac{Pt_m^2}{\mu T^3} = \frac{P}{\rho T^3}\frac{1}{\nu}t_m^2 \propto \frac{\varepsilon}{\nu}t_m^2 = \frac{t_m^2}{t_K^2}, \tag{2}$$

where t_K is Kolmogoroff's time scale of turbulence. In this way mixing time is measured in units of the smallest turbulent time scale. The values of this group range from 10^4 to 10^8, so t_m/t_K ranges from 10^2 to 10^4.

Zlokarnik (6) also used these groups, and added a new relationship:

$$\frac{\rho^2 PT}{\mu^3} = f\left(\frac{\mu t_m}{\rho T^2}\right). \tag{3}$$

In this case, the group on the left can be interpreted as:

$$\frac{\rho^2 PT}{\mu^3} = \frac{P}{\rho T^3}\frac{\rho^3}{\mu^3}T^4 \propto \frac{\varepsilon}{\nu^3}T^4 = \frac{T^4}{l_K^4}, \tag{4}$$

where l_K is Kolmogoroff's turbulent length scale. In the turbulent region T/l_K also ranges from 10^2 to 10^4.

The large differences between the Kolmogoroff time scale and mixing times and between the Kolmogoroff length scale and vessel dimensions make it implausible that these scales govern the mixing process. The largest energy containing eddies are probably of more importance. The shear field near the impeller might also be important, especially where large values of μ_a/μ_0 are concerned, for in that case turbulent eddies will not be strong enough to deform the added liquid.

The groups used by (5) and (6) can be used to choose a stirrer and to predict the power needed for homogenisation. To do this, however, t_m should be known in advance. It is doubtful, therefore, whether these groups can provide a better understanding of the mixing process, which is the major object of the present study.

In the literature mentioned above no effect of viscosity or density differences has been taken into account. Rielly & Pandit (7) examined the mixing of two layers of different viscosity, initially stratified as a result of a density difference. The mixing time could be correlated by:

$$N \cdot t_m \cdot \frac{D^3}{V} = f(\mathrm{Ri}); \qquad \mathrm{Ri} \equiv \frac{\Delta\rho \cdot g \cdot H}{\rho_L N^2 D^2}, \tag{5}$$

where D is the stirrer diameter, V the liquid volume in the tank, H the liquid height, ρ_L the density of the less dense liquid. The ratio D^3/V is added to account for different values of D/T, although this ratio was not varied in (7). For mixing time measurements with one diameter ratio, however, the influence of liquid properties on the mixing time is clear from the dimensionless mixing time $N \cdot t_m$ alone. As in the present study D/T has not been varied, $N \cdot t_m$ will be used here to present the data obtained.

As stratified layers were used in (7), the resulting mixing times are not representative for situations where liquid is added in smaller quantities and to a vessel that is already being stirred. The object of the present study is to gather information about the influence of differences in viscosity and density on mixing times in those cases.

EXPERIMENTAL

The experiments were carried out in a 0.29 m diameter fully baffled vessel mounted with a downward pumping 45° inclined blade impeller with a diameter of 116 mm (=0.4·T) (see fig.1). The clearance between the impeller and the bottom of the tank was equal to the impeller diameter. Stirrer frequencies of 0.7 to 6 s^{-1} were used, yielding power inputs of $0.4 \cdot 10^{-3}$ to 0.4 W/kg, as determined from a Po versus Re correlation.

Mixing times were measured using a conductivity method. After the injection of a saline solution the electrical conductivity in the bulk was recorded as a function of time. The probe was placed either behind a baffle or in the centre of the circulation loop which is formed by the flow in the vessel. Experiments showed that in the turbulent region the probe position did not affect the measured mixing time. The conductivity probe is also shown in fig.1. The tracer liquid was injected either about 1 cm below the surface or in the impeller plane. Mixing time was defined as the time after which the conductivity fluctuations did not differ from the final level by more than Δ_c times the total conductivity step for that experiment. In this work Δ_c equalled 0.05. For every condition ten measurements were carried out, giving average values for $N \cdot t_m$.

The viscosity of the bulk liquid or the tracer liquid was increased by adding Poly-Vinyl-Pyrrolidone (PVP). The resulting solutions were approximately Newtonian. In the earlier experiments the density of the liquids was determined by the amount of PVP and NaCl only and no additional correction was made. For later experiments, in order to be able to vary the viscosity and the density independently, mixtures of water, ethanol and glycerol were used in which Poly-Vinyl-Pyrrolidone was dissolved to achieve the desired viscosity. In this way, liquids can be made with viscosities ranging from 1 mPa·s to 200 mPa·s and densities from 900 kg/m^3 to 1100 kg/m^3.

RESULTS

Two cases will be considered separately: 1: addition of liquid which is as dense as or denser than the bulk liquid; 2: addition of a buoyant liquid.

Case 1: $\Delta\rho \geq 0$

In the first experiments the difference between the density of the added liquid (ρ_a) and that of the bulk (ρ_0) was always positive or zero. This causes the tracer liquid to be drawn into the stirrer in all cases, so that injection of the tracer near the surface results in the same mixing times as injection near the stirrer, provided that the added liquid does not adhere to the tank. These experiments show (see table 1) that for μ^* ($\equiv\mu_a/\mu_0$) ranging from 0.009 to 11 this viscosity ratio has only little influence on $N \cdot t_m$, taking into account that the standard deviation of $N \cdot t_m$ is about 20%. Therefore, in those cases the correlation of $N \cdot t_m$=constant can be used. Combined with the results of (2) where, in addition to solutions of PVP, concentrated solutions of sugar were added to the vessel, this result can be extended to μ^*=1400. The results of (2) were obtained using a Rushton turbine, but the results are analogous to those in this work.

TABLE 1 - Dimensionless Mixing Times for μ^* ranging from 0.009 to 11.

μ^* [-]	Re [10^3]	$N \cdot t_m$ [-]
0.009	1.0... 1.5	28...30
0.014	1.5... 2.5	27...29
0.18	1.0...20	25...26
1.0	10 ...30	23...25
11	1.0... 1.6	32...33
11	4.0...10	27

Case 2: $\Delta\rho < 0$

a. Injection near the surface. When tracer is injected about 1 cm below the free surface the mixing time is strongly influenced by viscosity and density differences.

- When the stirrer frequency is high enough to produce a large-scale flow strong enough to overcome buoyancy forces, the added liquid is transported to the stirrer region where the drop is broken up. In this case, again, $N \cdot t_m$ is the same as in the case of liquids of equal densities and viscosities. The graphs of conductivity versus the dimensionless time $N \cdot t$ (an example of which is given in fig.2a) show an irregular form.
- Low stirrer frequencies cause all tracer liquid to rise to the surface. The liquid is then slowly pulled into the bulk by turbulent eddies. This results in mixing times which can be 10 times longer than when the stirrer frequency is high. The graphs of the conductivity measurements (example in fig.2b) have an exponential form which corresponds to the theory on surface renewal. According to this theory the mixing time can be written as $t_m = -\ln(\Delta_c)/f_s$.
- At intermediate stirrer frequencies part of the added liquid is drawn into the impeller while the remaining part rises to the surface. This is clearly reflected in the conductivity registration (example in fig.2c). Sharp peaks during the first seconds indicate the break-up of part of the added liquid. The smooth second part of the graph is caused by the mixing of the remaining tracer liquid by the surface renewal mechanism. In this regime it is difficult to predict $N \cdot t_m$ because the mixing time is determined by the amount of liquid that rises to the surface. This in turn is determined by the flow conditions at the moment of injection. Because the flow in the vessel is turbulent, these conditions change randomly. If a fraction α of the added liquid is drawn directly into the impeller it can be shown that

$$t_m = \frac{-\ln(\Delta_c/\alpha)}{f_s}, \tag{6}$$

so that the mixing time will be shorter than in the surface renewal regime.

In this intermediate regime both $\Delta\rho$ and μ^* strongly affect the mixing time.

- In fig.3 the dimensionless mixing time is plotted as a function of the Reynolds number for μ^*=1.0 to 1.8. This graph shows that as $|\Delta\rho|$ increases $N \cdot t_m$ increases as well, because more tracer fluid rises to the surface.
- Fig.4 shows $N \cdot t_m$ versus Re for $\Delta\rho$=-5 kg/m^3. It is clear that in this case the mixing time increases rapidly with increasing μ^*. Two effects may be

responsible for this. First, the high viscosity of the tracer may cause the tracer liquid to stick together, thereby increasing the net upwards force. Secondly, the turbulent eddies near the surface may be too weak to draw in the tracer liquid once this has arrived at the surface. Fig.5 shows the same variables for $\Delta\rho$=-100 kg/m^3. In that case for all viscosity ratios the mixing times are increased compared to when $\Delta\rho$=-5 kg/m^3.

In fig.6 $N \cdot t_m$ is plotted as a function of the absolute value of the Richardson number. For experiments in with μ_0=10 mPa·s this gives a satisfactory result (fig.8, circles): all measurements lie in an s-shaped band which has its steepest slope between $|Ri|$=1 and $|Ri|$=10.

In fig.7 the relative standard deviations (σ/t_m) of the sets of ten measurements are plotted as a function of $|Ri|$. For low values of $|Ri|$, i.e. for high stirrer speeds, σ/t_m is low because t_m is determined by the tracer break-up in the stirrer region, which is reproducible. For high values of $|Ri|$ all tracer liquid rises to the surface, which also results in a predictable mixing time. In the intermediate region of $|Ri|$ the relative standard deviation increases as both mixing mechanisms act together, the mixing time depending on the amount of liquid that rises to the surface.

When the values for measurements with other values of μ_0 are included in the graph of $N \cdot t_m$ against $|Ri|$, however, these points fall outside the s-shaped band, as is clear from fig.7. For this reason, Ri alone does not suffice to predict $N \cdot t_m$.

b. Injection in the stirrer plane. When liquid is added near the stirrer, in a region where both shear stresses and turbulence intensities are high, the dimensionless mixing time is low and practically constant and the relative standard deviation is small. Viscosity and density differences do not remarkably affect the mixing time. Examples are given in table 2. Again, the correlation of $N \cdot t_m$=constant can be used.

TABLE 2 - Dimensionless mixing times for various viscosity ratios and density differences; tracer injection near the impeller; Re=$1.6 \cdot 10^3$ in all cases.

$\Delta\rho$ [kg/m^3]	μ^* [-]		
	0.1	1.0...1.8	10...14
0	25	26	29
-20	24	28	27
-100	30	25	24

CONCLUSIONS

Both viscosity and density differences can, but do not always, play an important part when liquids are mixed. Their role depends on the way the energy input of the impeller is divided between large-scale flow and turbulence, as well as on how these are distributed in the vessel. Under certain conditions mixing time can be unpredictable because of the turbulent nature of the flow. This can result in mixing time standard deviations of over 50%.

When liquid is added near the impeller the mixing time is determined by the break-up of the added liquid by the impeller. This break-up is not affected by viscosity and density differences for the values of these quantities used in this work. In those cases the correlation of $N \cdot t_m$=constant predicts the mixing time reasonably well.

When a liquid less dense than the bulk liquid is added near the surface buoyancy effects can increase the dimensionless mixing time considerably. For high stirrer speeds the liquid is drawn into the impeller, which results in short mixing times with a small relative standard deviation. Low stirrer speeds yield long dimensionless mixing times, also with a small relative standard deviation. In the transition region both mechanisms are at work, resulting in a low mixing time predictability and a large relative standard deviation.

These investigations in the program of the Foundation for Fundamental Research on Matter (FOM) have been supported by the Netherlands Technology Foundation (STW). The additional financial support by Akzo International Research B.V. is gratefully acknowledged.

SYMBOLS

D = stirrer diameter (m)

f_s = rate of (1/s)

g = acceleration of gravity (m/s^2)

H = bulk liquid height (m)

$l_K \equiv (\nu^3/\varepsilon)^{1/4}$ = Kolmogoroff's length scale (m)

N = stirrer frequency (1/s)

P = power input (W)

$Re \equiv \rho_0 N D^2/\mu_0$ = bulk Reynolds number (-)

$Ri \equiv (\Delta\rho \cdot g \cdot H)/(\rho_0 N^2 D^2)$ = Richardson number (-)

T = tank diameter (m)

t = time (s)

t_c = circulation time (s)

$t_K \equiv (\nu/\varepsilon)^{1/2}$ = Kolmogoroff's time scale (s)

t_m = mixing time (s)

V = vessel volume (m^3)

α = fraction of liquid drawn into stirrer (-)

Δ_c = mixing time criterion (-)

$\Delta\rho \equiv \rho_a - \rho_0$ = density difference (kg/m^3)

ε = energy dissipation (W/kg)

μ_0 = bulk viscosity at beginning of experiment (Pa·s)

μ_a = viscosity of added liquid (Pa·s)

$\mu^* \equiv \mu_a/\mu_0$ = viscosity ratio (-)

ν = kinematic viscosity (m^2/s)

ρ_0 = bulk density at beginning of experiment (kg/m^3)

ρ_a = density of added liquid (kg/m^3)

σ = standard deviation of t_m (s)

REFERENCES

1. Gavlak, A.M. & Calabrese, R.V., 1989, poster presentation, Mixing XII EF Conference, Potosi, MO, USA.

2. Smith, J.M. & Schoenmakers, A.W., 1988, Chem.Eng.Res.Des.66, 602-609.

3. Danckwerts, P.V., 1951, Ind.Eng.Chem.43, 1460-1467.

4. Davies, J.T., 1972, "Turbulence Phenomena", Academic Press, London.

5. Hoogendoorn, C.J. & Den Hartog, A.P., 1967, Chem.Eng.Sci.22, 1689-1699; partly published earlier in De Ingenieur 77, 1965, No.17, pp.Ch37-42 (in Dutch).

6. Zlokarnik, M., 1967, Chem-Ing-Techn.39, No.9/10, 539-549 (in German).

7. Rielly, C.D. & Pandit, A.B., proceedings, 6th Eur.Conf. on Mixing, Pavia, Italy, organised by AIDIC.

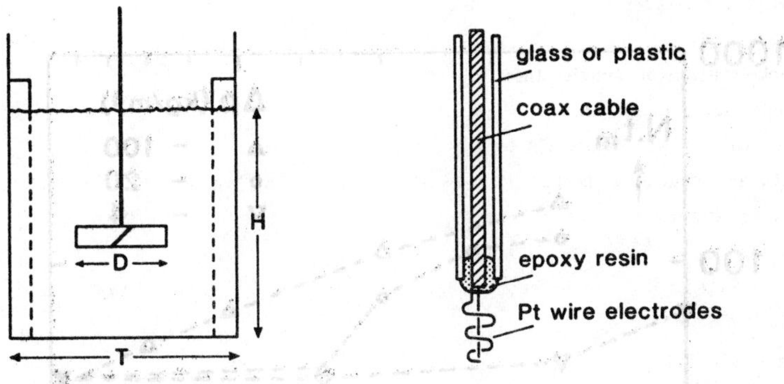

Figure 1. Mixing vessel and conductivity probe used for the experiments.

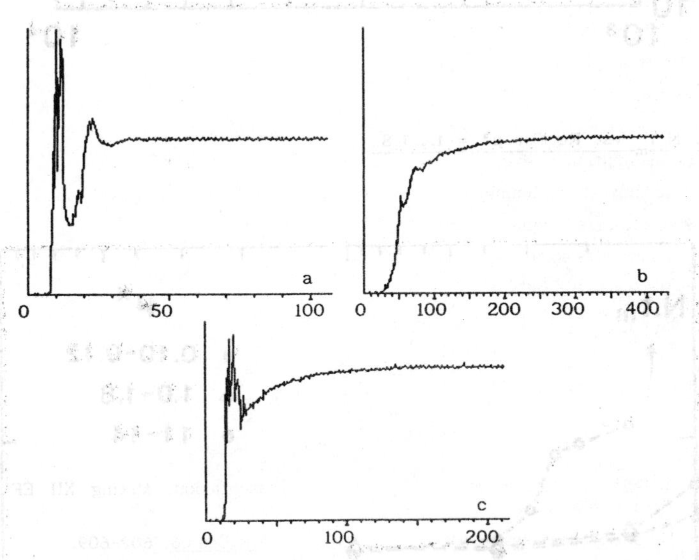

a) the tracer liquid is drawn into the stirrer and dispersed; $N \cdot t_m \approx 25$;

b) the tracer liquid rises to the surface and is dissolved by surface renewal; $N \cdot t_m \approx 150$;

c) part of the tracer liquid is drawn into the stirrer, while the remaining part rises to the surface; $N \cdot t_m \approx 80$.

Figure 2. Conductivity vs. N·t; conductivity in arbitrary units. Stirrer speed is not the same in the three cases; $\Delta\rho \leq 0$.

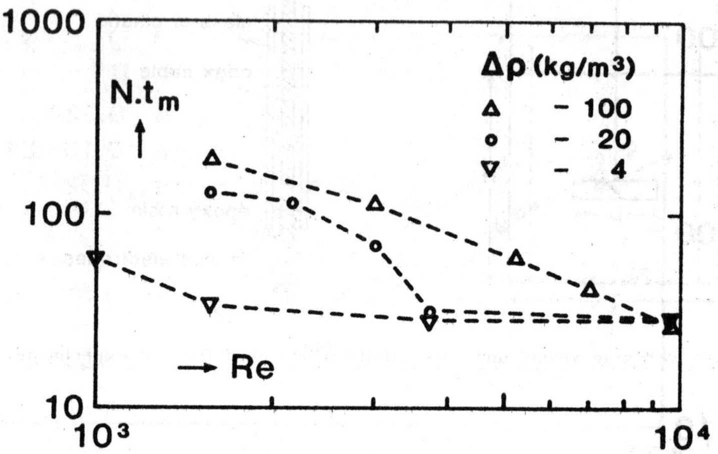

Figure 3. $N{\cdot}t_m$ vs. Re for $\mu^* = 1...1.8$.

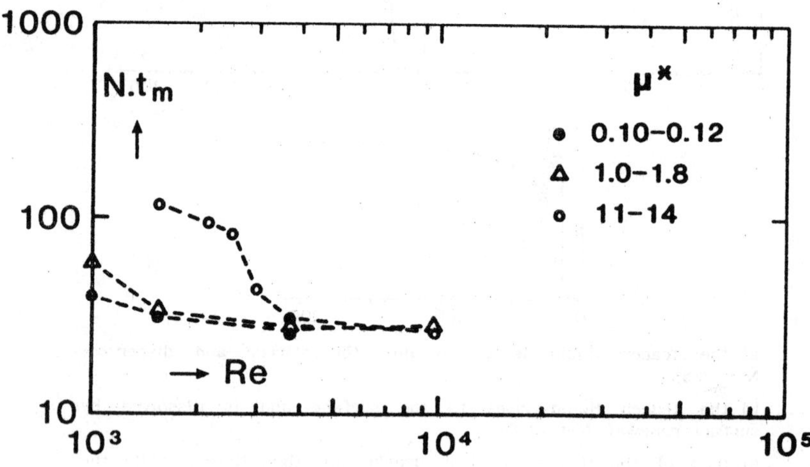

Figure 4. $N{\cdot}t_m$ vs. Re for $\Delta\rho = -5$ kg/m^3.

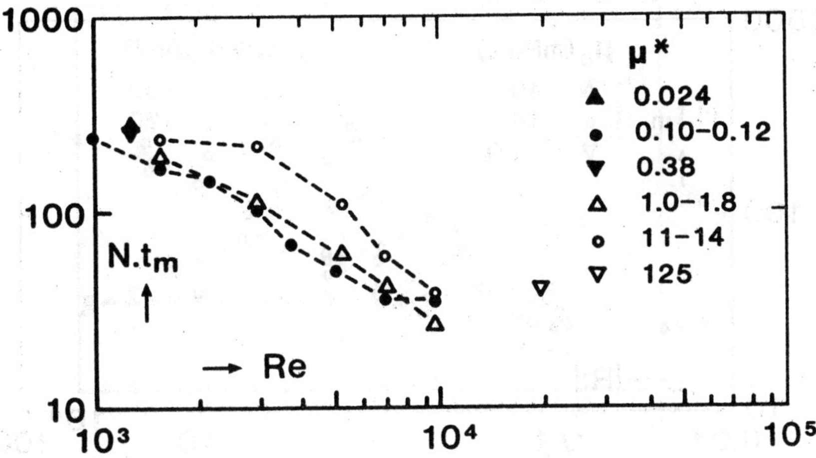

Figure 5. $N \cdot t_m$ vs. Re for $\Delta\rho$ = -100 kg/m^3.

(for figures 6 & 7 see next page)

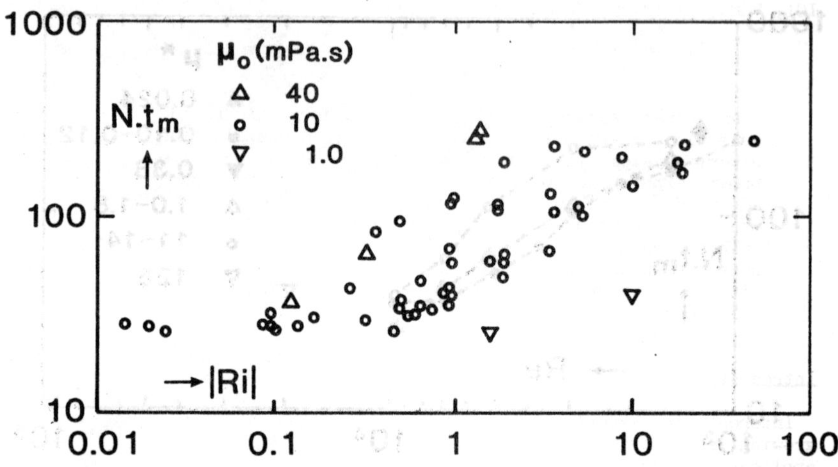

Figure 6. $N \cdot t_m$ vs. $|Ri|$ for all measurements with injection near the surface.

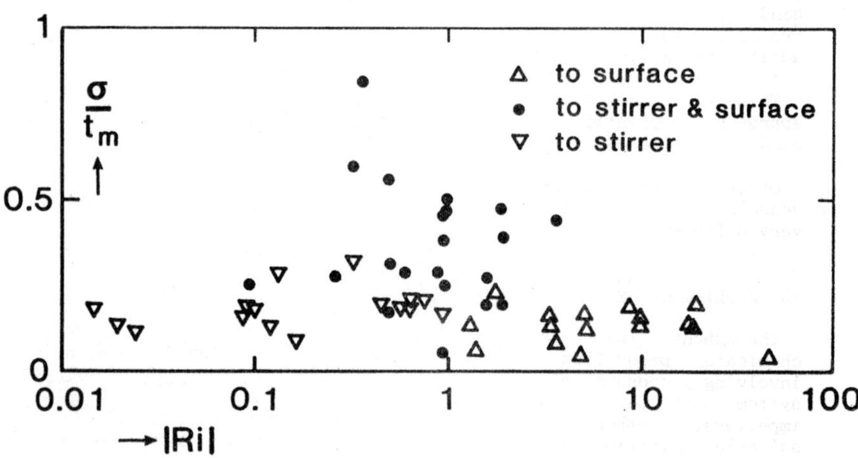

Figure 7. σ/t_m vs. $|Ri|$ for all measurements with injection near the surface.

AN EXPERT SYSTEM FOR THE SELECTION OF AGITATORS FOR STIRRED VESSELS

JW Morrison, PN Sharratt and KJ Carpenter
ICI, FCMO, Blackley, Manchester, M9 3DA

Summary

An expert system has been developed as an aid to the selection of agitation systems for stirred vessels. The expert system follows closely the method used by a human expert in solving the problem. A number of advanced logical and analytical techniques were used in the representation of the method.

The program uses the Prolog language for the expert system, which runs on a PC. The system is menu-driven, with context-sensitive help available to the user at all times.

Introduction

Interest in Expert Systems in the process industries has been growing steadily over the last ten years [1]-[6]. A wide variety of applications have been identified including control, on-line advice for plant troubleshooting, equipment selection, assistance in database searches, advice on calculation methods and even marketing aids.

The diversity of problems which may be tackled often leads to difficulties in the development of an expert system. If an expert tries to set up system using a general-purpose "shell", he may be unable to make the program solve the problem in the same way as he does. The result can be a cumbersome or slow system. On the other hand, programming in Artificial Intelligence languages (LISP or Prolog), to gain flexibility and speed, requires considerable additional skills and effort.

This paper describes an Expert System developed using the latter approach. Sophisticated mathematical and logical techniques have been used to mimic closely the problem-solving technique of an expert.

Other expert systems have been directed at the area of agitated vessels, for example [6]. In this, a rule based approach was used; very different to the method presented here.

The Problem of Agitator Selection

Throughout the process industries, and particularly in Fine chemicals production, stirred vessels are used for operations involving processing of liquids. For each stirred vessel, an agitation system must be designed. This usually consists of one or more impellers, together with baffles, if required. (Figure 1) The use of a suitable agitation system can often be critical to the success of a given application [7].

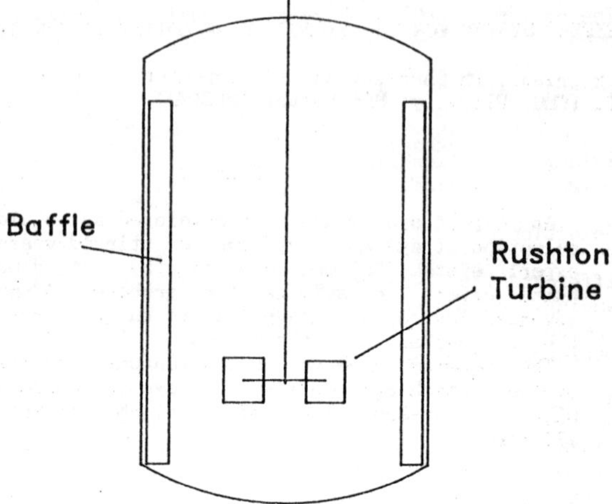

Typical Agitated Vessel

Figure 1

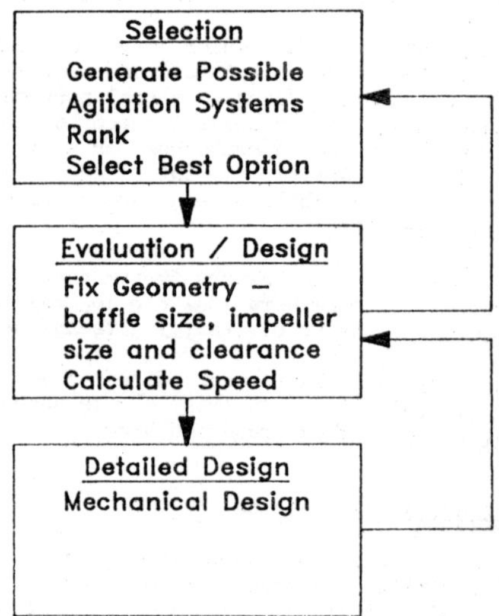

Agitator System Design

Figure 2

The design of agitation systems usually consists of three processes, as shown in Figure 2. Firstly, a suitable agitation system is selected - specifying the type and number of impellers, and the type and degree of baffling. Secondly, design calculations are carried out - for example, definition of the agitator speed required to perform the specified duties. In view of these calculations, the original selection may be modified, or replaced with a different configuration. Finally, detailed design is performed.

Selection of agitation systems for stirred vessels is both a complex and a difficult task. There is a wide range of both available systems (different impellers, types of baffles etc) and of duty performed (eg blending, solids suspension or heating). The selection of an optimal, or even feasible, system for a given duty requires considerable experience.

A duty may involve satisfying several different objectives simultaneously. A list of major "Agitation Objectives" was defined to include all of the duties normally performed in stirred vessels. This is shown in Table 1. For a duty with multiple objectives, the relative importance of the objectives must be assessed. To do this, an "Analytical Hierarchy Process" (AHP) was chosen. [9] This will be described in more detail later.

Table 1 - Major Agitation Objectives.

Liquid Blending
Gas Dispersion
Solids Dispersion
Immiscible Liquid/Liquid Dispersion
Heating / Cooling
Reactions

Each of the Agitation Objectives can be sub-divided into several secondary Objectives. For example, "Solids Dispersion" includes the suspension of sinking solids, incorporation of floating solids, high mass transfer and deagglomeration.

Note that a given duty might be served by a number of different agitation systems, each of which could be tuned (by choice of impeller speed, detailed dimensions etc) to perform adequately. Expertise must be used both to choose suitable systems and to provide an opinion as to which is likely to be best.

The chosen Algorithm

The selection algorithm follows closely the method by which an expert would solve the problem, and was divided into four parts -

i) The collection of data concerning the duty or duties to be performed by the agitator.

ii) Assessment of the relative importance of the components of the duty.

iii) Rejection of systems which are totally unsuitable for the specified duty.

iv) Ranking of the allowable systems on the basis of experience of

similar systems and their performance for the specified duty.
Note that the above does not include any design calculations. These would be carried out later in the process.

Knowledge Representation

An important consideration in the design of an expert system is the way in which the knowledge in the system is represented [8]. In this problem, two sets of knowledge must be stored - the requirements of the user and the capabilities of each of the agitation systems under consideration. The latter forms the "Knowledge base" of the expert system - supplied by human experts.

The user requirements consist of three elements -
i) The objectives and sub-objectives of the agitation system,
ii) The relative importance of the objectives in the overall duty,
iii) A number of additional properties - rheological type, approximate apparent viscosity and vessel aspect ratio.

For an agitation system, the following data are stored in the knowledge base -
i) Proscriptions, that is objectives and sub-objectives for which the system is unsuitable,
ii) Prescriptions, that is objectives and sub-objectives which must have been requested for the system to be specified,
iii) The range of viscosity over which the system can be used,
iv) The value of the system for each of its allowed sub-objectives, as judged by an expert, on a scale of 0 - 1 and
v) Other data - helpful additional information, including geometrical details and addition constraints.

With the exception of the "other data" (v), both sets of knowledge are stored as lists of ASCII strings. Information about forty distinct agitation systems is stored in the knowledge base. The requirements are assembled by the expert system as it interrogates the user. ASCII strings were used to facilitate updating of the knowledge base. The information about each system is easily read, and changes can be made simply using a normal file editor.

The "other data" are stored in a separate database of helptexts. Each item in the database is indexed by a single ASCII string which is one item in the list describing an agitation system.

Operation of the System

On entering the expert system, the user firstly selects one or more from a list of the major agitation objectives. If more than one has been selected, the Analytical Hierarchy Process (AHP) routine is invoked. Here, the user is asked to compare, pairwise, the importance of the objectives. The AHP routine generates weights corresponding to the user's opinions, and evaluates the consistency of his comparisons. Further details of the process are given by Saaty [9].

The user is then allowed to define more clearly his duty by selecting from among the sub-objectives corresponding to each chosen objective. He must also enter the rheological type of the fluid, its approximate apparent viscosity, the vessel aspect ratio, and the likely materials of construction, if known.

Once the list of user requirements is complete, the program moves to a search of the knowledge base. Each in turn of the agitation systems is compared with the list of requirements, and those which do not fit the objectives are rejected.

The comparison with the viscosity requirement is made by the use of "Fuzzy Logic". The range of allowable viscosities for each agitation system is represented as a "fuzzy set" [10,11]. This is a type of density function which is a relation between Grade Membership (goodness of fit) and viscosity. In this system, trapezoidal functions were used, as shown in Figure 3. Here, the system is a good fit to the viscosity of the duty between B and C, and the grade membership is one. On AB and CD, the system is able to perform the duty, but not as well. Over these regions, the grade membership varies between zero and one. Outside AD, there is no fit - the agitator cannot perform the duty and grade membership is zero.

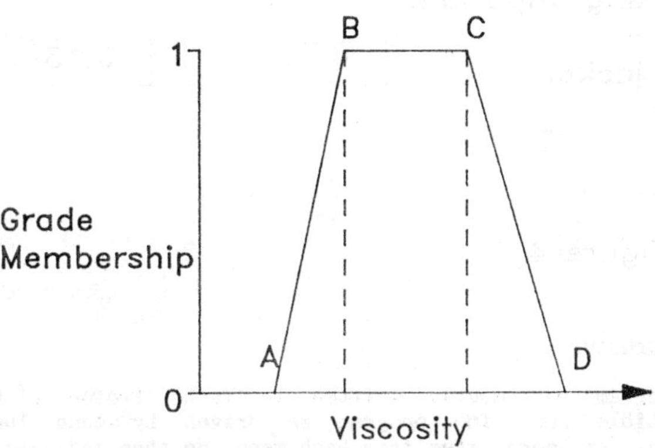

Figure 3

The system compares the apparent viscosity of the duty with the fuzzy set. It returns the grade membership as a measure of the goodness of fit of the system to the duty. This value is carried forward.

Once a list of feasible agitation systems has been identified, they are ranked. The "score" of a system is the scalar product of a vector of the weightings of the sub-objectives in the duty with a vector of capability of the agitator for each of those objectives. An example of this is shown in Figure 4. Here, a Rushton turbine has bee selected as one possible configuration for the agitation of a gas-liquid process with high gas rates. A second liquid is to be blended in, and good mixing is required. There is a need for heating supplied by an external jacket. These three objectives can be met by the Rushton turbine with values) 0.9, 0.5 and 0.7 respectively. Suppose that the relative importance of the objectives, as judged using AHP, are 0.4,0.45 and 0.15. The scalar product of the two vectors is 0.69. This is the score of the system. The suitable systems are ranked using the scores. The method is formally identical with the Decision Analysis methods described in [12].

	System Capability		Weighting of Objectives	
Gas Dispersion – high gas rates	0.9		0.4	0.360 +
Liquid Blending – good mixing important	0.5	∘	0.45	0.225 +
Heating – external jacket	0.7		0.15	0.105
				0.690

Figure 4

Program Structure

The program was mostly written in Turbo Prolog 2 for use on IBM-compatible PCs. The program is driven by menus. The user can select one or more item from each menu. He then indicates that the selection is complete, and the program proceeds to the next menu.

An important feature of the system is the availability of on-line help from any part of the program. The help is context-sensitive. If the user requests help, for example, in the menu for solids dispersion, he will receive general information about solids dispersion, together with specific advice on what the various menu items mean. The final, ranked list of possible agitators is presented as a menu. By selecting one of the menu items, the user may obtain details of the recommended configuration for that system and other relevant details.

The integration of the AHP routine (which performs a number of calculations) and the expert system was carefully designed. Prolog is a inefficient language for calculations, so the AHP code was written as a stand-alone program in Pascal. The expert system calls the AHP program by generating a command line containing the initial information required by AHP. After completion, the expert system receives the generated weightings which are appended to the list of requirements. This approach is both flexible and efficient.

Benefits

The system provides several benefits. A user inexperienced in agitator system design can specify systems for most duties which are normally encountered. For problematical duties, the user is referred to a human expert. This provides a screen which allows the expert to concentrate on the more serious problems. Because the expert system prompts the user for relevant data, it helps him to gather the appropriate data for design at an early stage. The program also has a strong educational element. The structured approach and helptexts can both increase the user's knowledge of the problem.

Future Developments

Separate programs are being developed for the calculations required during design of agitation systems. These will be called by the expert system in the same way as the AHP routine - either by a command line, or via a "mailbox" file.

Conclusions

In the construction of an expert system, care must be taken to understand the method by which the human expert solves the problem. By use of suitable analytical and logical techniques, his method can be mimicked by the expert system. The method by which knowledge is represented is also important, and can contribute to both efficiency and ease of maintenance. Integration with calculations can be achieved efficiently by use of separate routines written in languages designed for numerical computations.

References

[1] Chowdhury J and Caudwell R (1982) Chemical Engineering, Sep 20th, p45
[2] Chowdhury J (1985) Chemical Engineering, Jun 24th, p14
[3] Chowdhury J (1986) Chemical Engineering, Oct 13th, p14
[4] Basta N (1988) Chemical Engineering, Mar 14th, p26
[5] Andow P (1988) Process Engineering, Oct, p63
[6] Yue PL (1987) NW Branch IChemE, Symposium papers no 3, Personal Computing in the Process Industries", paper 9.
[7] Carpenter KJ (1988) Chem Eng Res & Des, 66, 2
[8] Ringland GA and Duce DA Eds.(1988) "Approaches to Knowledge Representation", Wiley
[9] Saaty TL (1980) "The Analytical Hierarchy Process", McGraw-Hill
[10]Dohnal M (1983) Computers in Industry,4,p341
[11]Dohnal M (1983) Computers in Industry,4,p347
[12]Kepner CH and Tregoe BB (1981) "The New Rational Manager", Princeton Research Press

ASSESSMENT AND DESIGN OF POWDER DEAGGLOMERATION IN FLUID AND PASTE MIXERS

D C-H Cheng*

This paper discusses a design procedure for the selection, operation and optimisation of mixers for powder deagglomeration duties. It incorporates a function describing the deagglomeration characteristic of the powder and a scale-up rule for the mixer, involving the mixer geometry and the mixture rheology. Practical application is considered. The determinations of the deagglomeration characteristic and the scale-up rule are illustrated by the dispersion of carbon black pigments in a range of high viscosity batch mixers. This serves to confirm the validity of the design procedure to a large extent. But the experimental verification is not entirely complete and further research requirement is pointed out.

INTRODUCTION

Mixing is a traditional process that has been developed by trial and error, and mixers for powder deagglomeration are even today selected and operated as a matter of art. Little science has entered into the subject, although some research into certain basic mechanisms has been carried out. This paper is an attempt to redress the situation. A design procedure is proposed involving a method for assessing the dispersibility of powders and one for studying mixer performance, leading to the establishment of scale-up rules. These form the basis for mixer selection and operation and this paper describes how they can be applied in practice.

It must be said that the design procedure is still being developed. This paper gives the theoretical basis of the methods and describes how far experimental verification has progressed to-date. It considers what additional research is needed in order to complete the development. It is usual that the development of an engineering design procedure required the input from many researchers over many years and the present case is no exception.

THEORETICAL CONSIDERATION

Deagglomeration in Uniform Shear Field

A mechanism for deagglomeration is the breaking of the bonds between ultimate particles in an agglomerate by abrasion and crushing. The forces

* Warren Spring Laboratory, Stevenage, Herts SG1 2BX

that bring this about can be derived from the shear stress and pressure generated when the suspending fluid is sheared. The literature provides some results on the forces that act on idealised bodies, and these give a clue to the mechanism of deagglomeration.

Theoretical results quoted by Krekel[1] and Reichert[2] show that the shear stress acting on a shear suspended in a shearing liquid varies with position on the surface from zero to a maximum,

$$\tau_{max} = 2.5\ \eta\dot{\gamma} \qquad (1)$$

Tadmor[3] calculated the force acting on a dumbbell made up of two spheres. The force varies with the orientation of the bond between the spheres with respect to the shearing direction. This passes through a maximum,

$$F_{max} = 3R_1R_2\ \eta\dot{\gamma} \qquad (2)$$

The rate of deagglomeration, ie the rate of bond breakage and the production of ultimate particles, can be assumed to be proportional to the maximum τ_{max} or F_{max}. Then for a given powder and liquid system,

$$\frac{dN}{dt} = P\dot{\gamma} \quad \text{or} \quad N = P\dot{\gamma}t \qquad (3)$$

where P is a factor that includes properties of the powder and liquid and the probability that the agglomerate would actually break up when the conditions are right. [Various assumptions are implied in eq (3) and the rest of the theoretical derivation of this paper. Not the least is that concerning the role of the strength of the agglomerates. These will be brought out and discussed when comparison with experiment is made below].

The effect of deagglomeration is to increase the mixture quality. This can have different attributes depending on the material being mixed, but the exact definition need not concern us here. If it is quantified, say by X, then it would be a function of N or, through eq (3), a function of total strain, $\gamma = \dot{\gamma}t$:

$$X = f(N) = X(\gamma) \qquad (4)$$

During a deagglomeration process, the variation of X with t would depend on the shear rate, $\dot{\gamma}$. But the different X(t) curves would collapse onto a master curve if X is plotted versus γ: (X,γ). That is, the degree of deagglomeration is a function of the total strain suffered by the powder. Experimental verification of this generalised idea is given below.

The exact variation of X on γ, eq (4), would depend on the nature of she powder, its composition by agglomerate size distribution and the strength of the agglomerates. For any powder, it should be possible to determine (X, γ) experimentally. The relationship is a measure of the deagglomeration characteristic of the powder.

Deagglomeration in Mixer

In a mixer, the degree of deagglomeration, X, is distributed with mean Xm and a standard deviation Sx. During mixing, both Xm and Sx would vary with time. The standard deviation depends on distributive mixing and tends to zero as the agglomerates become more uniformly distributed as time goes on.

The mean, Xm, on the other hand, depends on the progress of deagglomeration, ie dispersive mixing, and this is more complicated to enumerate. In a mixer, the shear rate varies from point to point and also the time each element of the mixture spends at each point in the mixer. The total strain suffered by each agglomerate would depend on its shear history and there will be a distribution of shear histories among the agglomerates. One can average the effect over all the agglomerates and consider an average shear rate for the mixer, $\dot{\gamma}_{av}$. Then the mean degree of deagglomeration would depend on the strain based on the average shear rate:

$$Xm = Xm(\dot{\gamma}_{av}.t) \tag{5}$$

Now, extending the idea of Metzner and Otto[4,5], the average shear rate is assumed to be proportional to the rotational speed of the mixer:

$$\dot{\gamma}_{av} = K\Omega \tag{6}$$

Then,

$$Xm = Xm(Nr) \tag{7}$$

where Nr = Ωt is the total number of revolutions undergone by the mixer.

The progress of deagglomeration in a mixer, Xm(t), would depend on the rotational speed, Ω. But the different Xm(t) curves would collapse onto a master curve if Xm is plotted against number of revolutions, Nr. This, (Xm, Nr), would be characteristic of the mixer, ie its geometry, specifically the shape and size. Experimental verification of the dependence of Xm on Nr is given below.

Development of a Design Procedure

The starting points for developing the design procedure are eqs (4) and (7). They are used to determine the effective strain in a mixer and to correlate it to the number of revolutions undergone by the mixture. Deagglomeration experiments are carried out in an ideal mixer (such as the cone and plate apparatus) and in a industrial mixer to determine the (X, γ) and (Xm, Nr) master curves (Figure 1). Corresponding γ_e and Nr are then obtained by equating X with Xm. The (γ_e, Nr) plot,

$$\gamma_e = \gamma_e(Nr) \tag{8}$$

is the effective strain characteristic of the mixer. It depends on the geometry of the mixer, ie the shape and size.

The next step in developing the design procedure is to correlate mixers of different sizes of the same shape. Experiments are carried out on different sizes of the mixer and we expect the (γ_e, Nr) curves to be different. But some geometric factor, Gf, may be found that will collapse them onto a master curve. This can be obtained by dimensional analysis and ad hoc correlation or more scientifically by considering the shear developed in the mixer.

For example, in the Sigma-blade mixer, intensive shear takes place between the tip of the rotor and the mixer wall. The shear rate can be estimated from $R\Omega/H$ and the strain is then $R\Omega t/H$. Thus replotting γ_e versus

Nr(R/H) would give a master curve. The geometric factor Gf = R/H.

In some other mixers, Gf = R^3/V where R is the rotor radius and V the mixture volume. This arises because the progress of mixing is determined by the probability that elements of the mixture are drawn into the mixing zone which has a volume proportional to R^3. The plot (γ_e, NrGf) would give a master curve. This is independent of mixer size, but is characteristic of the shape. Some experimental examples of such plots are given below.

Next, the (γ_e, NrGf) curve would vary with the non-Newtonian characteristic of the fluid, which governs the flow pattern in the mixer, which in turn determines the flow of the mixture into the actual mixing zone and hence the overall progress of mixing. The third step in the development is therefore to obtain a correlation with rheological property of the mixture.

To demonstrate how the master curve, (γ_e, NrGf), might depend on rheological property, consider the mixing of a power law fluid in the annulus between two coaxial cylinders in relative rotation. The shear rate is:

$$\dot{\gamma} = \frac{(2/n)\Omega}{1 - \beta^{(2/n)}}$$

where the radius ratio $\beta = R_1/R_2$. Thus the effective strain is

$$\gamma_e = \mathrm{Nr} \cdot f(\mathrm{Gf}, \mathrm{Rf}) \qquad (9)$$

where

$$f(\mathrm{Gf}, \mathrm{Rf}) = \frac{2/n}{1 - \beta^{(2/n)}}$$

The geometrical factor is Gf = β, and the rheological factor, Rf = n, is simply the power law index which describes the non-Newtonian viscosity characteristic. Other rheological factors may involve normal stresses or extensional viscosities, to cover the elastic properties of the fluid.

It is interesting to note that, whereas a master curve is obtained by multiplying Nr by Gf to cover the geometric effect (at least for the creeping flow regime), the rheological effect is more complicated than can be described by the triple product NrGfRf. There is interaction between Gf and Rf even in the simple example given and the product of Gf and Rf is found in the function f() of eq (9). For industrial mixers, more complicated functions of Gf and Rf might be expected.

In summary, the design procedure involves two relationships. One is the deagglomeration characteristic of the powder, eg (4); this can be determined using a ideal mixer, such as the cone and plate apparatus in which the shear rate is uniform. The other is the effective strain characteristic of the mixer, or scale-up rule, eq (9), which is determined by carrying out pilot scale mixing trials using powders of known deagglomeration characteristic and mixtures of a range of rheological properties, in mixers of varying sizes. It can be expected that the form of the scale-up rule would vary depending on mixer type. In the following sections, experimental verification of the procedure is given and a discussion on how the procedure may be applied.

APPLICATION OF DESIGN PROCEDURE

In this section, how the design procedure may be used in practical applications is discussed. As usual, there are two types of design problem to consider; the first concerns the design or selection of a new piece of equipment for installation and the other deals with the best use of an existing equipment. It is assumed that the scale-up rule, ie the [γ_e, Nr•f(Gf, Rf)] relationship, is already known for the mixer in question. Otherwise it has to be established as described earlier.

Design of a Mixer for New Installation

1. The first step in the design or selection of a mixer is to determine the deagglomeration characteristic of the powder, the [X, γ] relationship. This is unlikely to be known and in any case it will vary from batch to batch of powder. So it has to be measured on a sample of the powder in question.

   X is of course not restricted to the reflectivity method described in this paper. Depending on the mixture, some other convenient measure of mixedness would be used. The reader may find the review Reference 6 helpful.

2. We need also to know the desired degree of deagglomeration to be obtained in the mixer, expressed as a value of X. This would involve a separate study of the relationship between mixedness and X. Because of the wide range of mixing duties found in industry, no specific guidance can be given here on how to obtain this relationship. But it is a crucial step in the successful application of the design procedure.

3. From the desired X and the (X, γ) curve, we then determine the design total strain, γ_e, that we have to achieve in the mixer.

4. Then using the known scale-up rule, [γ_e, Nr•f(Gf, Rf)], we determine the corresponding Nr•f(Gf, Rf) as the design value.

5. The rheological property of the mixture has to be measured, to determine Rf.

6. The next step is to fix the design throughput and other constraints on the mixer parameters. Economic factors may be included here.

7. Then, from the design value of Nr•f(Gf, Rf), determine the mixer parameter(s) that the design exercise is to determine. Often the mixer parameters desired are the mixer size and speed, but it could equally well be the mixing time. The design may be to determine one or other of these, subject to constraints on the others. Again, it is not possible to generalise as to which parameter would be under constraint and which is to be determined. Each design exercise has to be considered separately.

8. The steps (4) to (7) may be repeated for different types of mixers. And so, subject to economic considerations if any, an optimum choice of mixer type may be made together with specific parameters.

Operation of an Existing Mixer

Often, an existing mixer is to be used for a new mixing duty. The following steps are then to be used.

1. Determine the deagglomeration characteristic of the powder, (X, γ).

2. From the desired degree of deagglomeration, determine γe required.

3. From the scale-up rule, [γ_e, Nr•f(Gf, Rf)], determine the design Nr•f(Gf, Rf) value.

4. Determine the rheological property of the mixture.

5. Fix throughput rate desired. The size of the mixer would be fixed and may be the speed. Determine any economic factors.

6. From the design Nr•f(Gf, Rf) value determine mixing time to achieve the desired γ_e. This should be a simple matter if the mixer speed is fixed.

However, the speed may be changed with the use of a new motor. This could shorten the mixing time and so reduce the production cost, but it would incur capital cost. The determination of the optimal speed and the corresponding mixing time would require a study taking into account economic factors, such as product value, motor cost and power and labour costs. Again, each design exercise is to be considered separately. The design procedure of this paper provides a basis for this to be done.

EXPERIMENTAL VERIFICATION

Determination of Deagglomeration Characteristic

Experimental verification of eq (4), which shows that the kinetics of deagglomeration is given by the total strain suffered by the mixture, has been given by a number of workers.

Krekel[1] studied the break up of calcium carbonate powder agglomerated by sodium chloride bridges and suspended in a silicone oil. The stressing unit was a coaxial cylinder apparatus in which the rotation shear was applied impulsively for short periods. The agglomerates were sized by microscopy and the average was used in defining the degree of pulverisation, V. It was found that V correlated with $\dot{\gamma}t$. it was independent of shear stress up to some critical value, but above this V decreased further indicating additional break up of the agglomerates.

Reichert[3] studied the deagglomeration of two commercial pigments in dimethyl silicone oils using a modified version of the Krekel apparatus. The white light extinction cross-section of the agglomerates, Av, was determined. this was an integrated measure of the mean size of the agglomerates and therefore an indication of the degree of deagglomeration. Again, it was found that Av was a function of $\dot{\gamma}t$ and not dependent on shear stress over the ranges used.

Kao and Mason[7] studied the removal of individual particles from an aggregate of PMMA spheres in a shear field generated between counter-rotating coaxial cylinders. They found that the break-up is given by the rate equation, $Ro^3 - Rt^3 = k_m\dot{\gamma}t$, where $k_m = 0.007\ mm^3$.

These results confirm the validity of eq (4).

Extensive verification of the equation had also been carried out at Warren Spring Laboratory from 1976 to 1982 under the aegis of the Fluid and Paste Mixing Co-operative Project (later renamed Thick Liquids and Pastes: Processing and Flow Programme). The experimental details, with some results, have previously been described[8,9]. This paper presents additional data. Briefly, the cone and plate apparatus, commonly used for viscosity measurement, was used as an ideal mixer to generate constant shear rates. Values between 0.58 and 113/s were used.

Two commercial carbon black pigment powders were studied. The first was Regal TRF-S: this was fluffy and soft. The other was Black Pearls 1300: this was a beaded pigment which was harder and stronger than the Regal.

They were dispersed in various white pastes made up of a chalk whiting (Snowcal 3ML) in either dibutyl phthalate or Hyvis polyisobutene. Model D was titanium dioxide in Hyvis 10. The pastes had different non-Newtonian viscosity characteristics. The flow curves, determined using cone and plate viscometer at 25°C, were fitted to the Herschel-Bulkley equation and gave the following flow curve parameters.

Model	Yield Stress	Consistency Coefficient	Power Law Index	Liquid Component
B	61.9 Pa	175 $Pa.s^n$	1.55	DBP + oleic acid
C	25.3	67.3	0.809	Hyvis 10
C1	8.2	33.1	0.646	Hyvis 3
C2	132	207	0.687	Hyvis 10
C3	58.3	77.8	0.517	Hyvis 3
D	-	17.5	1.00	see text
E	-	0.019	0.964	DBP
F	-	0.402	0.769	Water + glycerol + CMC
G	-	3.36	0.913	Hyvis 3
H	-	15.0	0.900	Hyvis 10

As deagglomeration progressed, the paste became first grey and then darkened to black. A reflectivity method was used to monitor the change in colour. The output reading was an arbitrary scale X, with high values corresponding to white and low to black.

Periodically during a test, the cone and plate were separated and the reflectivity of the exposed mixture was measured. Then the cone and plate were brought together and the shearing continued for another period of time. Previously published (X, γ) data show the experimental scatter obtained[8,9]. Figure 2 and 3 summarise the results for the two powders. They show that the master curve for a given powder in a particular paste was independent of the shear rate, thus confirming that eq (4) applies.

However, the master curve was dependent on the paste. The data were not fully consistent; witness the results for Black Pearls in models G and H. But on the whole, they show that a high viscosity paste tended to give lower X at the same strain. The reason for this behaviour may lie in the agglomerate strength distribution. It had been hypothesised[8] that the carbon black strength distribution might be bimodal. Then, for those pastes giving shear stresses that fell between the two modes, only the weaker agglomerates would break up and so the same (X, γ) curve would be obtained. If a higher viscosity paste is used to generate higher shear stresses, the stronger agglomerates could now break up, giving lower X and a new master curve. The three master curves on Figures 2 or 3 for each powder suggest that the strength distribution could be trimodal even.

The Black Pearl agglomerates were harder and stronger then Regal. This is reflected in the (X, γ) curve for Black Pearl lying above that for Regal where the same paste (B or C) was involved. That is, at the same shear rate and shear stress, Regal was broken down more than Black Pearl.

These results confirm eq (4) but only within certain provisos. It is clear that more work needs to be done on powder deagglomeration even in the simple cone and plate apparatus.

Establishment of Scale-up Rule

A limited experimental verification of eq (9) was carried out by Reichert and Ruhmling[10] using Werner and Pfleiderer Sigma kneaders, types LUK1/0.75 [0.75 l] and LUK3-III-2 [1 l], to mix a commercial pigment suspended in a dimethyl-polysiloxane liquid. The extinction cross-section Av of samples was measured as before. This was found to correlate with Ωt.Gf where $Gf = \pi(R_a + R_i)/b$, R_a and R_i were the radii of the kneading trough and hook and b the gap width of the kneader. However, different curves were obtained for different values of Ω.Gf (taken to be an average shear rate in the kneading gap), with higher Ω.Gf giving a higher degree of deagglomeration. Reichart and Ruhmling did not convert their results to effective strain.

In the FPM/TLP programme at WSL already mentioned, a much wider range of industrial mixers was used. It included both batch and continuous mixers, for low as well as high viscosity materials. In this paper, the high viscosity batch mixers are considered. They include the following:

- Maxta mixer[6]

  (Babcock Gardener laboratory scale mixer [1 l] and a 25 l mixer). The smaller mixer was also used with rotors that had been perforated to give by-pass of the mixture to relieve the pressure generated.

- planetary mixer

  (Collette MP900 mixer with bowl 2 [20 l] and bowl 5 [50 l])

- twin wing-blade kneader

  (Baker Perkins 20HS [20 l] and 200 HS [200 l] Dough mixers)

- Z-blade mixer

  (Brabender Plasticorder unit [0.6 l], Winkworth 8Z with serrated blade [4.5 l], and Baker Perkins mixer [60 l])

Again, the experimental details have been described before[9]. Briefly, the same carbon black powders and the higher viscosity pastes were used. During a test, samples of the mixture were taken from randomised positions within the mixer at different times of mixing and examined using the reflectivity probe. The readings were averaged to give Xm(t), as function of time, from which the (Xm, Nr) plot was obtained. This was then compared with the (X, γ) curve for the carbon black and paste in question, as described in the theoretical section above, to derive the (γ_e, Nr) curve. For each type of mixer, a number of (γ_e, Nr) curves were obtained for different sizes of the mixer, different carbon black grades and different rheological properties. These were used to assess the effect of geometry and rheology in eq (9).

Figure 4 shows the (γ_e, Nr) plot for the Maxta mixers. The Regal carbon black was dispersed in three pastes. The results were relatively independent of speed, size of mixer and rheology of the paste, falling on one master curve. This showed that both Gf and Rf were unity. The behaviour could be due to the unusual operation of the Maxta mixer described in Reference 9.

Figure 5 shows the results for the planetary mixers. The Regal carbon black was dispersed in two pastes. The different (γ_e, Nr) curves were collapsed quite well using Gf = $R/V^{1/3}$. The reason for this could be because mixing took place within the volume swept out by the rotor, and the rate of mixing depended on the probability of the mixture being drawn into this volume. The ratio of the volumes were R^3/V, hence Gf as given. But the results for the two pastes were distinct. Clearly a rheological factor Rf is needed to produce the master curve. We need to study the flow pattern of the paste in the mixer to gain some insight into what form Rf would take.

Figure 6 shows the results for the Z-blade mixer. Both Regal and Black Pearls carbon black were used, in a range of pastes. The two powders appeared to behave the same here and it is perhaps surprising that there was no significant effect of scale, ie Gf = 1. The difference in behaviour due to rheology, as previously observed[8], turned out to be somewhat more complicated. (In order to reduce confusion, the results for pastes C to C3 are represented by the average curves). It seemed that there was a difference between pastes B and C at low numbers of revolutions, but the difference was reduced as Nr increased. The difference was attributed to the operation of the Z-blade in cutting and displacing the mixture and the way the different pastes flowed or broke away in lumps over the blades[9]. The other pastes had different rheology and their results complicated the plot.

Figure 7 shows the results for the twin wing-blade kneader. There was not effect of size (Gf = 1), but the dependence on carbon black strength and rheology was complex. The kneader arm of the mixer had a complicated geometry.

Further Research Needs

The experimental work on powder deagglomeration in the cone and plate apparatus and in high viscosity batch mixers showed that theoretical ideas are applicable to a limited extent. The lack of success is very probably because the dependence on geometry is complicated, so that Gf is complex and not easily identified. Also, the rheological factor has yet to be investigated in a systematic way. However, the results suggested that the formation of the master curve by using Nr, Gf and Rf is possible. But, in general Gf and Rf, and the function f(Gf, Rf) may not be simple in form. They may be found by modelling of the mixing mechanism of a mixer. Or they may be found by trial and error. This is of course, the subject for further research.

CONCLUDING REMARKS

This paper describes a design procedure for the selection, operation and optimisation of mixers for powder deagglomeration duties. It involves (i) the deagglomeration characteristic of the powder, (X, γ), and (ii) the scale-up rule for the mixer, [γ_e, Nr•f(Gf, Rf)]. The discussion of applications, for the design or selection of a new mixer and for the operation of an existing mixer, shows that the procedure can have practical use.

The experimental determination of the deagglomeration characteristics and the scale-up rule is illustrated by results in the literature and WSL research on the dispersion of carbon black pigments in a wide range of mixers. These serve also to confirm the validity of the design procedure. It is seen that the experimental verification is not entirely complete. The geometric factor, Gf, has yet to be obtained for some mixers and the rheological factor, Rf has to be properly investigated. But, the results are very promising and, given the practical usefulness of the design procedure (it will put the design of mixers on a sound scientific basis), it can be recommended that further research should be carried out. The approach is also applicable to other mixing processes, provided of course that the degree of mixedness may be determined quantitatively. This in itself is another inducement for research because the achievement of objective measurement of mixedness would be a giant step forward in the study of mixing operations. It is in this spirit that I offer this paper to chemical engineering.

NOTATION

Av	white light extinction cross-section
Gf	geometric factor
H	mixing gap
K	constant relating shear rate to mixer speed
n	power law index
N	number of fundamental particles split from agglomerate
Nr	number of revolutions
R	radius of mixing arm
R_1, R_2	radii of coaxial cylinders or spheres in dumbbell
Ro, Rt	radius of agglomerate, initial and at time t
Rf	rheological factor
Sx	standard deviation of X
t	time
V	volume of mixer
X	measure of mixedness
Xm	mean value of X
β	$= R_1/R_2$
$\dot{\gamma}$	shear rate
γ	shear strain
γ_e	effective strain
η	viscosity
Ω	rotational speed

REFERENCES

1. Krekel J. 1966. Chemie-Ing Techn. 38, 229-234.

2. Reichart H. 1973. Chemie-Ing Techn. 45, 391-395.

3. Tadmor Z. 1976, Ind Eng Chem Fundam. 15, 346-348.

4. Metzner A B And Otto R E. 1957, AIChE Journal. 3, 3-10.

5. Nagata S. 1975, "Mixing Principles and Applications", New York: Wiley.

6. Anon. "Assessment of Mixture Quality", in Manual for Seminar on Mixing Technology, April 1988, Stevenage. Organised by Warren Spring Laboratory, Stevenage, Herts, UK.

7. Kao S V and Mason S G. 1975, Nature. 253, 619-621.

8. Schofield C and Stewart I W. 1980, Chem Engr, 486-489.

9. Schofield C and Stewart I W. 1982, Proc 4th European Conf. Mixing, Cranfield, Bedford: BHRA Fluid Engineering, pp 157-171.

10. Reichart H and Ruhmling K. 1976, Chemie-Ing Techn MS 371/76; Chemie-Ing Techn, 48, 559.

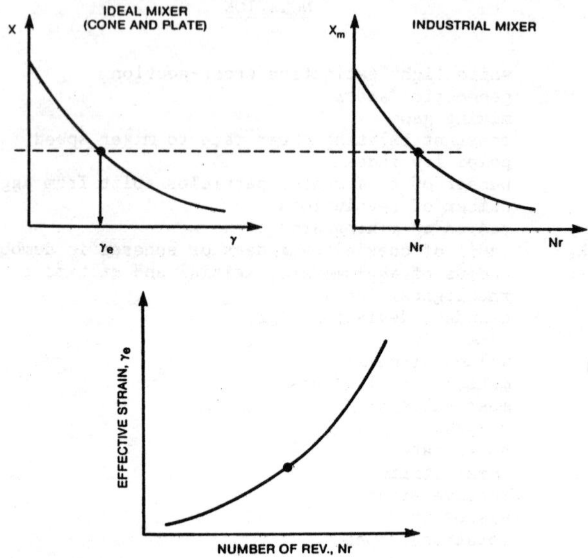

Figure 1 Determination of effective strain in industrial mixer

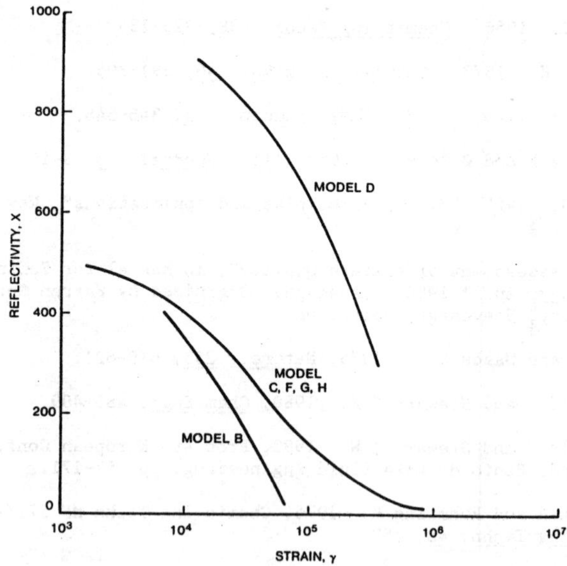

Figure 2 Deagglormeration characteristic of Regal carbon black

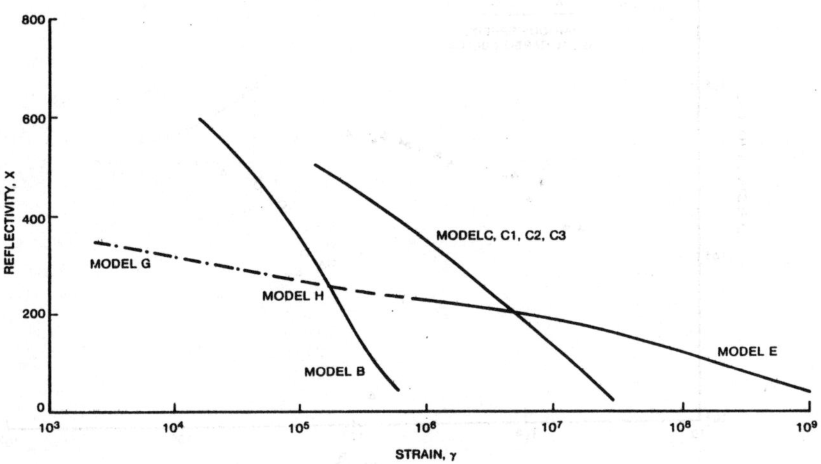

Figure 3 Deagglomeration characteristic of Black Pearls carbon black

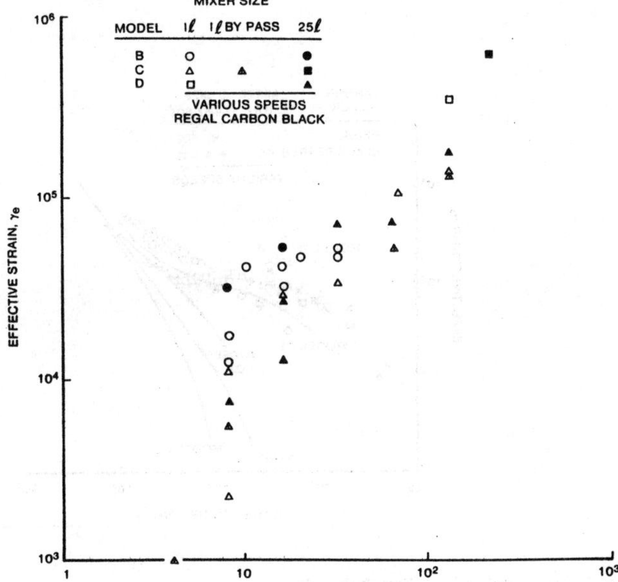

Figure 4 Scale-up rule plot for Maxta mixer

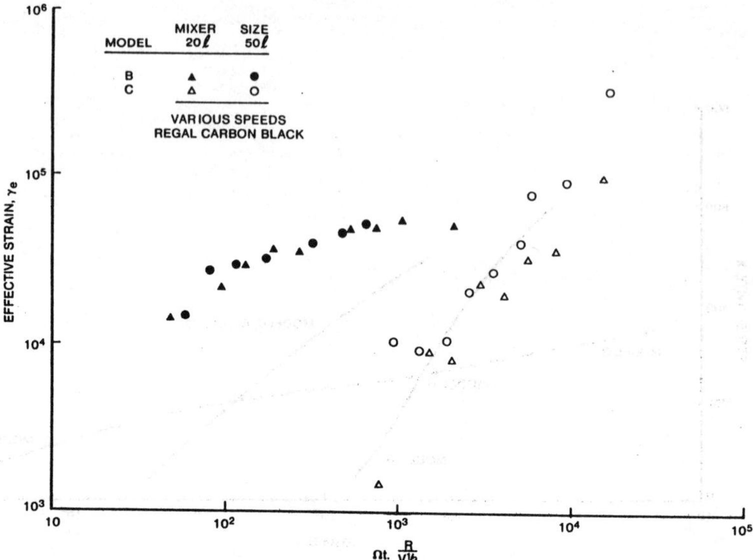

Figure 5 (γ_e, Nr.Gf) plot for planetary mixer

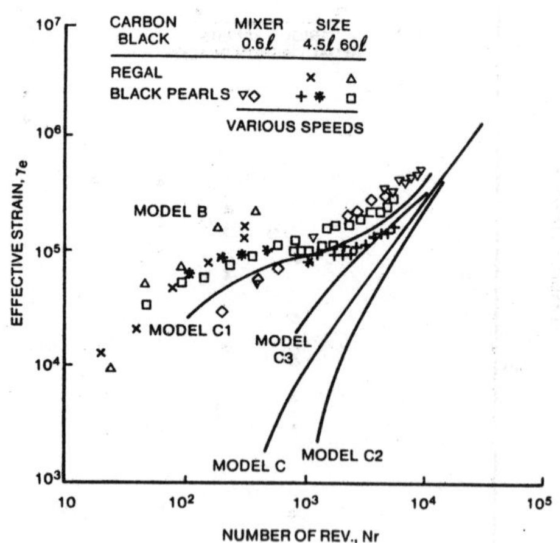

Figure 6 (γ_e, Nr) plot for Z-blade mixer

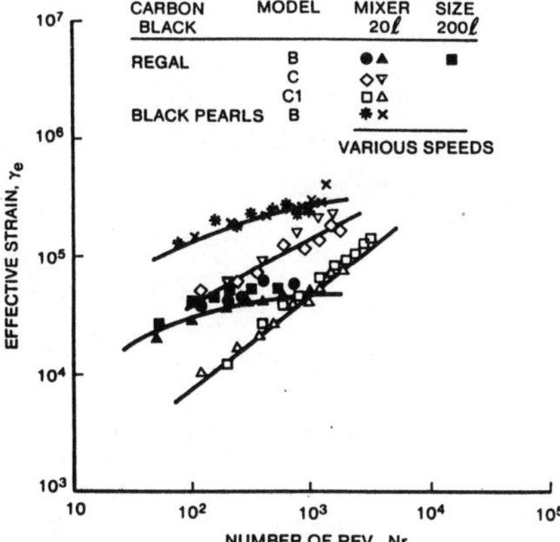

Figure 7 (γ_e, Nr) plot for twin wing-blade kneader

A STUDY OF THE PARTICLE RESIDENCE TIME DISTRIBUTION IN CONTINUOUS FLUIDISED BEDS

A.C. Hoffmann[1] and H. Paarhuis[2]

The particle residence time distribution in continuous bubbling fluidised beds has been related to the bed hydrodynamics. Residence time distributions have been simulated on basis of assumed mixing processes and compared with experimental data available in the open literature. The parameters involved have all been estimated from published correlations. Agreement is satisfactory in view of the simplifying assumptions made in the model, confirming that the assumed processes in the main are responsible for the solids mixing.

INTRODUCTION AND BACKGROUND

The subject of solid residence time distribution (RTD) in gas fluidised beds has until now been approached in a largely empirical and descriptive manner in the research literature. In the present work, this problem is attacked from a different angle. Specific particle mixing and transport processes have been proposed based on a study of other branches of the fluidisation literature. To test whether these transport processes can in fact account for the observed RTD's in fluidised beds, a numerical simulation of them has been implemented and the resulting RTD's qualitatively and as far as possible quantitatively compared with experimental data available in the literature. Below a brief rewiew of the existing approaches to the problem is given.

The literature concerned with the RTD of the particles in continuous vertically moving fluidised beds is extensive. In almost all studies attempts have been made to describe the process using the conventional tools of residence time theory.

Series of ideal mixers have been used in many studies (e.g. Gilliland and Mason (1), Klose and Herschel (2) and Pudel et al (3)). The number of mixer stages is in this approach used as an adjustable parameter. Also plug flow with superimposed axial dispersion has often been used (e.g.

[1]Dept. of Chemical Engineering, University of Groningen, Nijenborgh 16, 9747 AG Groningen, The Netherlands. (Author to whom correspondence should be addressed).

[2]Present Address: Keuken and de Koning BV, Korvezeestraat 88, 2628 DD Delft, The Netherlands.

Massimilla and Bracale (4), Weber and Rose (5) and Tripathy et al (6)). Where this model is used, the dispersion coefficient plays the role of adjustable parameter. Both of these approaches are often classified as 'one parameter models', and many workers test both against their experimental results. Another less used one parameter model is one based on the contention that the RTD of fluidised solids is to be attributed to the formation of a velocity profile (similar to that in a fluid flowing through a pipe) in the flowing solids (Morris et al (7)). In this approach, the value of the ratio of the maximum to the mean velocity in this profile was used as an adjustable parameter.

More complicated models have also been proposed. Series of ideally mixed tanks with reverse flow (Whittmann et al (8)), combinations of ideal mixers in series and in parallel (Heertjes et al (9)), combinations of mixed and stagnant zones with short circuiting (Krishnaiah et al (10)) and models based on gross solid circulation loops (Bernutti et al (11)). Reviews of the modelling efforts can be found in Verloop et al (12) and in (10). We shall return to the predictive performance of a few of these approaches in the discussion.

Common to all these modelling efforts is that they attempt to model the complex behaviour of continuous fluidised beds using the traditional methods of residence time theory. However they largely ignore the physical mechanisms governing the mixing process in the bed. One model which is related to the properties of fluidisation bubbles is that of Haines and King (13). They used the plug flow with axial dispersion model, but augmented the dispersion coefficient of the particles (assumed to be caused by random collisional movement of the individual particles in the bed) with an extra term attributed to their displacement when disturbed by a rising bubble.

A thorough investigation of the the mixing processes causing the spread in particle residence times in a continuous fluidised bed is called for.

PARTICLE MIXING AND TRANSPORT PROCESSES AND THEIR QUANTIFICATION

A set of mixing processes are proposed below based partly on a study of other branches of the fluidisation literature. Essentially this can be seen as an extension of ideas first put forward by Rowe and Partridge (14), and their application to the problem of the RTD of the solids in continuous beds. All the physical parameters necessary to quantify these processes are also estimated below from data and relations available in the published literature.

The following particle mixing and transport processes will be considered:

1) Transport of material in the wakes of rising fluidisation bubbles, accompanied by a volumetrically equal transport of material downwards in the bulk of the bed.

2) Dispersive mixing of the bed material by the stirring action of fluidisation bubbles.

3) Transport due to the continuous removal of product from the bottom of the bed and the addition of charge material at the top.

Process 1) was proposed to account for the mixing of the solids in fluidised beds in (14), these ideas will be seen in the Conclusions essentially to be borne out. The same transport and mixing processes were formulated in more detail by Gibilaro and Rowe (15) for the purpose of modelling the mixing/ segregation behaviour of batch fluidised beds consisting of binary mixtures of particles of different properties (size and/or density). This work has been continued by other workers, among others Chiba et al (16), Naimer et al (17), Ho et al (18) and Garcia-Ochoa et al (19).

In selecting the above processes, we are ignoring others which might also conceivably contribute to the spread in particle residence times, notably: a) a velocity profile in the flowing solids, b) gross solid circulation in the bed in connection with the flow of fluidisation bubbles and c) the existence of 'dead space' or stagnant regions within the bed. Indirect information as to the importance of these processes will of course be gained from knowledge of the degree to which the processes which are considered suffice in accounting for the mixing.

One effect which is not taken into account here is the possibility of exchange of material between the wake of a rising bubble and the surrounding wake, all the material once caught in the wake is assumed to be brought to the top of the bed. It is the present authors' experience from work with mixing and segregation in batch fluidised beds of equal density binary mixtures, that the effects of such exchange in beds of moderate height is likely to be very little.

Quantification of the processes

In the so-called bubbling fluidised beds, some of the fluidising gas penetrates the bed in fluidisation bubbles, and some flows interstitially in the 'bulk phase' (see Figure 1). As fluidisation bubbles rise through the bed, they carry with them a wake of material as sketched in the figure. The wake material is deposited on top of the bed by each bubble as it bursts at the bed surface. This causes mixing of the fluidised solids as particles a stochastically varying number of times are caught up in wakes and deposited on the surface of the bed, from where they must again penetrate the bulk wherein they are subjected to dispersion.

The 'wake fraction' is defined as the ratio: (wake volume)/(volume of wake and bubble), and increases with the size of the bubble itself. In order to calculate the wake fraction as a function of the size of the bubble, the 'wake angle' (see Figure 1) must be known. The following relation is used herefore, this relation is a best fit to the data of Hoffmann (20):

$$\Phi_w = 1.4 - 1.4*\exp(-55*D_B) \quad (1)$$

The fluidisation bubbles are formed at the gas distributor plate, and as they rise they coalesce and therefore grow in size. The size of the bubbles at formation and their rate of growth are estimated from the relation of Geldart (21):

$$D_B = \frac{1.3}{g^{0.2}} \left(\frac{U-U_{mf}}{1000} \right)^{0.4} \left(\frac{1}{1-f_w} \right)^{0.33} + 2.05\ h\ (U-U_{mf})^{0.94} \qquad (2)$$

The total flow of gas penetrating the bed in the form of bubbles is estimated from the so-called 'two phase theory'. This says that the gas flowing interstitially in the 'bulk phase' is just enough to keep the solids fluidised (Toomey and Johnstone (22)) i.e.:

$$\frac{Q_B}{A} = U - U_{mf} \qquad (3)$$

The total volumetric flow of bubble void is therefore axially constant, but as the bubbles grow with axial position due to coalescence, so does the wake fraction and therefore the total flow (upwards) of wake material.

Processes 1) and 3) can be described fully using these relations. It remains to estimate the degree of dispersion of the particles in the bulk phase due to the stirring action of the rising bubbles. For this the detailed measurements of Tanimoto et al (23) may be used. These workers placed particles in a horizontal layer in a quiescently (just fluidised, no bubbles) fluidised bed, and measured their position after the passage of one injected bubble. Their results fell (after normalisation w.r.t. the bubble radius) scattered around the curve sketched in Figure 2. This curve is in the present simulation approximated with the indicated straight lines. The nett flow of particles due to dispersion alone must of course be zero.

The development of a computer code wherein these processes are modelled numerically provides detailed information about the effect of each of them on the particle RTD. Details of the code is available from the authors on request. Below, the simulated RTD's are studied and compared qualitatively and in some cases quantitatively with experimental data available in the research literature.

DISCUSSION AND COMPARISON WITH LITERATURE

Figure 3 shows simulated F-curves ($F(\theta)$ is under steady state conditions simply the probability of a particle having a residence time less than $\theta \equiv T/\tau$ (Danckwerts (24)) under three different operating conditions. Comparing these curves with those given in (7) or Krishnaiah et al (10), it is evident, that the shape of the simulated curves qualitatively agree with those obtained experimentally in fluidised beds. The shape of the curves is also similar to those obtained in systems with stagnant zones. We shall briefly return to the features of the curves giving a physical interpretation of them in light of our knowledge of the particle mixing processes generating the curves.

It can also be seen in the figure, that increasing the gas flowrate keeping all else constant has the effect of improving the mixing of the solids. This is due to an increase in both of the mixing processes of transport in bubble wakes and dispersion in the bulk. Decreasing the solids flowrate also has the effect of enhancing the mixing, the particles are subjected to the dispersion in the bulk phase for a longer time interval, and they are more likely to be caught up in a bubble wake before they leave the bed.

Comparison with the results of Tailby and Cocquerel (25)

This work was performed under conditions where the particles in the beds were very close to being well mixed. They investigated systems with both co- and countercurrent gas and solids flow. They described their results in terms of 'hold-back' (HB) and 'segregation' (SG) for easy comparison of their results with ideal mixing and plug flow respectively. HB is simply the integral of the F function from $\theta=0$ to $\theta=1$.

They also worked with very high beds of a very modest diameter, and relation (3) predicts bubble diameters in the upper at least half of even the lowest bed comparable to the bed vessel diameter for all of their runs. The growth of bubbles would therefore be greatly suppressed in the upper part of their beds and it is likely that the flow there would be of a slugging character. The simulations might therefore be expected to exhibit a higher degree of mixing than the experimental data. Two of their runs were simulated, the co-current ones at the lowest bed height: 71 cm and the lowest value of $U-U_{mf}$: 0.038 m/s and the solid velocities of $7.62*10^{-4}$ and $3.82*10^{-4}$ m/s. The values of HB of the simulations were 0.520 and 0.441 respectively and in the experimental data they were 0.423 and 0.354 respectively. The discrepancy in the absolute values indeed indicates that the mixing in the simulations is better than that in the experimental data. The trends, however, agree, and the envisaged mixing processes clearly suffice in accounting for the mixing.

In their conclusions Tailby and Cocquerel state that increasing the solid throughput causes a tendency towards plug flow, while increasing the fluidisation velocity causes better mixing. We have seen in Figure 3 and in the HB data above, that both of these trends are present also in the simulated RTD's. They also conclude, however, that increasing the bed aspect ratio (defined as H_{bed}/D_{bed}) causes a tendency towards plug flow. This is not so in the simulated profiles. Three possible causes for this discrepancy are: a) wall effects in the upper part of the beds of Tailby and Cocquerel as we mentioned above, b) the fact, that material exchange wake/bulk is unaccounted for in the present simulations or c) deviations in the beds from the 'two phase theory' (see below).

Comparison with the data of Morris et al (7)

Many of the data of Morris et al were generated in beds aerated at rates close to, at, or even under the minimum fluidisation velocity of the powder. To use the present model, we must compare with data obtained at conditions, under which the bed is well fluidised. Again we use data generated in beds which are low. Figures 4 and 5 show comparisons of simulated and experimental F-curves of two of the runs of Morris et al. It is evident, that there is quite good agreement between the simulated and the experimental data in both cases.

Morris et al found that they obtained F-curves in moving aerated *packed* beds qualitatively similar to the curves obtained in fluidised beds. The reason for the observed solid RTD in the packed beds is the existence of a solids velocity profile in the flowing solids. Morris et al held that the same effect must be at work also in fluidised beds. It would seem probable, however, that in bubbling fluidised beds the fluidisation bubbles provide so much radial exchange of momentum, that a velocity profile does not occur. It seems highly unlikely in fact, that a solids velocity profile of the order of a fraction of a millimeter per second should not be completely swamped by effects of the bubble flow. Moreover, a velocity profile would not explain the variation of the RTD either with the gas or the solid flow rate.

Morris et al found a profound influence of the height of the bed upon the RTD of the solids in the bed. A higher bed under otherwise unchanged conditions caused a tendency towards plug flow. The trend shown in (7) is qualitatively the same, but much stronger than that in the work of Tailby and Cocquerel discussed above. Morris et al worked at very low values of $U-U_{mf}$, and it would seem very likely that under these conditions, and in beds as high as those used by them, deviations from the 'two phase theory' would be significant with bubble flow less than that predicted by Equation (3) in the lower part of the beds (Werther (26), Yacono (27)).

Comparison with the work of Verloop et al (12)

Verloop et al represented the residence time distribution data obtained from fluidised beds in the form of 'intensity curves', I_t vs. θ (not to be confused with the 'internal age distribution', which is also often denoted by an I). I_t is under conditions of steady flow defined in terms of $C(\theta) \equiv dF(\theta)/d\theta$ and $F(\theta)$ as:

$$I_t(\theta) = \frac{C(\theta)}{1 - F(\theta)} \qquad (4)$$

It turns out that the intensity curves obtained from fluidised beds are of a very characteristic shape, exhibiting a local maximum at some value of θ less than 1. This representation is therefore very useful for testing the merits of models predicting the RTD of solids in fluidised beds.

Verloop et al showed that only one of the many models that they treat in their extensive review can account for this characteristic shape. This is a model wherein the total particle stream is split in several fractions, each fraction going through its own series of stirred tanks, the different series being of varying lengths. From what went before in the description of the mixing processes it is not difficult to see why this approach should work.

In one simulation a large number of particles were used, sufficient to generate the relatively smooth curve shown in Figure 6. There is no mistaking the local maximum at a value of θ of approximately 0.35. The inference is that such a maximum, when seen in RTD data stemming from fluidised beds, not necessarily should be attributed to the presence of stagnant regions in the bed, as is mostly done, but can be explained by the fact that some of the particles repeatedly are caught up in bubble wakes and therefore remain circulating in the bed for very long.

CONCLUSIONS

The particle mixing and transport processes 1) to 3) can in themselves account for the mixing observed in bubbling fluidised beds of a moderate diameter, no other mixing mechanisms need be taken into consideration.

Not only will these processes account for the degree of mixing, but RTD data simulated on basis of them exhibit in details the same characteristics as experimental data. The trends in the experimental residence time distributions with varying operating conditions are correctly present in the simulated data except the tendency towards plug flow in higher beds.

Simulated residence time distributions have shown quantitative agreement with experimental data even though no adjustable parameters were used for the simulation.

On basis of the knowledge of the transport mechanisms responsible for the simulated RTD's, the following conclusions can, among others, be drawn: a) the characteristic shape of the intensity curves can be attributed to the repeated transport of particles in bubble wakes. This shape does not necessarily imply that there are stagnant regions in the bed. b) The initial sharp rise of the F-curve often observed in beds of a moderate value of $U-U_{mf}$ is due to the arrival in the bottom of the bed of the (smeared out through dispersion) front of those particles, which penetrated the bed without being caught up in a bubble wake. c) The transport of particles in bubble wakes is the main factor determining the shape of the obtained residence time distribution curves, bearing out the ideas of Rowe and Partridge (14).

LITERATURE

1. Gilliland, E. R., and Mason, E. A., 1952, Ind. Eng. Chem. 44, 218.

2. Klose, E., and Herschel, W., 1985, Chem. Techn., Heft 4.

3. Pudel, F., Strümke, M., and Sündermann, U., 1986, Wiss. Z. der Techn. Hochsch. Magdeburg 30, Heft 6, 51.

4. Massimilla, L., and Bracale, S., 1957, Ricerca Scientifica 27, 1509.

5. Weber, B., and Rose, K., 1970, Chem. Techn., Heft 10, 594.

6. Tripathy, G., Pandey, G.N., and Singh, P.C., 1971, Indian Journal of Technology 9, 281.

7. Morris, D.R., Gubbins, K.E., and Watkins, S.B., 1964, Trans. Inst. Chem. Engrs. 42, T324.

8. Whittmann, K., Wippern, D., Schlingmann, H., Helmrich, H., and Schügerl, K., 1983, Chem. Eng. Sci. 38, 1391.

9. Heertjes, P.M., de Nie, L.H., and Verloop, J., 1967, "Proceedings of the International Symposium on Fluidisation", Neth. Univ. Press, Amsterdam, 476.

10. Krishnaiah, K., Pydisetty, Y., and Varma, Y.B.G., 1982, Chem. Eng. Sci.37, 1371.

11. Berrutti, F., Liden, A.G., and Scott, D.S., 1988, Chem. Eng. Sci. 43, 739.

12. Verloop, J., de Nie, L.H., and Heertjes, P.M., 1968, Powder Technol. 2, 32.

13. Haines, A.K., and King, R.P., 1972, AIChE Journal 18, 539.

14. Rowe, P.N., and Partridge, B.A., 1962, in: "The Interaction Between Fluids and Particles", Rottenburg, P.A. (Hon. Ed.), London: The Institution of Chemical Engineers.

15. Gibilaro, L.G., and Rowe, P.N., 1974, Chem. Eng. Sci. 29, 1403.

16. Chiba, S., Chiba, T., Nienow, A.W., and Kobayashi, H., 1979, Powder Technol. 22, 255.

17. Naimer, N., Chiba, T., and Nienow, A.W., 1982, Chem. Eng. Sci. 37, 1047.

18. Ho, Tho-Ching, Kirkpatric, M.O., and Hooper, J.R., 1987, AIChE Symp. Ser. 83, 42.

19. Garcia-Ochoa, F., Romero, A., Villar, J.C., and Bello, A., 1989, Powder Technol. 58, 169.

20. Hoffmann, A.C., 1983, Ph.D. Thesis, Dept. of Chemical and Biochemical Engineering, University College London.

21. Geldart, D., 1972, Powder Technol. 6, 201.

22. Toomey, R.D., and Johnstone, H.F., 1952, Chem. Eng. Progress 48, 220.

23. Tanimoto, H., Chiba, S., Chiba, T., and Kobayashi, H., 1981, J. Chem. Eng. of Japan 14 273.

24. Danckwerts, P. V., 1953, Chem. Eng. Sci. 2, 1.

25. Tailby, S. R., and Cocquerel, M. A. T., 1961, Trans. Instn Chem. Engrs. 39, 11.

26. Werther, J., 1975, in: "Fluidisation Technology", Vol 1, Keairns, D.L. (ed.), Hemisphere Publ. Co., p. 215.

27. Yacono, C. X. R., Ph.D. Thesis, 1975, Dept. of Chemical and Biochemical Engineering, University College London.

SYMBOLS

A = cross-sectional area of fluidised bed (m^2)

$C(\theta)$ = the probability of a particle having non dimensional residence time between θ and $(\theta+d\theta)$ (-)

D = diameter (m)

$F(\theta)$ = the probability of a particle having non dimensional residence time less than θ (-)

f = fraction (-)

g = acceleration due to gravity (m/s^2)

H, h = height of bed and height in bed respectively (m)

$I_t(\theta)$ = intensity function, definition in Equation (4) (-)

Q = volumetric Flowrate (m^3)

T = time (s)

U = superficial gas velocity (m/s)

v = solid velocity (m/s)

Greek:

ϕ = angle (-)

τ = mean residence time (s)

θ = $\equiv T/\tau$ (-)

Subscripts:

B = bubble

mf = at minimum fluidisation conditions

w = wake

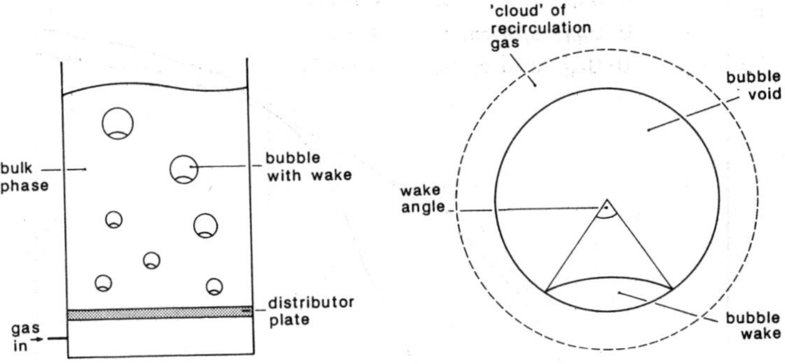

Figure 1 Schematic of bubbling fluidised bed and fluidisation bubble.

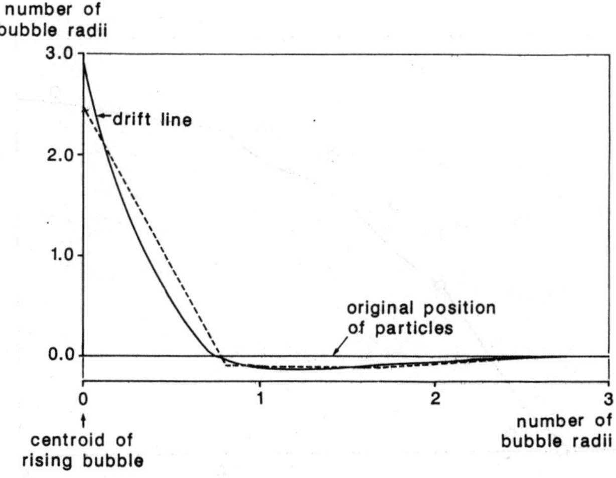

Figure 2 Drift of solids after the passage of a fluidisation bubble (after Tanimoto et al (23)).

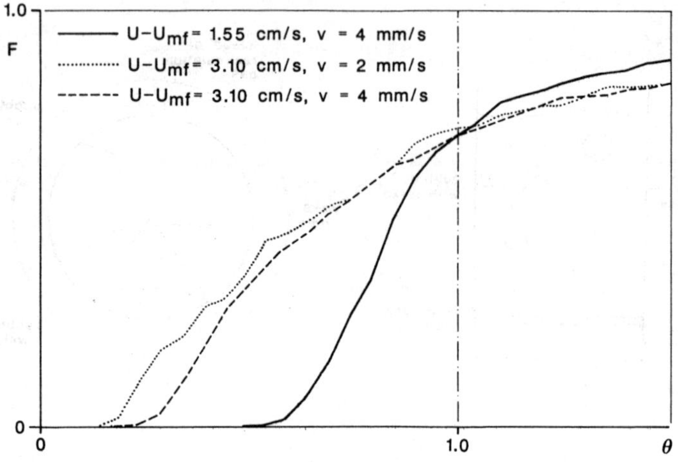

Figure 3 Simulated F-curves. Bed height: 23 cm.

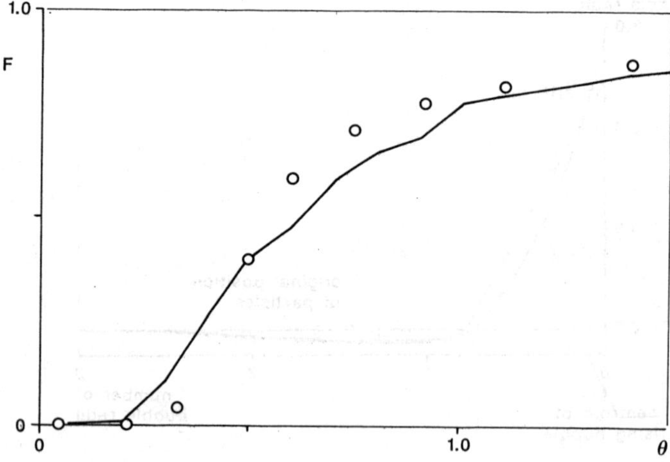

Figure 4 Comparison of a simulation (line) with experimental data in (7). Bed height=23 cm, $v=4.29*10^{-4}$m/s, $U-U_{mf}$=0.0155m/s

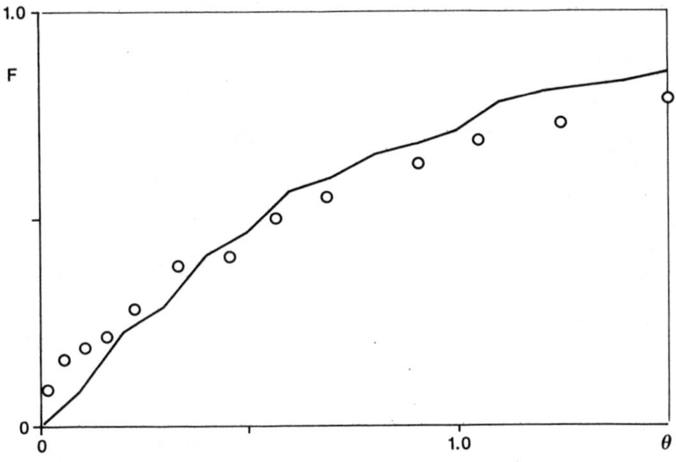

Figure 5 Comparison of a simulation (line) with experimental data in (7). Bed height=23 cm, $v=4.29*10^{-4}$m/s, $U-U_{mf}=0.0235$m/s.

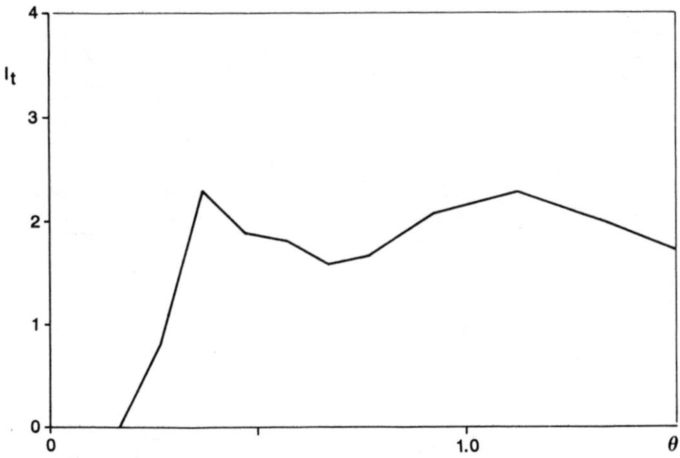

Figure 6 Simulated intensity curve. Bed height=10 cm, $U-U_{mf} = 3.50*10^{-4}$m/s, $v=4*10^{-3}$m/s.

LASER ANEMOMETRY STUDY OF SHEAR THINNING FLUIDS AGITATED BY A RUSHTON TURBINE

E.Koutsakos,* A.W.Nienow* and K.N.Dyster.

Mean velocity profiles and turbulence parameters were measured in a stirred vessel equipped with a Rushton turbine using a laser Doppler velocimeter. The working fluids were water and a range of non-Newtonian shear thinning fluids with rheological properties typical of those found in fermentation processes. The power curve of the system showed that a distinct change of the hydrodynamic regime occurred at Re≈60 which can be related to the change in fluid flow characteristics. A simple relationship has been used to correlate the mean velocity profiles along the centre-line of the impeller. The ratio of the centre-line mean radial velocity to impeller tip velocity was found to be directly proportional to the flow number, Fl. It was generally shown that for Re≤60 then Fl∝Re and the mean velocity profiles were strongly dependent on impeller speed, where as for Re ≥ 60, then $Fl \propto Re^{0.2}$ and the velocity profiles became relatively independent of impeller speed.

INTRODUCTION

Fermentation industries have traditionally used mechanically agitated vessels to ensure intimate contact and good mixing of the stirred phases. The design, development and scale-up procedures of such industrial size mixing vessels depend on representative and reliable information obtained from small-scale experiments. Recently developed techniques such as the laser doppler anemometry (LDA) can be used to provide detailed profiles of the complex fluid velocity distribution in such vessels.

Though LDA is being increasingly applied to stirred vessels, an up to date literature survey reveals that no LDA measurements have been reported using viscous or non-Newtonian fluids. All the previous studies are concerned with measurements of velocity distribution in water, i.e. simulating agitation of low viscosity fluids. However during a typical fermentation process the rheological character of the biomass can change from Newtonian to non-Newtonian and its viscosity increase by 100-fold or more.

* *Centre for Biochemical Engineering, School of Chemical Engineering University of Birmingham, P.O.Box 363, Birmingham* B15 2TT, UK.

Most commonly, fermentation broths acquire a shear thinning or pseudoplastic character, i.e. a relatively low apparent viscosity in the high shear rate region near the rotating impeller. However, away from the impeller zone, high apparent fluid viscosities are observed [1,2], due to the existence of low shear rate regions. Thus high gradients of velocity are formed with increasing distance away from the impeller which lead to poor mixing or even regions of stagnant fluid within the agitated vessel [3].

The change of rheological properties of the fermentation broth with the passage of time can cause significant changes in the power drawn by the agitator [4,5], the liquid flow patterns [1,3], the gas dispersion [2] and consequently the mass transfer rate [6,7]. It is therefore important to investigate the effects of the change of rheological fluid properties on the hydrodynamics of the agitated system. For this purpose a research programme has been set up at the *Centre for Biochemical Engineering* at Birmingham University, to investigate the effects of rheological properties of fluids on the mean and fluctuating fluid velocity distribution measured using LDA. Impeller power consumption was also measured. Some of the early results of this research project are presented here.

EXPERIMENTAL

Laser Doppler Velocimeter:

Instantaneous measurements of radial and tangential velocities were measured by a TSI laser doppler velocimeter (LDV), Fig.1a, which was equipped with a frequency shifter so as to eliminate directional ambiguity. The system consisted of a laser gun capable of producing a 32mW He-Ne laser beam of wavelength, 632.8nm. A rotatable beam splitter attached at the exit point of the laser gun allowed the measurement of two velocity components. The crossing point of the focused laser beams (i.e the sampling volume) was accurately positioned inside the stirred vessel at a pre-specified location (see Fig.2). The orientation of the laser beams for measuring tangential and radial velocities is shown in Fig. 1b.

The frequency of the scattered light produced by the passage of minute particles (present in the stirred liquid) through the measuring volume was converted by the photodetector into an electrical signal: the **doppler signal**. An oscilloscope was used to monitor the quality (signal to noise ratio) of the doppler signal. The photodetector was equipped with a photomultiplier used for signal amplification; this was proved particularly useful in the backward scatter mode where the signal is about 100 to 1000 times weaker than the forward scatter signal. The frequency of the doppler signal was calculated by the signal processor (TSI model 1980B) over eight doppler periods; this corresponded to five cycles within a burst. A 3% comparison criterion was set on the counter-signal processor to check the repeatability of the time for each cycle within a burst. A high rate of sample capture was used (2048 to 4096) with a sample interval of 100 to 500μs. The frequency of the doppler signal was converted by the data acquisition system into a mean velocity, $\bar{V}$. The velocity measurements were obtained as ensemble averages in each location in the vessel, irrespective of the impeller blade position.

A statistical analysis of the doppler signal by the data acquisition system determined the fluctuating velocity component, V', which was the sum of the periodic component caused by the passage of the impeller blades and the random fluctuating component at the point of measurement.

Solutions used:

Distilled water and a range of concentrations of CMC (Carboxy-methyl-cellulose) and Natrosol (Hydroxy-ethyl cellulose) were used; these fluids exhibit pseudoplastic, shear thinning behavior (Table: 1) and were found to obey a Power Law ($\tau = K \dot{\gamma}^n$) relationship.

Table: 1 Rheological properties of fluids used.

FLUID	K	n
Distilled Water	1×10^{-3}	1.00
0.14% CMC	0.237	0.58
0.28% CMC	0.451	0.57
1 % CMC	9.120	0.42
1 % Natrosol	1.214	0.56
2% Natrosol	19.84	0.37

Agitated vessels:

The experiments were carried out in a standard, flat bottom glass vessel shown in Fig. 2. The cylindrical glass was inserted inside a square glass vessel filled with water. More details of the experimental set-up are given in a previous publication by Koutsakos and Nienow [8]. The temperature of the stirred tank was maintained constant at 21 ± 1°C.

Agitation was provided by a Rushton turbine located at mid-height of the quiescent liquid height. The geometrical configuration of the turbine and its dimensions are shown in Fig. 3 and Table 2, respectively.

Power measurement:
An instrument employing a patent gear mechanism (designed by ChemLab Scientific Products Ltd) to accurately measure the torque in laboratory scale apparatus was used to determine the impeller power consumption. The instrument uses several transducers to measure the movement of a gear caused by torque applied on the shaft of a rotating impeller.

RESULTS AND DISCUSSION

Power curve of the system:
The power curve (Po vs. Re) for the range of fluids (Table 1) and the agitated vessel used, Fig. 2, is shown in Fig. 4. Also superimposed on Fig. 4, are the power curves reported by Bates et al [11] for Newtonian fluids and that of Metzner et al [4] obtained for a range of non-Newtonian fluids. As described in a previous publication by Koutsakos and Nienow [8] the Metzner and Otto [10] method was used for shear thinning fluids to relate the average shear rate to the impeller speed via the proportionality constant, k_s=11.5, which was proposed for Rushton turbines [10].

The apparent viscosity for the non-Newtonian fluids was then calculated by Eqn. 1a and the Reynold's number by Eqn. 1b.

$$\mu_a = K \ (k_s N)^{n-1} \quad \{1a\}$$

$$Re = \rho N^{2-n} D^2 / K \ (k_s)^{n-1} \quad \{1b\}$$

The experimental Power curve in Fig. 4 show that a slope of -1 is obtained in the laminar region which corresponds to Re numbers of up to ~20. A comparison of the power curves in the laminar region clearly indicates that, using the Metzner and Otto approximation, although a slope of -1 is obtained, the actual values of Po for these non-Newtonian fluids are on average 67% higher than those obtained for Newtonian fluids [11]. This agrees favorably with the data of Nienow et al [5] who reported a maximum of 60% increase in Po for Re numbers up to ~10 for fluids which are moderately viscoelastic. It is evident that measurable deviations from the -1 slope occur at Re>~60, which marks the beginning of the transition region. In this region the experimental non-Newtonian data for, Po, are at first, slightly lower than the Newtonian curve. The sharp minimum of the power curve described by Metzner [4] was less pronounced in this investigation. Previous workers [5] have also reported a smoother change between the laminar and transition region. Metzner and Otto [9] postulate that this reduction in Po, is due to the depressive effect of pseudo-plasticity upon turbulence.

It is observed that by increasing the Reynold's number beyond Re>~60, the power curve becomes relatively flat; thus large increases in Reynold's number causes only small changes in the power number. Therefore this value of Re, within the transition region should also represent an important transition point for the velocity distribution. For the range of 100<Re<1000, the Newtonian power curve tends towards the lower experimental values of Po for non-Newtonian fluids. However, the experimental data of Po are in good agreement with the non-Newtonian power curve reported by Metzner and Otto [10]. The slight curvature of the power curve observed for 100<Re<1000, was considered by Nienow et al [5] to be related to the change of angle of discharge from the impeller blade.

The Power number at turbulent conditions (Re> 10^4), for the Rushton turbine under investigation was found to be, Po=5.6. This value can be argued to be sightly higher than previously reported values [17]. This may be due to the slight differences in the impeller/vessel configuration, vessel size or possibly due to the different technique used to measure the torque on the rotating shaft of the impeller. However, the trend of the power curve is evidently in excellent agreement with previously published data.

Outlet velocity profiles:

For a given radial distance away from the impeller tip the mean radial velocities were measured at several axial locations, as shown in Fig. 2. The results are expressed in the form of the dimensionless ratio, $\overline{V_r}/\pi ND$ vs. z(axial sampling location) for a given radial position. These so called *outlet flow profiles* were measured at 2,4,9,18 and 25mm away from the impeller tip and at different impeller speeds. The results obtained for water are shown in Fig.5. Some data were omitted in order to make the diagram clearer to the reader. It is evident from Fig. 5 that the outlet velocity profiles are independent of the impeller speed; these speeds corresponds to Re numbers in the turbulent, flat part of the power curve. Previous studies have shown [11,12] this to be true in the vicinity of the impeller; Fig. 5 extents this to locations much further from the impeller tip.

The results obtained for 0.28% CMC, Fig. 6, indicate that the outlet velocity profiles in the immediate vicinity of the impeller remain almost independent of impeller speed, from 300 to 800rpm; these speeds correspond to Re values above the critical value of Re≃60, which distinguishes between the relatively flat and sloped part of the power curve. Fig.6 also shows that the mean velocities at the centre-line of the Rushton turbine are only weakly affected by speed at a given radial position. Findings reported in a previous publication by Koutsakos and Nienow [8] with different non-Newtonian fluids (Table 1) also corroborate that for Re>60 the outlet velocity profiles change very little with impeller speed for a given fluid in this range.

A study on non-Newtonian fluid as early as 1960 by Metzner and Taylor [16] using visual observation and "streak" photography reported a change in flow fields as one moves from laminar into the early transition range as defined by the power curve. These observations agree with the qualitative measurements reported above which distinguish between the mean radial velocity profiles of the lower/curved region (Re<~60) of the transition regime (near the laminar region) and its flat section (Re> ~60) towards the turbulent regime.

The pumping capacity of the impeller, Q, was calculated using Eqn. 2 shown below which utilises the experimental outlet velocity profiles measured with the LDV.

The dimensionless discharge or Flow number, Fl (which is a measure of the pumping capacity of the impeller) was calculated using Eqn. 3 which incorporates the experimentally determined value of Q.

$$Q=\pi\ (D+2s) \int_{-w/2}^{+w/2} \overline{V}_r\, dz \qquad (2)$$

$$Fl = Q / N D^3 \qquad (3)$$

The calculated Flow number for the results in Fig. 5 for water remained constant with increase in impeller speed. The value of Fl=0.78 measured at 2mm away from the impeller blade is in agreement with the recommended mean value of Fl=0.75, obtained for turbulent conditions by Revill [18] after a critical review of the literature. The calculated values of Fl, for radial positions away from the impeller tip (r-R>2mm) remained fairly constant with a value of Fl=0.79±0.02 [8]. Cooper and Wolf [19] reported the value of Fl=0.86 in water and 0.68 in air for Rushton turbines; this highlights the effect of fluid viscosity on the velocity distribution around an impeller. Günkel and Weber [12] and Desouza and Pike [20] proposed the value of 1.0 and 0.95 respectively. However, unlike other studies, the latter investigations [12,20] determined the value of, Fl, by integrating Eqn. 2 for the limits greater than ± w/2.

By increasing the apparent fluid viscosity this has the effect of reducing the mean velocity at the centre-line of the Rushton turbine and this causes a flatter outlet velocity profile [8]. The effect of the apparent fluid viscosity, μ_a (Eqn. 1a), on Flow number is highlighted in Fig. 7. The results show that there is a distinct change between the relation of Fl and μ_a, at Re≃60, as follows,

for Re < 60 $$Fl \propto (\mu_a)^{-1} \tag{4a}$$

$$\text{or } Fl \propto Re \tag{4b}$$

for Re > 60 $$Fl \propto (\mu_a)^{-0.2} \tag{5a}$$

$$\text{or } Fl \propto Re^{0.2} \tag{5b}$$

The value of Fl (i.e. the pumping capacity of stirrer) increases relatively steeply with reductions in μ_a until Re≃60. Further reduction of μ_a (e.g. by increasing impeller speed) results in relatively smaller increases in the value of Fl. Thus, the change at Re≃60, (which corresponds to $\mu_a \simeq 0.2 Nsm^{-2}$ for this scale of operation) between the flat and sloped part of the power curve has a definite effect on the velocity distribution around the rotating impeller. At low values of μ_a approaching the viscosity of water, the Flow number approaches the expected value of 0.78 which designates a fully turbulent condition. Further reductions in μ_a, or increases in Re (i.e. with increase in scale) are not expected to change the value of Fl (see dotted line in Fig. 7).

The (total) r.m.s fluctuating radial component, V_r', was also measured and presented in the form of an outlet velocity profile. Fig. 8 shows the data obtained at 300rpm for different fluids; the effect of increasing the fluid's viscosity is to reduce the fluctuating radial velocity component particularly at the centre-line of the impeller. This coincides with the flattening of the mean velocity profiles [8] and consequently the reduction of the Flow number with increases in μ_a. Evidence is therefore provided for the suppression of both the mean and fluctuating radial velocity component with increase in pseudoplaticity and viscosity of the fluids.

A modified Flow number, Fl', is introduced in this study which is obtained by substituting the mean radial velocity in Eqn. 2 by the fluctuating radial velocity component thus,

$$Fl' = \frac{\pi(D+2s) \int_{-w/2}^{+w/2} V_r' \, dz}{N \ D^3} \quad \{6\}$$

The dimensionless value of Fl', is a measure of the velocity fluctuations in the discharge stream of the impeller. Table 3 below shows the values of Fl' (from Fig. 8) for different fluids at N=300 rpm. This is an alternative way

Table:3 Values of Fl' for different fluids at N=300rpm. (data from Fig. 8)

Fluid	Water	0.14% CMC	0.28% CMC
Fl'	0.455	0.127	0.072

of showing the reduction of turbulence in the impeller discharge stream due to increase in the fluid's viscosity and pseudoplasticity. Similar data have been obtained for viscous Newtonian fluids in order to determine the effect of fluid viscosity alone on turbulence. These will be reported at a later date.

It is interesting to note from Fig.9 that, the V_r' velocity profiles in the case of 0.28% CMC increase with impeller speed even when Re>60. The largest velocity fluctuation occurs at the centre-line of the impeller. Similar observations where reported with 0.14% CMC solution [8]. Thus the conclusions drawn from measurements of the mean velocity outlet profiles are not necessarily applicable for the fluctuating velocity. However, experiments carried out in fully turbulent conditions with water agree with previously reported data [21] that the outlet velocity profiles of V_r' are virtually independent of impeller speed.

Centre-line velocity distribution:

The radial and tangential velocities at the center-line of the Rushton turbine (z=0 in Fig.2) were measured. The data were graphically presented in the form of the dimensionless plot of, $\overline{V}/Vtip$ vs. r/R, referred to as the *centre-line velocity profile*.

In Fig. 10 the results obtained for water show that the centre-line profile for the mean radial velocity is not affected by the impeller speed. The correlation proposed by Van der Molen [9] shown in Eqn. 7 below, where ξ=1.0, was found to describe the data reasonably well. The parameter ξ was introduced in this study as a measure of deviation form the fully turbulent well mixed condition described by Van der Molen's correlation.

$$\overline{V}_r \;/\; V_{tip} = 0.85\ \xi\ (r/R)^{-7/6} \qquad \{7\}$$

Slight disagreement were noted in Fig. 10 near the impeller tip; these could be due to the differences in the impeller geometry (Fig. 3) which have been reported to cause measurable effects in the fluid flow patterns [15] and power drawn by the impeller [17]. Van der Molen Eqn. 7 for mean velocity distribution is also independent of impeller diameter and vessel size. This findings have important design implications with regards to scale-up rules.

A comparison of the centre-line velocity profiles at N=300rpm for different fluids is shown in Fig. 11. It is clear from the results that for a given radial position the mean radial velocity decreases with increasing fluid viscosity. However, the centre-line profiles seem to follow an exponential decrease. To test this, the proposed Van der Molen's Eqn. 7 was employed to correlate the data by adjusting the value of the parameter, ξ, to account for changes in fluid viscosity. The calculated values of ξ for each fluid (from Fig. 11) are plotted against Reynold's number in Fig. 12. This simple correlation method appears to describe the data with reasonable accuracy for data where Re>60. However, for cases where Re<60, e.g. 1 % CMC in Fig. 12, although the above suggested correlation works well for locations in the immediate vicinity (r/R<≃1.1) of the impeller, the formation of a *cavern** (i.e. fluid velocity reduced to zero at r/R≃1.2) causes large

---

* *Cavern:* a well mixed region; outside its well defined boundaries the fluid is stagnant [1].

deviations to occur at radial positions further than the impeller blade ($r/R > 1.1$) [8]. In a previous study by Koutsakos and Nienow [8] the cavern boundary, for Re<≃60, was characterized by a fluctuating mean velocity region, around the value of zero (see dotted line in Fig. 12) and this was attributed to the periodic passage of the impeller blades.

The distribution of the centre-line velocity for 1% Natrosol shown in Fig. 12 (Re≃60) begins to deviate from the proposed correlation at locations much further than the impeller tip, $r/R > 1.8$. The velocity tends towards zero and this causes the formation of a *pseudo-cavern*: this is a cavern with its boundaries not so well defined between stagnant and well mixed regions. This phenomenon might be due to the low shear regions formed away from the impeller zone which subsequently would form an increase in the apparent viscosity of the fluid and hence lower the effective value of Re below the critical value of Re≃60. Thus, the assumption usually made, that the viscosity of a non-Newtonian fluid throughout the agitated vessel can be represented by an *apparent fluid viscosity*, may not be strictly correct. This finding supports previous observations by Metzner and Taylor [16] that, it is possible to have both laminar and turbulent regions in a mixing tank at the same time.

A previous investigation with non-Newtonian fluids [8] has shown that for cases where Re is well above the value of sixty, i.e. Re≥170, then Eqn. 7 can be successfully used to correlate the centre-line velocity profiles without deviations. The results of a series of experiments are summarised in Fig. 13 in the from of the correlating parameter ξ vs. Re (ξ is a measure of the effect of fluid viscosity on the centre-line velocity). The results are described by the following equations,

for $Re < 60$, $\xi \propto Re$, (8a)

for $60 > Re > 10^4$, $\xi \propto Re^{0.2}$ (8b)

for $Re > \sim 10^4$ $\xi = 1.0$ (8c)

Eqns. 8 show that for Re>60 the value of ξ is less affected by the Reynold's number (impeller speed or μ_a) than for cases where Re<60. Thus the effect of fluid viscosity on the centre-line velocity profiles is much more prominent for Re<60 (i.e. at the lower end of transition region) than for Re>60 (i.e.

the flat section of the power curve). Cavern formation was observed for cases where Re<60, particularly for fluids exhibiting high pseudoplasticity. Pseudo-caverns appear in the region of Re≃60, but, they tend to disappear with increase in Re. The value of ξ in turbulent conditions (i.e. Re> 10^4) measured in water, reaches the expected constant value of unity as predicted by Eqn. 7.

A comparison of Eqns. 4 and 5 with Eqns. 8, reveals that for the range of experimental parameters studied the Flow number, Fl, is directly proportional to the correlating parameter, ξ. This suggests that the distribution of the fluid velocity around the impeller is directly related to the centre-line velocity. It is also postulated that the change in slope of the power curve from $Po \propto Re^{-1}$ to Po≃constant at Re≃60 is due to the fact that the orientation of the fluid velocity distribution around the impeller becomes relatively unaffected by increases in Reynold's number above this Re value. This is supported by a previous publication by Koutsakos and Nienow [8] which reported that the tangential velocity distribution at the centre-line of the Rushton turbine was also only weakly dependent on Re for Re>60.

Further work is carried out at Birmingham University to establish the effect of fluid rheology on the hydrodynamics of stirred vessels for a wider range of system parameters, e.g. viscous Newtonian fluids.

CONCLUSIONS

The outlet profiles of the mean radial velocity were shown to be only weakly dependent on the impeller speed for conditions where Re>60. This flow regime corresponds to the flat section of the power curve where further increases in the Re number cause relatively little effect on the Power number. However, the outlet velocity profiles of the fluctuating radial velocity increase with impeller speed until Re approaches the fully turbulent condition. The effect of the increase in fluid viscosity and pseudoplasticity is to reduce turbulence.

The Fl number of the Rushton turbine at fully turbulent conditions was found to be 0.78. It was generally shown that for Re<60 then $Fl \propto Re$ and for Re>60 then $Fl \propto Re^{0.2}$. Cavern formation was observed when Re<60. Pseudo-caverns with less well defined boundaries were described at Re≃60 but these disappeared with further increases in Re number.

The centre-line profiles of the mean radial velocity for non-Newtonian fluids were reasonably correlated by introducing the parameter, ξ, in Van der Molen's Eqn.7. The general result obtained in this study that, $Fl \propto \xi$ reveals that the overall fluid motion in the impeller zone is proportional to the centre-line mean velocity of the Rushton turbine.

NOMENCLATURE

D	impeller diameter, [m]
Fl	Flow number, $Q/N D^3$
Fl'	fluctuating Flow number, Eqn. 6
K	power law, fluid consistency in Eqn. 1b, [Pa s^n]
k_s	proportionality constant in Eqn. 1a, [-]
N	impeller rotational speed, [rps]
n	flow behavior index in Eqn. 1a, [-]
P	Power drawn by impeller, [W]
Po	Power number, $P/\rho N^3 D^5$ [-]
R	impeller radius, [m]
r	radius of sampling location, [m]
Re	Reynold's number, $\rho N D^2/\mu$ [-]
s	distance between impeller tip and velocity measurement [-]
T	tank diameter, [m]
V	velocity, [m/s]
V_r	radial velocity, [m/s]
V_{tip}	impeller tip speed, πND [m/s]
w	width of impeller blade, [m]
z	axial distance from impeller centre-line (z=0)

Greek symbols:

$\dot{\gamma}$	shear rate, ($\dot{\gamma}=k_s N$) [s^{-1}]
μ	fluid viscosity, [Nsm^{-2}]
μ_a	apparent fluid viscosity, Eqn. 1a., [Nsm^{-2}]
ξ	Van der Molen's correlating constant, Eqn. 7, [-]
π	mathematical constant, [3.142]
ρ	fluid density, [kg/m^3]
τ	shear stress, ($\tau = K \dot{\gamma}^n$), [Pa]

Subscripts:

a	apparent
r,z,θ	radial, axial and tangential components

Superscripts:

' root mean square value

- time or space average

REFERENCES

1) Elson, T.P., Proceedings of 6th European Conference on Mixing, Pavia, Italy, 1988, p. 485.

2) Solomon, J., Ph.D. Thesis, University College London, 1980.

3) Elson, T.P., Cheesman, D.J., Nienow, A.W., Chem.Eng.Sci., 1986, **41**, No.10, p. 2555.

4) Metzner, A.B., Feehs, R.H., Ramos,.H.L., Otto, R.E., Tuthill, J.D., A.I.Ch.E.J., 1961, **7**, No.1, p. 3.

5) Nienow, A.W., Wisdom, D.J., Solomon, J., Machon, V., Vlcek ,J., Chem.Eng.Commun., 1983, **19**, p. 273.

6) Baker, M.R., Emery, A.N., Nienow, A.W., BHRA 2nd International conference on Bioreactor Dynamics, 1988, paper B3, p.79.

7) Cooke, M., Middleton, J.C., Bush, J.R., BHRA 2nd International conference on Bioreactor Dynamics, 1988, paper B1, p.37.

8) Koutsakos, E., Nienow, A.W., Proceedings of the 1989 annual conference of the British Society of Rheology, Sept. 1989 (in print).

9) Van der Molen, K., Van Maanen, H.R.E., Chem.Eng.Sci., 1978, **33**, p.1161

10) Metzner, A.B., Otto, R.E., A.I.Chem.J., 1957, **3**, pp.3-10.

11) Bates, R.L., Fondy,.P.L., Corstein,.R.R., Ind.Eng.Chem.(Proc.Des.Dev.), 1963, **2**, p. 310.

12) Gunkel,L.A., Weber, M.E., A.I.Ch.E.J., 1975, **21**, Part:1, p. 931.

13) Costes, J., Couderc, J.P., Chem.Eng.Sci., 1988, **43**, No.10, p.2751.

14) Cooper, R.G., Wolf, D., Can.J.Chem.Eng., 1968, **46**, p. 94.

15) Armstrong, S.G., Ruszkowski, S., . Proceedings of 6th European conference on Mixing, Pavia, Italy, 1988, p. 1.

16) Metzner, A.B., Taylor, J.S., A.I.Ch.E.J., 1960, **6**, No. 1, p. 109.

17) Bujalski, W., Nienow, A.W., Chem.Eng.Sci., 1987, **42**, No.1, p. 317.

18) Bartels, P., Ph D thesis, University of Delft, 1987.

19) Cooper, R.G., Wolf,D., Can.J.Chem.Eng., 1968, **46**, p.94.

20) Desouza,A., Pike,R,W., Can.J.Chem.Eng., 1972, **50**, p.15.

21) Wu,H., Patterson,G,K., Chem.Eng.Sci., 1989, **44**, No.10, p.2207.

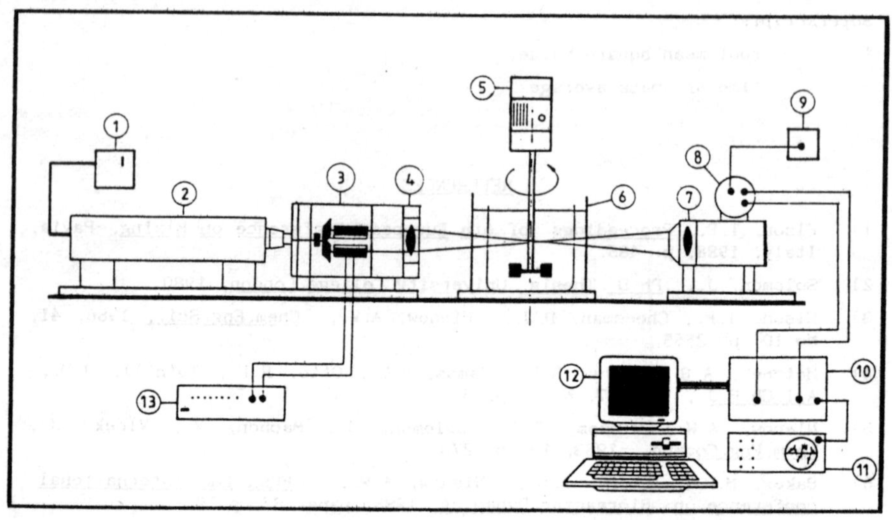

1. POWER SUPPLY
2. LASER GUN
3. ROTATABLE BEAM SPLITTER, (BRAGG CELL)
4. FOCUSING LENS
5. POWER MOTOR
6. SQUARE TANK
7. LIGHT COLLECTING LENS
8. PHOTODETECTOR
9. PHOTOMULTIPLIER
10. SIGNAL PROCESSOR
11. OSCILLOSCOPE
12. MICROCOMPUTER
13. FREQUENCY SHIFTER

(a) The LDV instrumentation

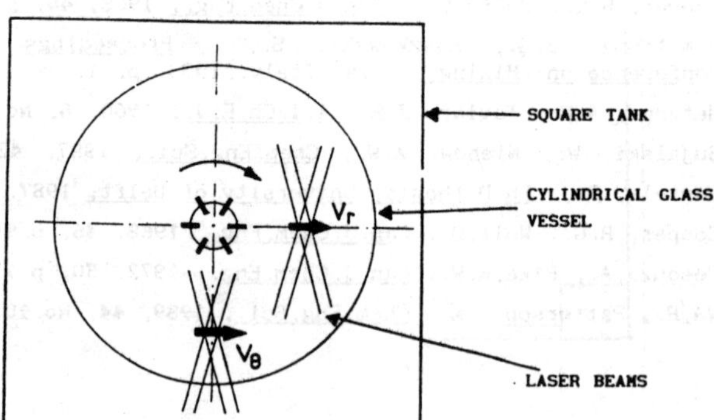

(b) Orientation of Laser beams for measurement of V_θ and V_r.

Figure: 1 Schematic diagram of the set-up for the LDV system.

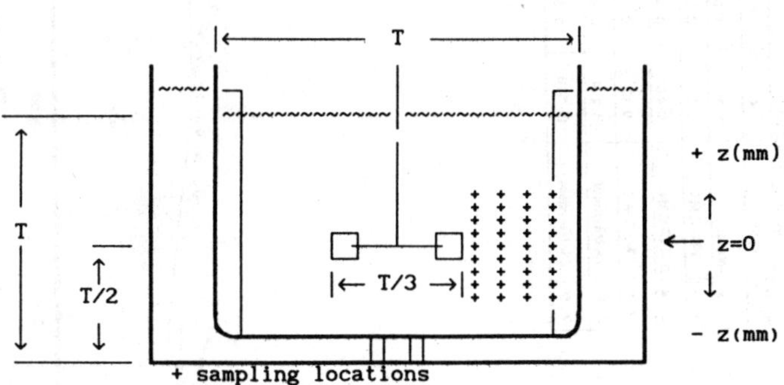

Figure: 2 The stirred vessel with relative dimensions (T=0.15m)

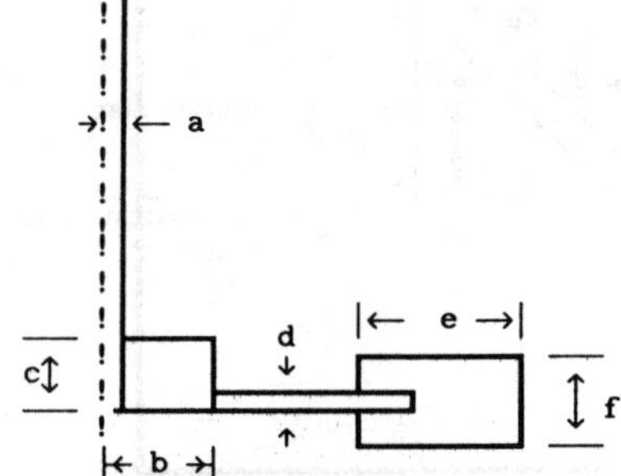

Figure: 3 Geometry of Rushton turbine used.

Impeller diemnsions		
	Fig. 3	ref[9]
a	0.06D	0.075D
b	0.16D	0.125D
c	0.20D	0.25D
d	0.03D	0.038D
e	0.25D	0.25D
f	0.20D	0.20D
w	0.03D	0.038D

w = width of impeller blade
D = impeller diameter

Table: 2 Dimensions of Rushton turbine used compared to Van d'Molen's impeller [9].

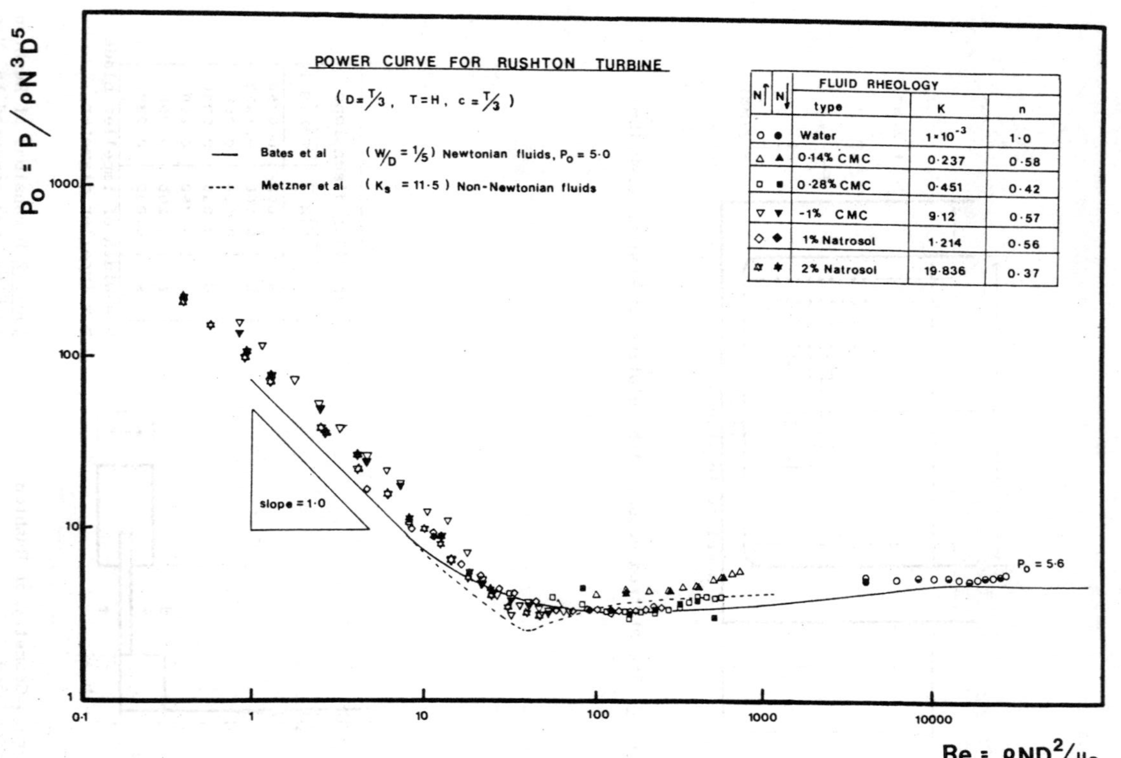

Figure 4. Power curve for Rushton turbine (D=T/3, c=T/3)

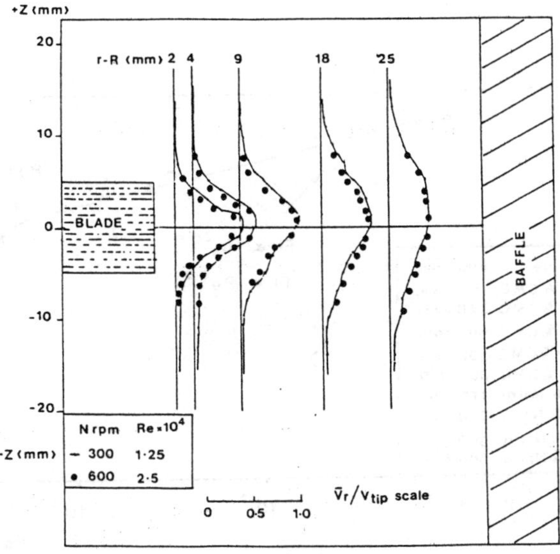

Figure 5. Normalized outlet velocity profiles in water at N=300 and 600rpm.

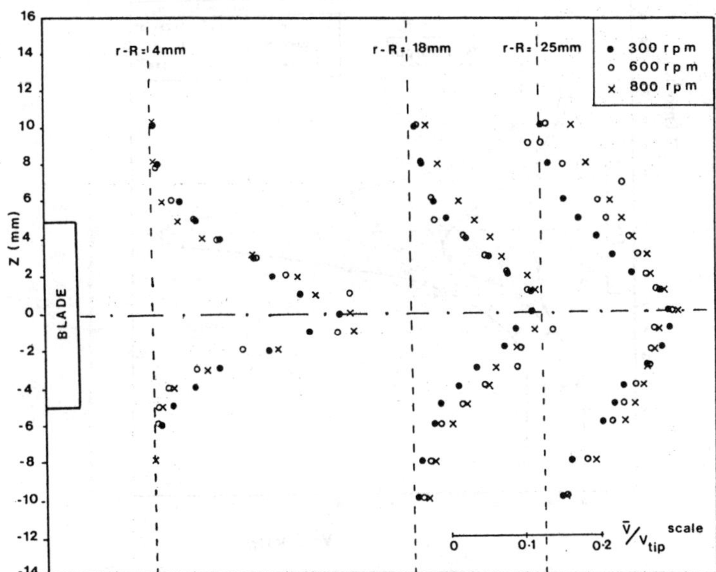

Figure 6. Normalized outlet velocity profiles of $\bar{V}_r/V_{tip}$ for 0.28% CMC at different impeller speeds (Re>60).

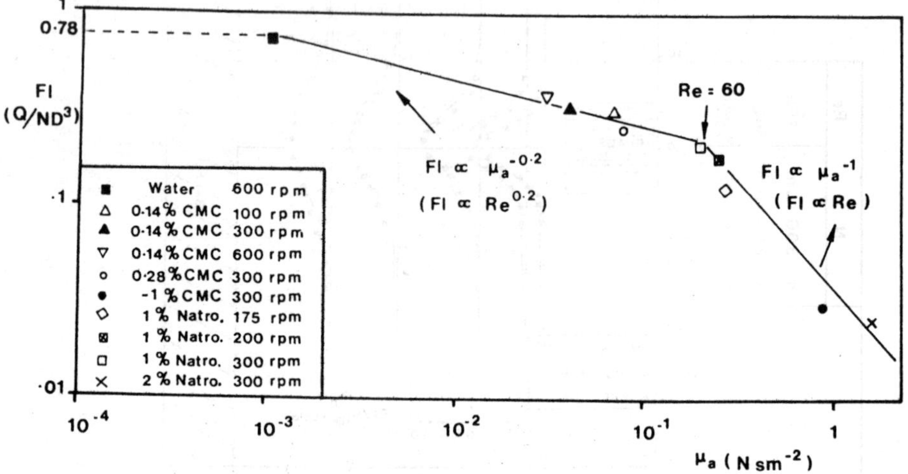

Figure 7. Effect of apparent fluid viscosity, μ_a, on Flow number, Fl.

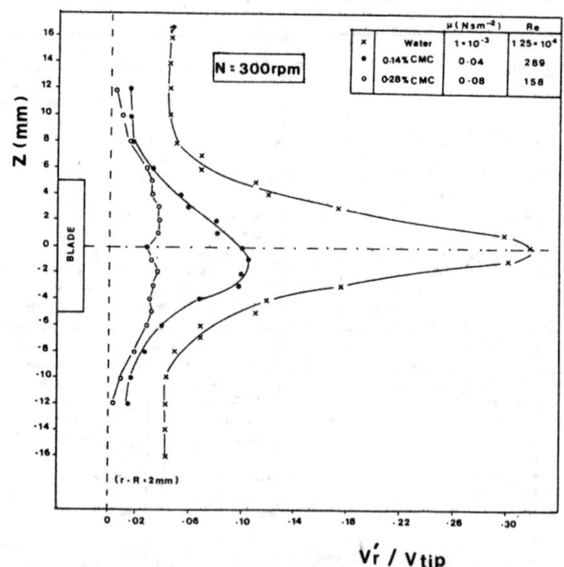

Figure 8. Effect of fluid viscosity on normalised fluctuating outlet velocity profiles of Vr'.

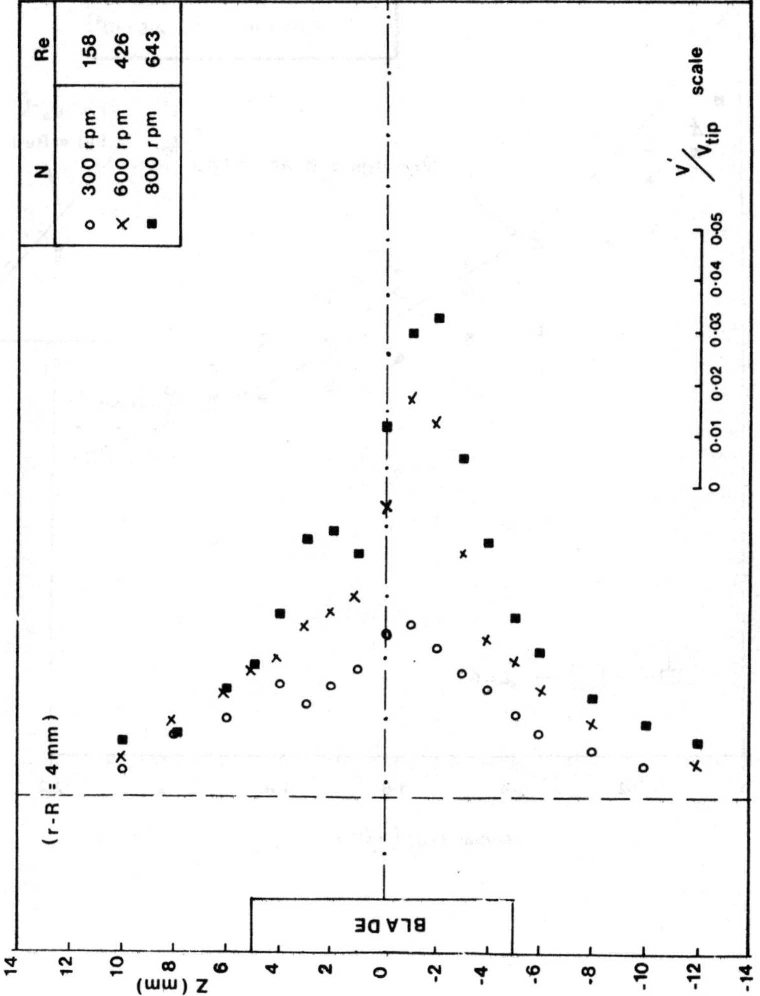

Figure 9. Distribution of V_r'/V_{tip} for 0.28% CMC at various impeller speeds.

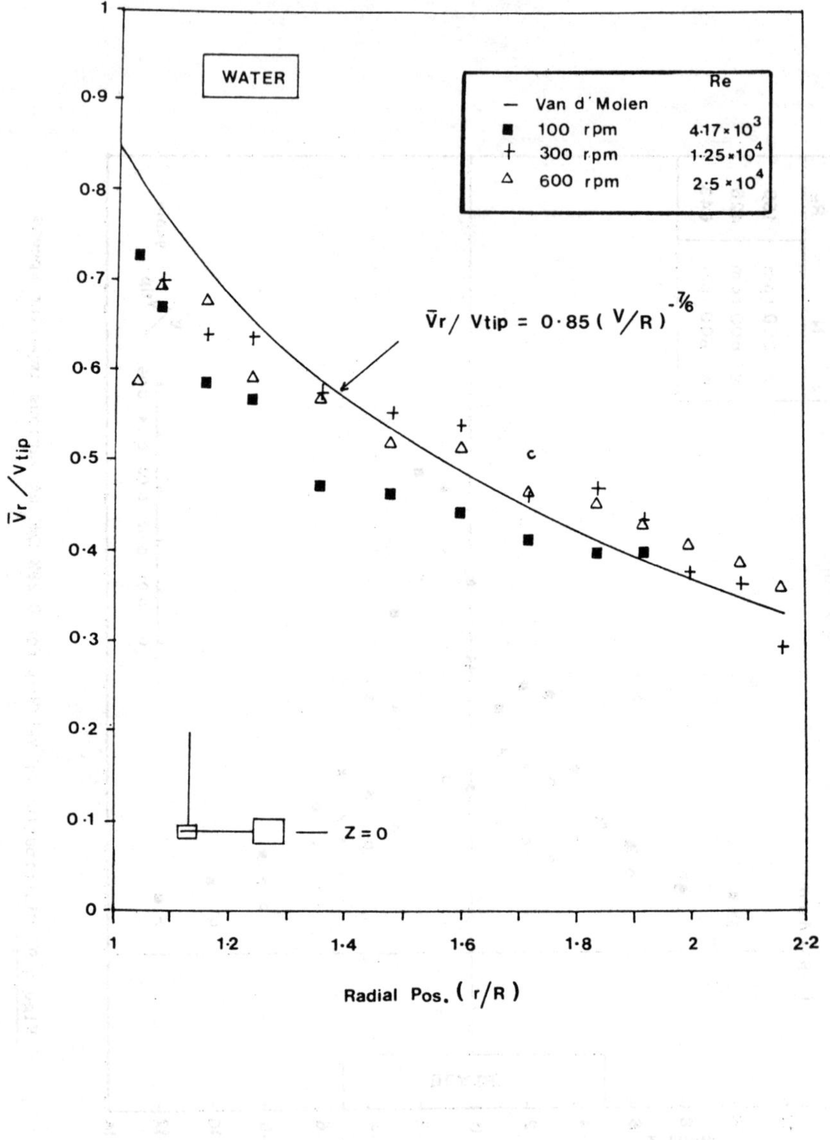

Figure 10. Mean radial velocity profile at z=0, for different impeller speeds

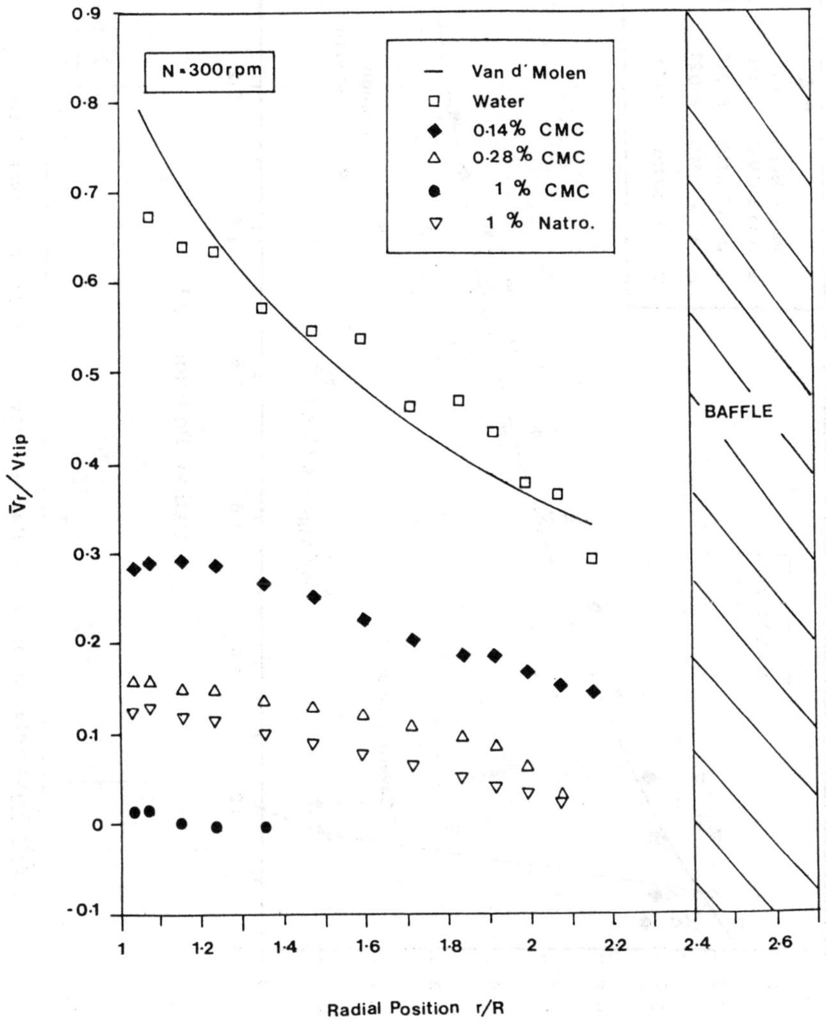

Figure 11. Radial decrease of the mean radial velocity at z=0 and N=300rpm for different impeller speeds.

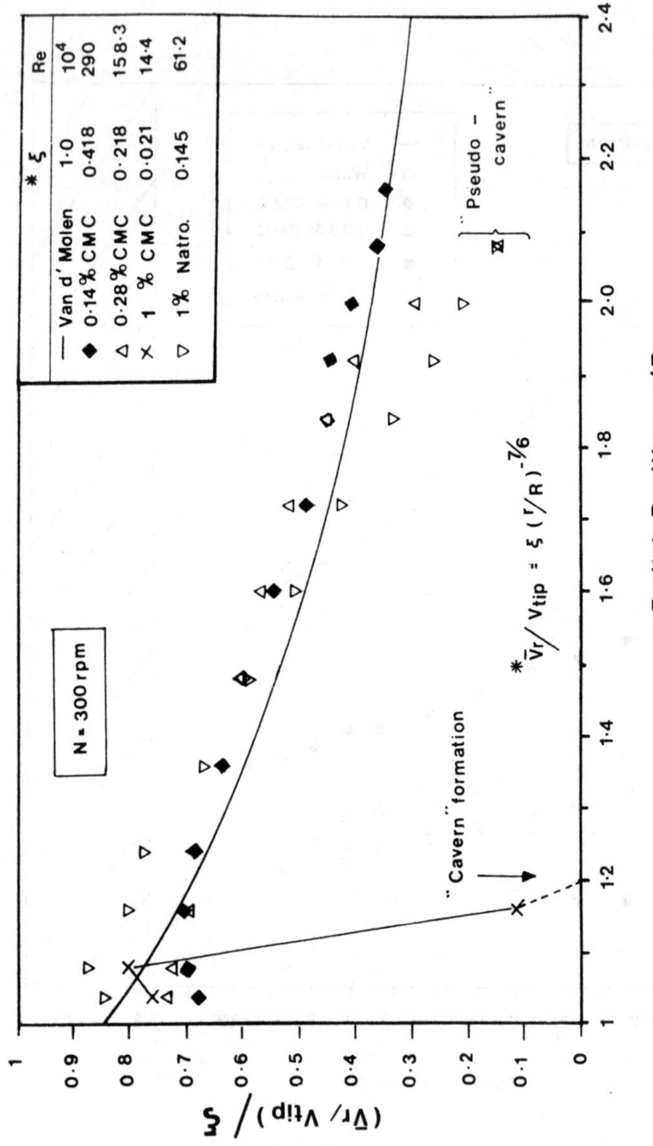

Figure 12. Correlation of mean radial velocity at z=0, for different fluids

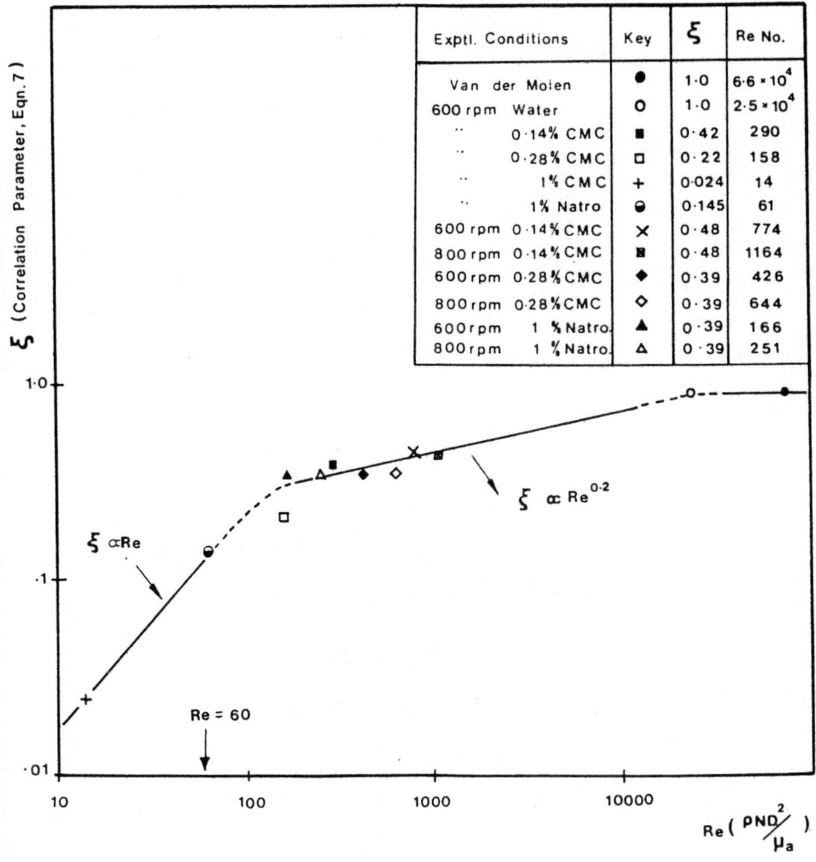

Figure 13. Graph of ξ values as a function of Re number.

MIXING AND CIRCULATION TIMES IN RHEOLOGICALLY COMPLEX FLUIDS

E. Brito-De la Fuente*, J.C. Leuliet**, L. Choplin* and P.A. Tanguy*

For an helical-ribbon impeller, power consumption and both mixing and circulation times have been studied during the mixing of Newtonian and elastic fluids, in the laminar regime of flow. An unique correlation accounting for the elastic effects is proposed in terms of the Weissenberg number, for the power consumption of both Newtonian and elastic liquids. No significant difference was detected in mixing and circulation times for the two types of fluids. The results have been compared with those reported in the literature and the role played by the rheological parameters has been analyzed.

INTRODUCTION

Mixing rheologically complex fluids is increasingly important since those fluids are encountered in a great diversity of processes and industries. Several rheological complexities can be found in those fluids like a shear-dependent viscosity and a viscoelastic behavior and their effects on mixing performance have been the subject of only few studies in the last years. Although it has been reported in the literature that elasticity affects mixing by altering power requirements and by modifying flow patterns and thereby mixing and circulation times, however, more experimental data are needed in order to clarify and quantify these effects on mixing performance.

Mixing performance has been usually described by measuring the power requirements and in some cases, by using the concepts of both the mixing and circulation times. Power consumption determinations are always related to the economy of mixing, as well as, to the design of mixing equipment. However, power is also closely related to flow pattern, therefore, it is expected that rheological complexities may alter power consumption. In the viscous regime of flow, the liquid's elasticity has been reported to have little influence on the torque [1]. However, the level of elasticity has not been discussed.

Several definitions of mixing times can be found in the literature [2], but almost all of the authors agree in that the mixing time is the time needed to reach a specified degree of homogeneity or the time required for a tracer to

* Department of Chemical Engineering, Université Laval, Québec, QC, G1K 7P4, CANADA.

** Permanent address: I.N.R.A., L.G.I.A., 369 Jules Guesde, 59650 Villeneuve d'Ascq, FRANCE.

disperse. It should be pointed out that the mixing time concept as well as the classical techniques available to measure it, are used to describe macro-mixing. When dealing with rheologically complex fluids, the experimental techniques are not straightforward neither the interpretation of the experimental results. As a consequence, literature data on mixing times show a considerable amount of scatter which makes the comparison of experimental values very difficult.

Circulation time is related to the pumping capacity of the impeller. This time can be measured from the periodicity of the response curve obtained in the mixing time measurements, if probe methods are used. The use of flow followers to measure circulation times is also a common practice. Although it is known that elasticity increases circulation time, however, the increasing factor depends on the experimental technique used, as well as, on the geometrical configuration of the mixing system.

Close clearance impellers are widely used for mixing high-viscosity liquids. Among these, helical ribbon agitators have found widespread industrial applications and they have been the focus of research in recent years. As it has been shown in the literature [1,3], the helical ribbon agitators provide a shorter homogenization time (mixing time) when compared with impellers like turbines. Although it is clear that helical ribbon impellers require much more power than turbines at the same Reynolds number, the energy required to reach a specified level of uniformity or homogenization, is considerably less.

Several reviews dealing with the mixing of non-Newtonian fluids have been published recently [1,4]. The role played by elasticity on power consumption and mixing and circulation times has been recently studied using different types of impellers [5,6]. Collias and Prud'homme [5], using a Rushton turbine found that elasticity increases the torque required for mixing and also increases the mixing time. The authors found that ten circulation times were necessary to homogenize the elastic fluids and only four for Newtonian fluids. Carreau *et al.* [6], using an Archimedes' screw rotating inside a coil of tubing, reported a value of three for the ratio: mixing time/circulation time, for Newtonian fluids. The same authors mentioned that elasticity increased the mixing and circulation times by a factor of two to three.

In this study, the effect of elasticity on both power consumption and mixing time is investigated using an helical ribbon impeller. The effect of elasticity on circulation time is derived from mixing time measurements. In order to separate the effects of shear-dependent viscosity from the effects caused by the elasticity, an elastic fluid having a constant viscosity is used.

MATERIAL AND METHODS

Material

Equipment. The geometrical parameters of the helical ribbon impeller used in this study are shown in Fig. 1. The helical ribbon was built in polish stainless steel. The mixing vessel is a flat-bottomed glass cylinder of 0.210 m of diameter and 0.435 m of height (New Brunswick Scientific Co.). The ratio: height of the liquid/height of the impeller was kept constant and equal to 1.14.

The impeller is driven by a system: motor-speed control (Mod. 8293 Pacific Scientific - Mod. KBMD-240D, KB Electronics, Inc.), which allows to have a variable speed, as well as, clockwise and counter-clockwise direction of rotation. The speed of the motor shaft is measured by using a digital stop-watch. This last technique was chosen because the maximum speed reached in the tank was 60 rpm. Such a lower speed is due to the high viscosities of the fluids studied.

The torque is measured by a non-contact strain gauge torquemeter (Mod. MCRT 24002T, S. Himmelstein Co.), in the range from 0 to 2.8 N m. The temperature of the fluid is measured by a thermistor thermometer (Mod. 5831, Omega Engineering, Inc.), equipped with two thermistor probes (Mod. OL-703-PP, Omega Engineering, Inc.). The position of the two probes inside the mixing vessel is indicated in Fig. 2. The probe T_w (near to the vessel wall) was placed at 0.060 m from the free liquid surface, and the probe T_s (near to the impeller shaft) at 0.005 m above the agitator level.

Fluids. The Newtonian fluids studied are mixtures of polybutene (MW: 1300; Parapol 1300 - Exxon Chemical, Co.), and kerosene. Two Newtonian solutions were prepared and named solution No. 1 and solution No. 2, respectively. The composition of these solutions is shown in Table 1. The dependence of the viscosity with respect to the temperature for these two solutions is expressed by an Arrhenius type function:

$$\eta = A \exp\left[B \left(\frac{1}{T} - \frac{1}{T_0} \right) \right] \tag{1}$$

T is the absolute temperature and T_0 is a reference temperature (T_0 = 298 K).

The elastic fluid (solution No. 3) is obtained by adding 0.05% (w/w) of polyisobutylene (MW: 1.2 x 10^6; Vistanex L-120, Exxon Chemical, Co.), to solution No. 2.

Rheology. The rheological properties of the three solutions were determined on a Rheometrics Inc. System IV rheometer with a fluid transducer using a cone-and-plate geometry. Steady-shear viscosity tests have shown that the viscosity of the three solutions is independent of the shear rate, $\dot{\gamma}$. Furthermore, the viscosity of solution No. 3 (the elastic fluid) is equal to the viscosity of solution No. 2, as it is expected.

The primary normal stress difference, N_1, for solution No. 3, can be considered a quadratic function of the shear rate, in the range from 3 to 60 s^{-1}, as it is shown in Fig. 3. In the same figure, the values of N_1 at different temperatures are also presented. From Fig. 3, N_1 is correlated to $\dot{\gamma}$ by:

$$N_1 = \alpha \dot{\gamma}^2 \tag{2}$$

In equation (2), α is an Arrhenius type function of the temperature which can be expressed by:

$$\alpha = A' \exp\left[B' \left(\frac{1}{T} - \frac{1}{T_0} \right) \right] \tag{3}$$

Table 1 resumes the values of the constants A, B, A' and B' in the above equations. Fig. 4 shows the experimental data as well as the representation of equations (1) and (3). It should be noted that N_1 is even more sensitive to the temperature than the viscosity. The viscosity decreases by a factor greater than two when the temperature increases from 20 to 30°C.

Methods

Power Consumption. For Newtonian fluids, torque measurements as a function of rotational speed, are transformed into dimensionless power number, Np, and Reynolds number by:

$$Np = P/\rho N^3 D^5 \qquad (4)$$

where

$$P = (2\pi N/60)\ C \qquad (5)$$

and

$$Re = \rho N D^2/\eta \qquad (6)$$

In order to compare our experimental results with some models reported in the literature, the correlation originally proposed by Chavan and Ulbrecht [7,8], was chosen. This last correlation has been developed for helical ribbon agitators and it is based on a Couette flow analogy. Furthermore, the correlation takes into account the more important geometrical variables found in a mixing system with an helical ribbon impeller. Skelland [4] and Ulbrecht and Carreau [1], have recently reviewed Chavan and Ulbrecht's correlation. For a Newtonian fluid, this correlation takes the form:

$$Np = 2.5\pi B\ a'_s \left(\frac{d'_e}{D}\right)\left(\frac{D_T}{d'_e}\right)^2 \left\{\frac{4\pi}{(D_T/d'_e)^2 - 1}\right\} Re^{-1} \qquad (7)$$

or

$$Np = Kp \left(\frac{\rho N D^2}{\eta}\right)^{-1} \qquad (7a)$$

where

$$\frac{d'_e}{D} = \frac{D_T}{D} - 2\frac{W}{D} \Big/ \mathrm{Ln}\left\{\frac{(D_T/D) - 1 + 2(W/D)}{(D_T/D) - 1}\right\} \qquad (8)$$

and

$$a'_s = \frac{(H/D)(S/D)}{3\pi}\left[\frac{\pi\{(S/D)^2 + \pi^2\}^{0.5}}{(S/D)^2} + \mathrm{Ln}\left\{\frac{\pi}{(S/D)} + \frac{[(S/D)^2 + \pi^2]^{0.5}}{(S/D)}\right\}\right]$$
$$\{1 - [1 - 2(W/D)]^2\} \qquad (9)$$

B = number of helical ribbon blades and S = pitch of the impeller.

For elastic fluids, if the power number and the Reynolds number are defined according to equations (4) and (6), {the constant viscosity of the fluid makes no necessary the generalization of the Reynolds number, and this is one of the main advantages in studying Boger's fluids (Boger [9,10])}, then it is necessary to define a dimensionless number to characterize the fluid elasticity. We used the Weissenberg number, W_i, which is defined in this case as:

$$W_i = N_1/\tau_{12} \qquad (10)$$

where τ_{12} is the shear stress.

It should be noted that several definitions of the Weissenberg number exist in the literature. One may cite, for example, the definition adopted by Collias and Prud'homme [5]:

$$W_i = N_1 \; N/\dot{\gamma}^2 \; \eta \tag{11}$$

or the one taken by Carreau *et al.* [6]:

$$W_i = N_1/2 \; K_s{}^2 \, \eta_e N \tag{12}$$

in which η_e is the viscosity calculated at an effective shear rate $\dot{\gamma}_e$.

Although the above definitions may only differ by a multiplying factor, however, we consider that only the main rheological parameters should be kept in the Weissenberg number definition. It should be noted, however, that in the case of elastic-constant viscosity fluids with N_1 proportional to $\dot{\gamma}^2$, equation (10) reduces to:

$$W_i = \frac{\alpha}{\eta} K_s \; N \tag{13}$$

whereas equation (11) reduces to:

$$W_i = \frac{\alpha}{\eta} N \tag{14}$$

The equations (13) and (14) differ then by a multiplying factor equal to K_s. When N_1 is not proportional to $\dot{\gamma}^2$, a value of K_s has to be used, no matter what definition of W_i is chosen. In this paper, we decided to work with equation (13) in order to show the approach that one may follow if a value of K_s is needed to define the W_i number when working with elastic constant viscosity fluids, for which N_1 is not proportional to $\dot{\gamma}^2$.

In this case, for the elastic fluid studied here, τ_{12} is proportional to $\dot{\gamma}$ (because the fluid has a constant viscosity), and if we consider equation (2), it may be seen that for a fixed temperature, W_i is proportional to the shear rate existing in the mixing vessel. Then, a shear rate representing the experimental mixing conditions should be determined. In order to estimate this shear rate, the following considerations have been made:

i) Chavan and Ulbrecht [7,8], proposed a correlation very similar to equation (7), but for pseudoplastic fluids:

$$Np = 2.5\pi B \; a'_s \left(\frac{d'_e}{D}\right)\left(\frac{D_T}{d'_e}\right)^2 \left\{\frac{4\pi}{n[(D_T/d'_e)^{2/n} - 1]}\right\}^n \left(\frac{\rho N^{2-n} D^n}{K}\right)^{-1} \tag{15}$$

where n and K are the flow and consistency indexes, respectively;

ii) Metzner and Otto [11] proposed an effective shear rate, $\dot{\gamma}_e$. If this shear rate is estimated according to the mixing conditions existing in the vessel, then the resulting power curves for Newtonian and pseudoplastic fluids coincide. Therefore, $\dot{\gamma}_e$ should have a value such that:

$$Np = Kp\left(\frac{\rho ND^2}{\eta_e}\right)^{-1} = Kp\left(\frac{\rho ND^2}{K\,\dot{\gamma}_e^{\,n-1}}\right)^{-1} \qquad (16)$$

Then, equating (15) and (16) and substituting Kp by equation (7a), it can be easily shown that:

$$\dot{\gamma}_e = Ks\ N \qquad (17)$$

where

$$Ks = 4\pi\,\frac{\{(D_T/d'_e)^2 - 1\}^{1/n-1}}{\{n[(D_T/d'_e)^{2/n} - 1]\}^{n/n-1}} \qquad (18)$$

The dependence of Ks with n for the impeller used in this study is shown in Fig. 5. It should be noted that Ks is not a strong function of n: Ks decreases from 27.0 to 23.9 when n changes from 0.1 to 1.

Because in the case of an elastic fluid a Ks value should be defined in order to estimate $\dot{\gamma}_e$ (for W_i determination), and because the Ks function in equation (18) is not defined for n = 1.0, we consider appropriate to calculate $\dot{\gamma}_e$ according to equation (17) with a Ks value given by:

$$Ks = \lim_{n \to 1.0} Ks(n) = 23.91 \qquad (19)$$

Mixing and Circulation Times. Mixing times are measured using the thermal method. This method employs a change of temperature as the property to be measured. Fluctuations in temperature after the injection of a small amount of heated fluid (2% of same fluid) are monitored by using two thermistors (T_s and T_w in Fig. 2) and a data acquisition interface. The temperature of the hot injection fluid was 60°C and the initial temperature in the mixing vessel was in the range from 20 to 25°C. During all experiments, the injection point and the position of the probes were kept constant (Fig. 2).

Several advantages are found with this technique. It allows for an unlimited number of experimental runs with the same fluid. The thermal technique does not change drastically the rheological properties of the fluids, even though in our case, the fluids are thermo-sensitive. The small decrease in viscosity, due to a global increase of temperature always less than 0.7°C, was taken into account in all calculations.

An example of an experimental probe response is shown in Fig. 6. Several times can be identified in Fig. 6: a dead time t_d, between the injection of the hot fluid and its first detection at the thermistor probe; a circulation time, t_c, time between two successive peaks. It should be mentioned that for Reynolds numbers < 1.0, it was not possible to detect t_c from the probe response, because only one clear peak was obtained.

Several definitions for mixing time exist in the literature [2], but for the thermal technique, the one proposed by Hoogendoorn and den-Hartog [12] seems to be the most widely used. For the above authors, in order to have an homogenization of 90%, the mixing time, t_1, should be the time at which:

$$\forall\ t \geq t_1 \ :\ 0.9 \leq \frac{\theta(t) - \theta_i}{\Delta\theta_\infty} \leq 1.1 \qquad (20)$$

$\theta(t)$ is the temperature measured at time t and $\Delta\theta_\infty$ is the difference between the final temperature θ_f and the initial temperature θ_i.

Many difficulties are encountered in the application of the above definition, mainly because the temperature approaches θ_f in an asymptotic manner and the end point of the experiment is difficult to detect with precision, making $\Delta\theta_\infty$ too much imprecise. This particular subject will be discussed separately in a future communication.

Therefore, we have chosen another criteria for homogenization which is less dependent on the experimental imprecisions. Homogenization is considered to be accomplished when the mixing time, t_2, is such that:

$$\forall\, t \geq t_2 \quad : \quad |\theta(t) - \theta_f| \leq \Delta\theta_\infty/2 \qquad (21)$$

The difference between the temperature of the hot injection fluid and the initial fluid temperature inside the mixing vessel, was always in the range from 35 to 40°C. Then, this last definition means that, in our case, one may consider the mixing vessel homogeneous, when the absolute value of the difference between the equilibrium temperature and the measured temperature, is systematically less than 0.35°C.

RESULTS AND DISCUSSION

Power Consumption

Newtonian Fluids. Figure 7 shows the experimental results for power consumption for the three solutions studied. For the Newtonian fluids, the power number is given by:

$$Np = 135\ Re^{-1} \qquad (22)$$

The Chavan and Ulbrecht correlation [equations (7) to (9)], predicts for our geometry:

$$Np = 147\ Re^{-1} \qquad (23)$$

Comparing the results expressed in both equations (22) and (23), it can be seen that a difference less than 9% exists between the Kp values. Then, the experimental values for the Newtonian fluids are in good agreement with those predicted by the Chavan and Ulbrecht correlation. This suggests that the prediction given by equation (15) concerning the pseudoplastic fluids, may be equally good. This means that we can assume that the Ks value derived from equation (15) and calculated by equation (18), may be not far from its true experimental value. Some experiments are underway using pseudoplastic fluids and will answer how far from the experimental value is the prediction of equation (18).

Elastic Fluid. For the elastic fluid, the power consumption increases compared with that needed to mix Newtonian fluids at the same Reynolds number. This augmentation is observed for Reynolds numbers higher than 0.4; below this value, no effect on power consumption due to elasticity was observed. In order to quantify the increase in power consumption, a parameter β was used, which is defined by:

$$\beta = \frac{(Np)_{elastic}}{(Np)_{Newtonian}} - 1 \tag{24}$$

A plot of ß vs. the Weissenberg number defined in equation (10) is shown in Fig. 8. It should be pointed out that the dimensionless number $(Np)_{Newtonian}$ was calculated by using equation (22) and the experimental values of the Reynolds number. The data shown in Fig. 8 are well correlated by:

$$\beta = 6.802\ W_i^{1.727} \tag{25}$$

Equation (25) is represented in Fig. 8 by a solid continuous line. The dotted lines represent the values predicted by using equation (25) with ± 5% of difference from the experimental Np values.

The results obtained in Fig. 8 are used to shift the elastic data on the Newtonian curve. Then, an unique power consumption curve as a function of the Reynolds number is obtained if the function: $Np\ (1 + 6.802\ W_i^{1.727})^{-1}$ is plotted vs. Re. This curve is shown in Fig. 9. Therefore, the final correlation is:

$$Np = 135\ Re^{-1}\ (1 + 6.802\ W_i^{1.727}) \tag{26}$$

The comparison of our experimental results with those reported in the literature is not straightforward. This is mainly due to the existing contradictions regarding the influence of the elasticity on the power consumption. For example, Chavan and Ulbrecht [13] studying the performance of some off-centered helical agitators and some other close clearance impellers [7], concluded that the elasticity did not affect the power consumption in the laminar region. Skelland [4], citing several studies on a great diversity of agitators (screws, helical-ribbons, turbines, paddles, etc.) came to the same conclusion.

On the other hand, there are several authors that have reported a decrease on power consumption in the laminar region, due to elastic effects. This decrease may be small for Rieger and Novak [14], who worked with helical-ribbons and anchor impellers, or it may be quite significant for Kelkar *et al.* [15], and Oliver *et al.* [16], who studied the elasticity effects using a Rushton turbine.

By the contrary, Prud'homme and Shaqfeh [17], and Collias and Prud'homme [5], using a Rushton turbine, have reported that elasticity increases the torque and therefore the power consumption required for mixing in the laminar region. These authors have proposed a correlation based on a dimensionless number, the elasticity number, which is equal to the ratio W_i/Re, to quantify the effects of elasticity.

Although it has been concluded by Nienow and Elson [18] that the effect of the elasticity on the mixing is far from being clear, however, for Ulbrecht and Carreau [1], the contradictions found in the literature can be explained by considering that the experimental conditions between researchers are not always the same.

If the main effect of the elasticity is the modification of the secondary flow pattern, then it seems imperative to quantify the balance between inertial and elastic forces, i.e. the ratio W_i/Re. For comparison purposes, only one definition of the W_i number should be considered. For this discussion, we adopted the definition given by Collias and Prud'homme [5] [equations (11) and

(14)]. Then, based on the work of Walters and Savins [19] concerning the flow around a sphere rotating in a viscoelastic liquid, if the ratio W_i/Re is smaller than 0.042, then inertial forces dominate the whole flow field, and if the ratio is larger than 0.125, then the elastic forces determine the flow field. It should be noted, however, that the above limits have been reported erroneously as 1/6 and 1/2 respectively by Ulbrecht and Carreau [1]. Then, it seems that the ratio W_i/Re is a key parameter to explain the contradictions found in the literature regarding the elasticity effect. For example, Oliver et al. [16] worked with a ratio W_i/Re close to 7 (for Re = 1.0); on the other hand, Collias and Prud'homme [5], worked within a range of W_i/Re where the maximum value was 0.06. Although in both cases the geometry was the same (Rushton turbine), however, the reported effects on power consumption due to the elasticity are quite different, as it has been mentioned previously.

In our case, the experimental values of W_i/Re [according to equation (11)] are in the order of 0.0046. Considering the results given by Walters and Savins [19] for a sphere, this would mean that we are in the region where inertia dominates the flow field and the elastic forces have not influence. However, it must be remembered that in this work we are dealing with flow patterns produced with an helical-ribbon impeller. More experiments are being carried out at our laboratory to study the performance of helical-ribbon impellers, at values of W_i/Re above and below the limits of 0.042 and 0.125. These studies will better permit to analyse the role played by the elasticity on the power consumption.

Mixing and Circulation Times

Mixing Time. In Fig. 10, the experimental results of mixing time as a function of Reynolds number are shown. It should be noted that these results are presented in dimensionless form, and that they were calculated using the experimental probe response curves and the mixing time criteria established in equation (21). The data can be fitted by:

$$N\,t_2 = 55 \pm 22 \qquad (27)$$

Equation (27) has the traditional form found in the literature for mixing time, in the laminar region. Although the scatter of the data may be considered important, however, it may be explained on the basis of the imprecisions found in measuring the temperature of equilibrium. At very low Reynolds, the end point of the experiment is difficult to detect with precision and this is reflected on the values of $\Delta\theta_\infty$. Considering the low values of Re covered [Re < 6], the fit is reasonable.

Using the criteria proposed by Hoogendoorn and den-Hartog (equation (8), Ref. [12]), the scatter of the data is greater and the correlation obtained is:

$$N\,t_1 = 102 \pm 78 \qquad (28)$$

In the same Fig. 10, some experimental results obtained with solution No. 3 (elastic fluid), are also shown. These results suggest that elasticity does not modify mixing time, in the range of W_i/Re studied.

Circulation Time. Circulation times were calculated from the response curves obtained during mixing time experiments, only for Re > 1.0. The experimental relationship between mixing and circulation times is summarized in Fig. 11. Mixing time data in Fig. 11 were calculated by using both equations (20) and (21). The data can be correlated, according to the mixing time definition chosen, by:

$$n_{c1} = (N\,t_1)/(N\,t_c) = 4.4 \qquad (29)$$

and

$$n_{c2} = (N\,t_2)/(N\,t_c) = 2.4 \qquad (30)$$

Two important remarks can be derived from the results of figure 11: 1) There is no significant difference in the behavior of Newtonian and elastic fluids regarding the ratio n_c, and 2) More scatter in the data is obtained when the Hoogendoorn and den-Hartog [12] definition of mixing time is used.

Analysis and Discussion of Results. The analysis of the experimental data of mixing and circulation times, can be made by comparison with the results in the literature which are summarized in Table 2. The following remarks can be made:

1) Our results on mixing and circulation times for Newtonian fluids, are qualitatively in agreement with those reported in the literature for the same type of agitator (Hoogendoorn and den-Hartog [12], Guerin *et al.* [25], Ulbrecht and Carreau [1], and Rieger *et al.* [26]). From Table 2, the values of N t_m are in the range from 22 to 163, and in our case, whatever the definition taken for t_m [eq. (20) or eq. (21)], the experimental values fall within this range. A more precise comparison is difficult to make, because the influence of the impeller geometry is important (Rieger and Novak [26]), although to our knowledge, it has not been correlated to mixing time.

2) In the case of circulation times for Newtonian fluids, the values of the ratio n_c found in the literature, fluctuate from 3 to 4. Our experimental values are close to this range, independently of the mixing time definition chosen [eq. (20) or eq. (21)].

3) The role of elasticity on both mixing and circulation times is clear for most of the authors cited in Table 2. For them, elasticity increases the time required for mixing, as well as the circulation time. In our case, elasticity did not affect mixing time. This may be explained on the basis of the lower W_i/Re ratios used in this work. In most cases, the elasticity criteria has not been integrated in the reported correlations for mixing and circulation times. Exceptions to this are the correlations proposed by Ulbrecht [21], Ulbrecht and Carreau [1] and Carreau *et al.* [6], which are respectively:

$$(n_c)_{elastic}/(n_c)_{Newtonian} = (1 + 0.45\,W_i)^{0.5} \qquad (31)$$

$$(t_m)_{elastic}/(t_m)_{Newtonian} = (1 + 0.45\,W_i)^{0.3} \qquad (32)$$

and

$$(t_m)_{elastic}/(t_m)_{Newtonian} = 1 + 3.76\,W_i^{0.435} \qquad (33)$$

It must be noted that the above equations were developed for a mixing system having an helical-screw in a draught tube. The W_i number definition in eq. (33) is different from the one given by eq. (31) and eq. (32).

Even though our impeller geometry is quite different from the one used to develop equations (31) to (33), we have calculated the values of the ratios expressed in those equations, using our experimental W_i numbers. The results are: 1.003, 1.005 and 1.414, respectively. This may confirm the fact that, for our mixing system, elasticity does not have any significant effect on mixing and circulation times. The higher value obtained with eq. (33) may be explained by considering that this equation was developed for $W_i > 1.0$ [calculated by eq. (10)], and in our case, the maximum value of W_i was 0.25.

CONCLUSIONS AND PERSPECTIVES

The mixing performance of an helical-ribbon agitator has been studied by measuring power consumption, and both mixing and circulation times. From the known mixing time and the power consumption correlation, the relative efficiency of the agitator can be estimated. The experimental results have shown that, in the laminar region ($Re < 10$):

i) Power consumption for Newtonian fluids is predicted by the correlation proposed by Chavan and Ulbrecht [7,8], with a difference of only 10%;

ii) For the elastic fluid ($W_i \leq 0.25$), power consumption increases by almost 60% due to elasticity. Using a Ks value derived from both the principle proposed by Metzner and Otto [11] and the correlation proposed by Chavan and Ulbrecht [7,8] for pseudoplastic fluids, this augmentation of power consumption has been correlated with the Weissenberg number as follows:

$$Np = 135\ Re^{-1}\ (1 + 6.802\ W_i)^{1.727}$$

The difference between experimental data and predictions are less than 5%.

iii) Dimensionless mixing and circulation times are essentially constant. The ratio mixing time/circulation time is 2.4 or 4.4, depending on the definition chosen for mixing time calculations, but independent of the rheological properties of the fluid. It has been suggested that these results may be attributed to the "low elasticity" of the fluid.

These results have shown that the quantification of the elastic effects in the mixing performance of any agitated system is not straightforward. In the same direction, we have seen that the use of different criteria for mixing time calculations, produce a great diversity of results and conclusions which are very difficult to compare. The use of the Weissenberg number seems not sufficient to quantify the role of elasticity. By the contrary, we think that the ratio W_i/Re may be a better criteria in determining the impact of elasticity. Unfortunately, the range of W_i/Re covered in this study is very small to involve this number in any correlation.

Current studies include the use of other impeller geometries, pseudoplastic fluids having different values of the flow index (n). Other elastic fluids (Boger type) will be studied in order to increase the range of the ratio W_i/Re and other (less viscous) Newtonian fluids in order to determine the onset of the transition regime.

ACKNOWLEDGEMENTS

The authors are indebted to the "FCAR-Action spontanée" program of the Province of Quebec, to the National Council of Science and Technology of Mexico (CONACYT) and to the "Coopération France-Québec Biotechnologies" for financial support.

REFERENCES

1. Ulbrecht, J.J. and Carreau, P., 1985, in "Mixing of Liquids by Mechanical Agitation", Vol. 1, Chapter 4, J.J. Ulbrecht and G.K. Patterson ed., Gordon and Breach Publishers, N.Y.

2. Hiby, J.H., 1981, Int. Chem. Eng., 21, No. 2, 197-204.

3. Nagata, S., 1975, "Mixing: Principles and Applications", Kodansha Ltd. and John Wiley and Sons.

4. Skelland, A.H.P., 1983, in "Handbook of Fluids in Motion", Chapter 7, N.P. Cheremisinoff and R. Gupta ed., Ann Arbor Science, N.Y.

5. Collias, D.J. and Prud'homme, R.K., 1985, Chem. Eng. Sci., 40, No. 8, 1495-1505.

6. Carreau, P.J., Guerin, P. and Paris, J., 1986, Proceedings of the World Congress III of Chemical Engineering, Tokyo, Japan, 405-408.

7. Chavan, V.V. and Ulbrecht, J.J., 1973, Ind. Eng. Chem. Process Des. Dev., 12, 472-476.

8. Corrigenda to Ref. 7, 1974, I/EC Process Des. Dev., 13, 309.

9. Boger, D.V., 1977/78, J. Non-Newt. Fluid Mech., 3, 87-91.

10. Boger, D.V., in Rheology, Vol. 1: Principles, VIII Int. Congress on Rheology, Naples, Sept. 1-5 1980, G. Astarita, G. Marucci and L. Nicolais ed., Plenum Press, N.Y. and London, 1980.

11. Metzner, A.B. and Otto, R.E., 1957, AIChE J., 3, No. 1, 3-10.

12. Hoogendoorn, C.J. and Den-Hartog, A.P., 1967, Chem. Engng. Sci., 22, 1689-1699.

13. Chavan, V.V. and Ulbrecht, J.J., 1973, Trans. Inst. Chem. Engrs., 51, 349-354.

14. Rieger, F. and Novak, V., 1974, Trans. Inst. Chem. Engrs., 52, 285-286.

15. Kelkar, J.V., Mashelkar, R.A. and Ulbrecht, J.J., 1972, Trans. Inst. Chem. Engrs., 50, 343-352.

16. Oliver, D.R., Nienow, A.W., Mitson, R.J. and Terry, K., 1984, Chem. Eng. Res. Des., 62, 123-127.

17. Prud'homme, R.K. and Shaqfeh, E., 1984, AIChE J., 30, No. 3, 485-486.

18. Nienow, A.W. and Elson, T.P., 1988, Chem. Eng. Res. Des., 66, 5-15.

19. Walters, K. and Savins, J.G., 1965, Trans. Soc. Rheol., 9, 407-416.

20. Coyle, C.K., Hirschland, H.E., Michel, B.J. and Oldshue, J.Y., 1970, AIChE J., 16, 903-906.

21. Ulbrecht, J., 1974, The Chem. Engr., No. 286, p. 347-353 + 367, 1974.

22. Chavan, V.V. Arumugam, M. and Ulbrecht, J.J., 1975, AIChE J., 21, No. 3, 613-615.

23. Chavan, V.V., Ford, D.E. and Arumugam, M., 1975, Can. J. Chem. Eng., 53, 628-635.

24. Carreau, P.J., Patterson, I. and Yap, C.Y., 1976, Can. J. Chem. Eng., 54, 135-142.

25. Guerin, P., Carreau, P.J., Patterson, W.I. and Paris, J., 1984, Can. J. Chem. Eng., 62, 301-309.

26. Rieger, F., Novak, V. and Havelkova, D., 1986, Chem. Engng. J., 33, 143-150.

Table 1. Rheological parameters [in equations (1), (2) and (3)] and compositions for all tested fluids. PB: Polybutene; KER: kerosene and PIB: Polyisobutylene.

Solution	Density Kg/m³	Composition (w/w)	A Pa s	B K^{-1}	A' Pa s^2	B' K^{-1}
1	885	PB97%-KER3%	33.67	7744	-	-
2	872	PB91%-KER9%	6.693	6870	-	-
3	872	PB91%-KER9%-PIB0.05%	6.814	6776	0.0880	17170

Table 2. Mixing and circulation times for Newtonian, pseudoplastic and elastic fluids in mechanically agitated systems.

Ref.	Geometry	Results and comments
[12] 1967	Helical-ribbon with two ribbons	• For Newtonian fluids: $N\ t_{75\%} = 65$ • Determination of mixing time: thermal method
[20] 1970	Helical-ribbon-screw	• $(t_c)_{Newtonian} = (t_c)_{Shearthinning}$ • $(t_m)_{Newtonian} = (t_m)_{Shearthinning}$ • $n_c = (t_m/t_c) \sim 3$
[21] 1974	Helical screw with a draught tube	• Newtonian fluids, for Re < 10: $n_c \sim 7$ • $(n_c)_{elastic}/(n_c)_{Newtonian} = \sqrt{1 + 0.45\ W_i}$
[22] 1975	Helical-ribbon-screw	• Newtonian fluids: $N\ t_{90\%} \sim 100$ • $(N\ t_c)$ and $(N\ t_m)$ are independent of the shear-thinning behavior of the fluid • $(t_m)_{elastic}/(t_m)_{Newtonian} \sim 2$
[23] 1975	Helical screw with a draught tube	• Newtonian and Shearthinning fluids: $N\ t_m \sim 215$ • For different elastic fluids: $215 \le N\ t_m \le 640$
[24] 1976	Helical-ribbon	• For Newtonian fluids: $n_c \sim 3.5$ • n_c can be multiplied by 2 or 3 for a slightly elastic CMC solution • n_c can be multiplied by 7 for a highly elastic fluid
[25] 1984	Helical-ribbon of different geometries	$30 \le Re \le 250$ • Glycerol: $n_c \sim 3.5$ • 2% CMC: $n_c \sim 8.0$ • For all tested geometries: $22 \le N\ t_m \le 53$
[5] 1985	Rushton turbine	• For Newtonian fluids: $n_c \sim 4$ • For elastic fluids: $n_c \sim 10$ • For Re = 40: $(t_m)_{elastic}/(t_m)_{Newtonian} \sim 2$

Table 2. Continued

Ref.	Geometry	Results or comments
[1] 1985	Helical-ribbon of five different geometries	• Both t_c and t_m increase with elasticity, but not necessarily by the same factor • Glycerol: $25 \leq N\ t_m \leq 61$ • 2% CMC: $51 \leq N\ t_m \leq 189$ • 1% SEPARAN: $108 \leq N\ t_m \leq 163$ Conclude that t_m depends on both geometry and fluid
	Helical screw with a draught tube	Reference: Ford and Ulbrecht (1976) • $(N\ t_c)_{elastic}/(N\ t_c)_{Newtonian} = (1 + 0.45\ W_i)^{0.3} (\eta_0/\eta_e)^{0.3}$ • $(N\ t_m)_{elastic}/(N\ t_m)_{Newtonian} = (1 + 0.45\ W_i)^{0.8}$
[26] 1986	Helical-ribbon 8 different geometries	• Complete study for Newtonian fluids with: $0.3 \leq Re \leq 4\ 10^4$ • In the laminar region, for the 8 agitators: $29 \leq N\ t_m \leq 123$
[6] 1986	Archimede's screw inside a coil of tubing that function as a draft tube	• $10 < Re < 400$: end of the laminar regime and transition regime • Laminar regime with Newtonian fluids: $N\ t_m \sim 60$ • Laminar regime with Newtonian and shearthinning fluids: $n_c \sim 3$ • Elastic fluids: $(t_m)_{elastic}/(t_m)_{Newtonian} \sim 1 + 3.76\ W_i^{0.435}$

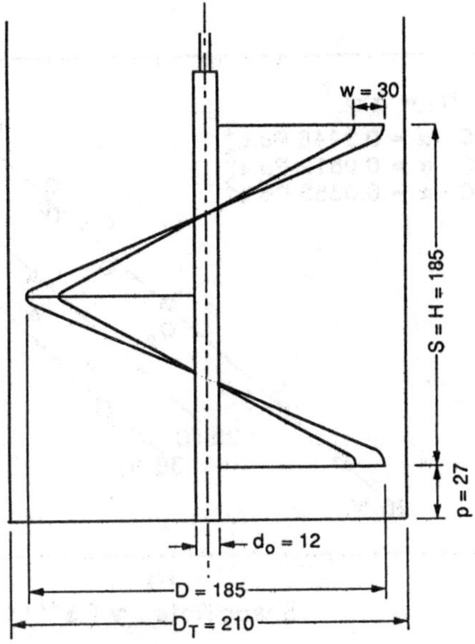

Figure 1. Helical-ribbon impeller (all geometrical dimensions are in mm.).

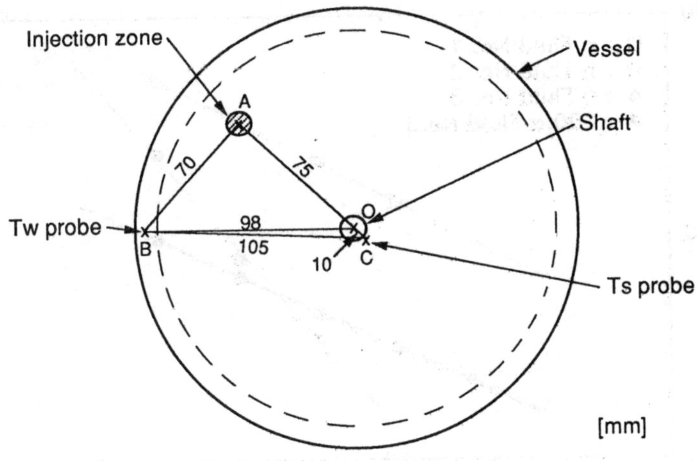

Figure 2. Position of the thermistor probes (T_s and T_w) and localization of the injection point during mixing time experiments.

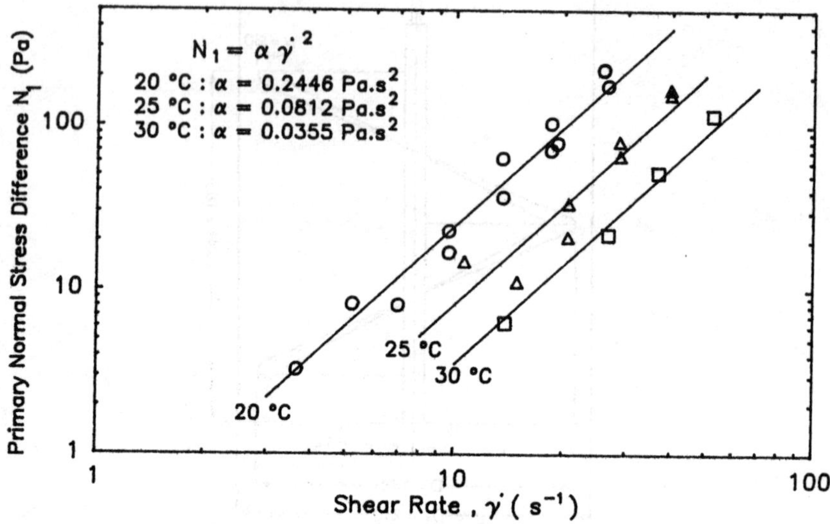

Figure 3. Primary normal stress difference as a function of temperature and shear rate for solution No. 3 (elastic fluid).

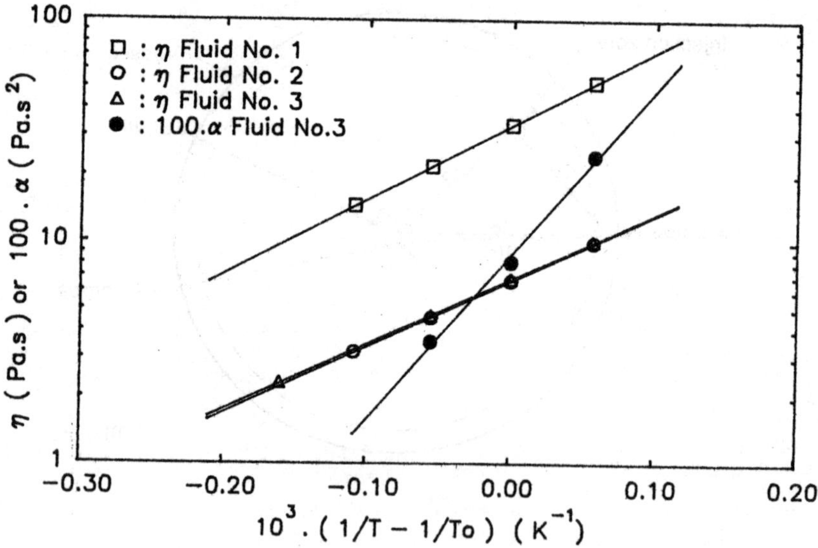

Figure 4. Viscosity and α parameter [defined in eq. (3)] as a function of temperature for the three fluids studied.

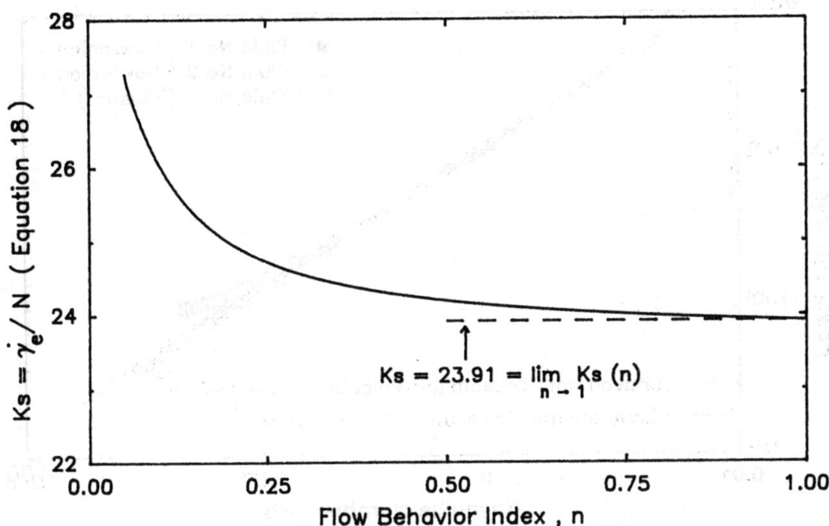

Figure 5. Prediction of the K_S value as a function of n using equation (18).

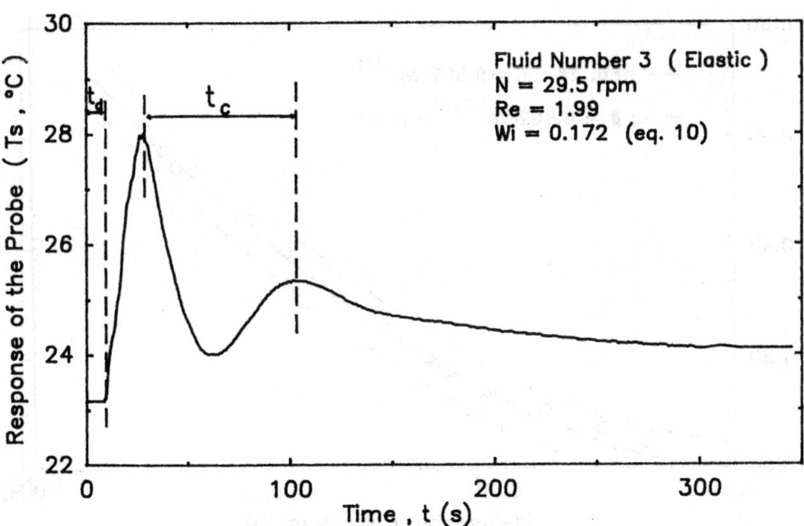

Figure 6. Example of a thermal probe response obtained during mixing time experiments.

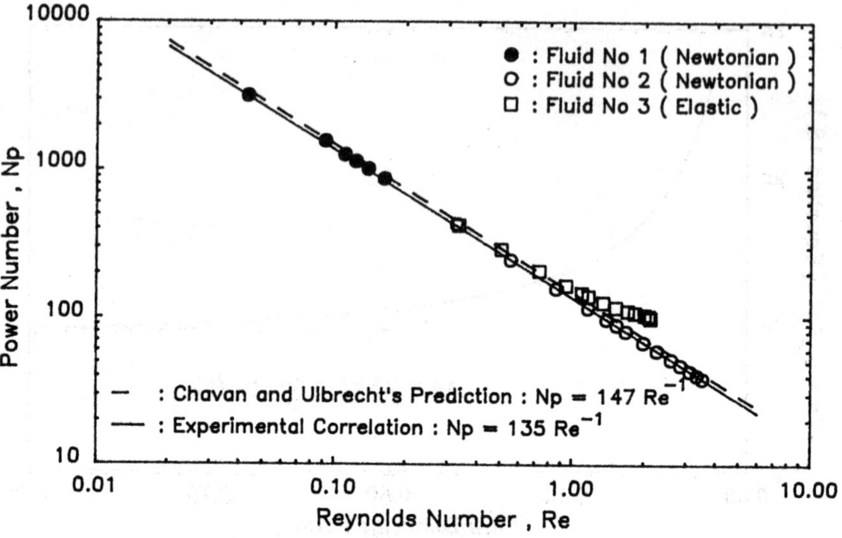

Figure 7. Power number as a function of Re for the three fluids studied.

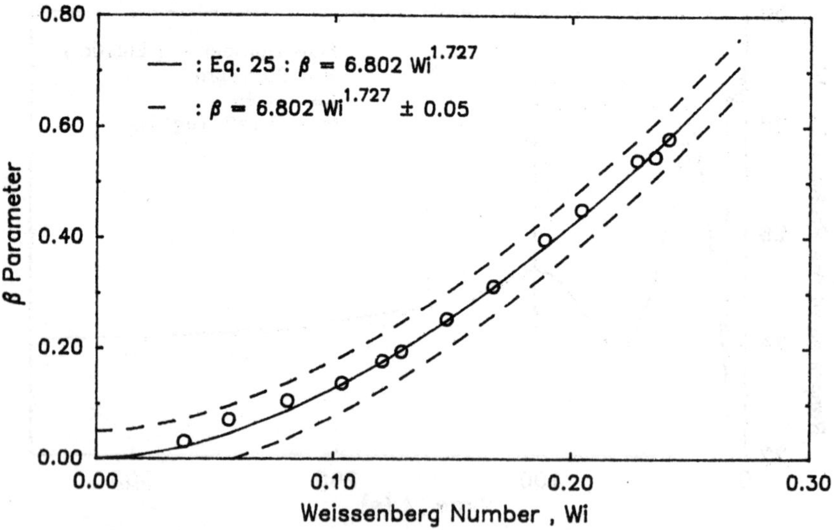

Figure 8. Effect of the Weissenberg number on power consumption. The β parameter is defined in equation (24).

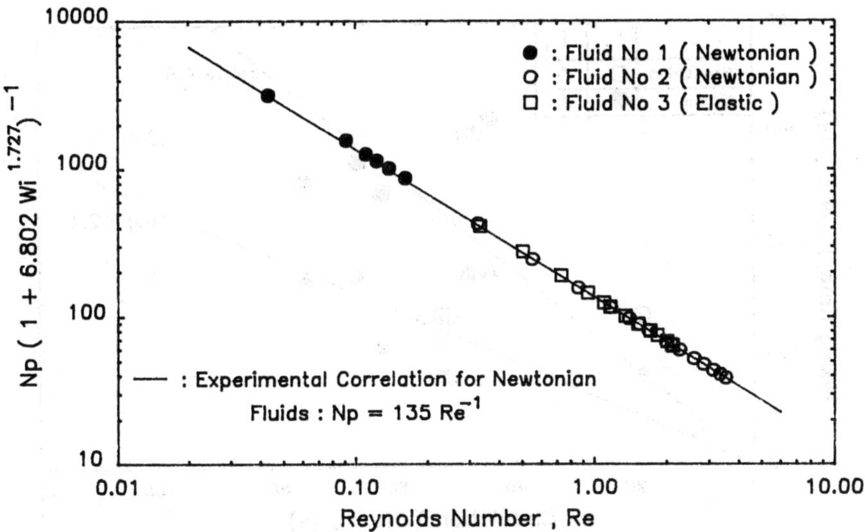

Figure 9. Proposed power correlation for Newtonian and elastic fluids.

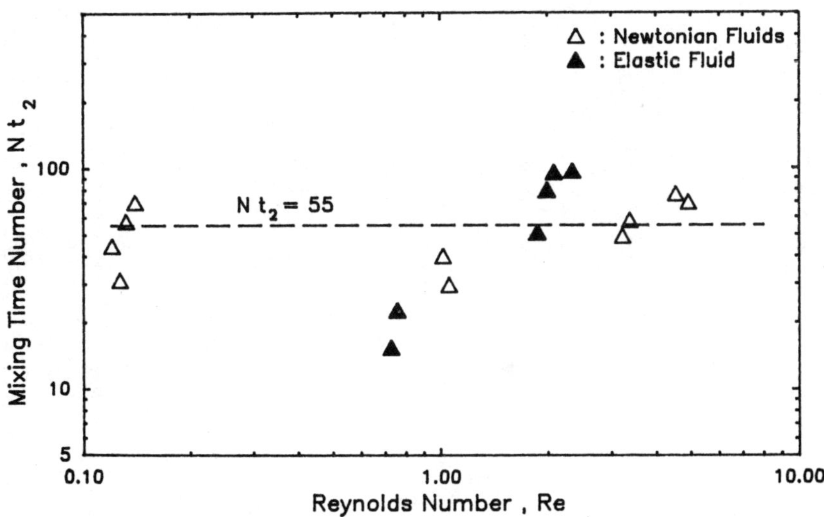

Figure 10. Dimensionless mixing time versus Reynolds number for Newtonian and elastic fluids.

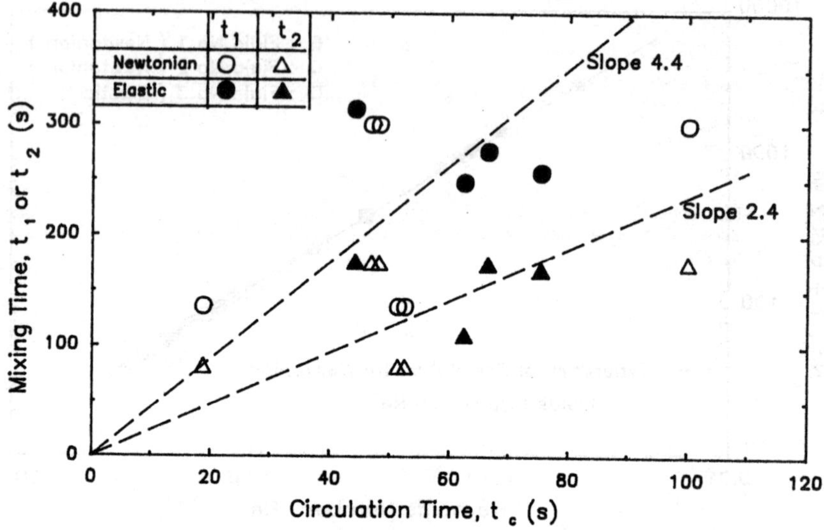

Figure 11. Mixing time versus circulation time for Newtonian and elastic fluids.

Three-Dimensional Modelling of the Flow Through a LPD Dow-Ross Static Mixer

P.A. Tanguy*, R. Lacroix* and L. Choplin*

The complex 3-D flow pattern inside a LPD Dow-Ross static mixer is investigated using the finite element method. The study is carried out with one and two mixing elements over a wide range of flowrates. Validation is made using results obtained on pressure drop. The agreement with published data is very good. The simulation work also allows the understanding of the layer-forming and entanglement mechanisms typical of the mixing process.

INTRODUCTION

The development of mixing process technology is an important field of investigation in chemical engineering due to its growing application in the chemical and food industries. Mixing can be achieved as a batch or a semi-batch operation, using some conventional agitation systems or increasingly as a continuous operation. In the latter case, motionless mixers or static mixers are generally selected (Pahl and Muschelknautz, 1982).

Static mixers consist of stationary rigid elements mounted in the flow channel (often a pipe) to guide the flow. These elements are used to split the fluid into streams which are then combined and split up again in such a way that a mixing effect occurs by entanglement of the fluid layers (Schott et al., 1975). Unlike classical agitation for which the mixing energy is supplied by the rotating impeller, static mixing gets its mixing energy from the fluid flow.

Mixing action is accomplished through several mixing elements, the degree of homogeneity increasing with each additional element. The shape of the elements plays an important role in the mixing mechanism and strongly effects the flow pattern across the pipe. Important factors that influence selection for a given application include desired degree of homogeneity, pressure drop, distribution of residence time and shear rate.

The characterization of parameters governing mixing is generally carried out experimentally (Kabatek and Ditl, 1988) and there is a great deal of data available on pressure drop (Cybulski and Werner, 1986; Lecjaks et al., 1987), residence time (Nauman, 1982; Kemblowski and Pustelnik, 1988) and heat transfer (Nauman, 1979; Genetti, 1982) for specific static mixers. No unified treatment has been published yet and the science of static mixing can be still considered in emergence. The recent progress of advanced numerical technology and the availability of fast computers could provide a powerful tool to better understand the mixing process (Shintre, 1988) and help develop a sound engineering basis for the development and screening of new designs, restricting experimental testing to a smaller number of prototypes.

* CERSIM, Dept. of Chemical Engineering, Laval University, Quebec City, QC, G1K 7P4, Canada

The objective of this paper is to demonstrate the capability of the numerical approach in this domain. More specifically, the purpose is to show that accurate predictions on the flow behaviour through static mixers can be obtained numerically by using state-of-the-art three-dimensional finite element techniques. The study is carried out in the case of a viscous Newtonian fluid flowing across a LPD Dow-Ross static mixer. Results on pressure drop and fluid layer-forming mechanism are presented and compared with available experimental data.

DESCRIPTION OF THE PROBLEM

We consider the static mixer shown in Fig. 1. This system is composed of a tubular housing of length L and radius R in which a series of semielliptical plates has been discriminately positioned. A single mixing element consists of two plates having a 45 degrees angle to the horizontal axis of the flow stream.

We wish to study the flow of a viscous fluid of constant viscosity at a low Reynolds number through this system. For this purpose, we consider the three follo
wing configurations:

one element, R = 0.0508 m and L = 0.508 m hereinafter referred to as SM1D2,
one element, R = 0.1524 m and L = 1.524 m hereinafter referred to as SM1D6,
two elements, R = 0.1524 m and L = 1.79 m hereinafter referred to as SM2D6.

The elements are located at mid-length of the housing and the radius to length ratio is chosen so as to ensure that the fluid is fully developed at the outlet.

MATHEMATICAL MODEL

Equations of change

The equations governing the flow of an incompressible, Newtonian fluid through a static mixer under isothermal conditions at steady state are the classical Navier-Stokes equations:

$$\rho(\mathrm{u.grad\ u}) + \mathrm{grad\ p} - \mathrm{div}\ (\mu\ \mathrm{grad\ u}) = 0 \tag{1}$$

$$\mathrm{div\ u} = 0 \tag{2}$$

where ρ is the fluid density, μ the Newtonian viscosity, u the velocity and p the pressure.

Due to the fully three-dimensional nature of the flow in such a mixer and the lack of symmetry conditions, the computational domain consists of the complete system described before. Cartesian coordinates in three dimensions x,y,z, where z is the main flow direction, are used to express the differential operators in (1) and (2). Boundary conditions are set as follows:

- fully developed flow at the inlet of the tube: $u_x = 0$, $u_y = 0$, $u_z = f(x,y)$ where f(x,y) is a paraboloid centred at the inlet plane whose volume is the flowrate,

- fully developed flow at the outlet i.e. no normal force in the z-direction, $u_x = 0$ and $u_y = 0$,

- no-slip condition at the wall and at the surface of the mixing elements: $u_x = 0$, $u_y = 0$ and $u_z = 0$.

Weak form

The finite element method is used for the solution of the above equations. Using variational calculus principles and the Galerkin method, it is easy to show (e.g. Tanguy, 1982) that equations (1)-(2) can be transformed into:

$$(u.grad\ u,\Phi) - 1/\rho\ (p,div\ \Phi) + \mu/\rho\ (grad\ u,\ grad\ \Phi) = 0 \tag{3}$$

$$(div\ u,\psi) = 0 \tag{4}$$

where Φ and ψ are the Galerkin test functions for the velocity and the pressure respectively and (,) the standard inner product in L^2.

In order to solve equations (3)-(4), it is necessary to discretise the weak form using a suitable approximation for the velocity and the pressure. In the present work, enriched tetrahedral elements $P_1{}^+$-P_0 (Figure 2) are used. The $P_1{}^+$-P_0 element which was introduced recently for fluid flow (Gadbois et al., 1990) is based on the classical linear tetrahedron P_1-P_0 (linear velocity, constant pressure) on which bubble functions have been added at the middle of each face, enabling the control of flowrate within the element. In order to deal directly with degrees of freedom representing velocities and not velocity corrections as it is usual with bubble functions, the shape functions of the enriched tetrahedral were made orthogonal by a Gram-Schmidt orthogonalisation process as explained in (Pelletier et al., 1989). Let us mention finally that this element which is the tetrahedral counterpart of the $Q_1{}^+$-P_0 hexahedral satisfies the Brezzi-Babuska compatibility condition at the element level (Gadbois et al., 1990), ensuring that correct velocities and pressure can be computed.

Solid model and finite element mesh

A major difficulty in the numerical modelling of the flow through static mixers is the generation of the mesh. In the present work, the generation of the solid model (geometry) and the mesh were carried out using PATRAN PLUS software from PDA Engineering. Two models were created, one with one mixing element and the second with two mixing elements.

Depending on the configuration (one or two mixing elements), it was necessary to create 30 to 50 hyper-patches. The fluid domain was meshed in a chunk-wise fashion, each chunk corresponding to one quarter of a portion of the tubular housing. The mesh was specified to be relatively coarse before and after the mixing elements and it was refined in the mixing region.

As PATRAN PLUS software could not deal with $P_1{}^+$-P_0 elements, linear elements were generated and additional degrees of freedom were added at the middle of each face through a customised program. The final mesh included 4016 elements yielding 24503 degrees of freedom for the mixer with one element, and 5536 elements yielding 33644 degrees of freedom for the mixer with two elements.

Numerical algorithm

After discretisation of equations (3)-(4), the final form of the system is obtained:

$$Av + B^Tp + C(v)v = 0 \tag{5}$$

$$Bv = 0 \tag{6}$$

where A is the diffusion (viscous) matrix, C(v) the advection matrix and B the divergence matrix.

The resolution of (5)-(6) by a direct method would require large computational resources due to the size of the matrix A. Indeed, for the two meshes considered, the storage of A is 78 megabytes and 122 megabytes respectively. In order to avoid the large cost of dealing with matrix A directly, we decided to use an iterative scheme, the incomplete Uzawa algorithm (Robichaud et al., 1990), which makes use of an incomplete factorization of A. This algorithm is based on a decoupled resolution of (5)-(6). The first step of the method consists of solving (5) by a preconditioned conjugate gradient method, p being given. The resulting velocity field is then projected onto a divergence-free subspace and the pressure is updated at that level. The advantage of the incomplete Uzawa algorithm over a direct method is twofold. Firstly the storage requirement is much smaller due to the use of an incomplete factorization. In the present work, 3.9

megabytes and 5.3 megabytes were necessary for the two meshes considered, respectively. Secondly, the cpu time to get the solution varies as $NEQ^{1.26}$ with the proposed method compared to $NEQ^{2.33}$ with a direct solver.

NUMERICAL TESTS

A series of numerical simulations (run 1 to run 5) was carried out for the three static mixing configurations presented before, to determine the pressure drop, the distribution of residence time and study the layer forming mechanism. The conditions of the runs were selected according to the set of experimental data available and are shown in Table I.

Mixer radius (m)	Flowrate (litre/min)	Viscosity (Pa.s)	Nb. elements.
0.0508	10-100	10	1
0.1524	10-400	10	1
0.1524	10-400	1100	1
0.1524	10-400	10	2
0.1524	10-400	1100	2

Table 1: Numerical and Experimental Tests Conditions

For all the tests, the value of the fluid density was taken as 1000 kg/m^3.

RESULTS

The first series of results deal with the pressure field in the case of one static mixing element. According to the manufacturer technical sheet, the total pressure drop through the static mixer, evaluated as the pressure drop per element times the number of elements, varies linearly with the flowrate on a Log-Log scale.

We present in Fig. 3, a comparison of the computed and experimental pressure drops versus the volumetric flowrate for runs 1-3. It can be seen that the agreement is very good. The deviation between the two curves varies from 1.1% to 8.5% in the range of flowrates considered. The linear relation existing between Log Q and Log ΔP is also found numerically.

The second series of results deal with the pressure field obtained with two mixing elements. We present in Fig. 4 the results for the pressure drop per element. The manufacturer specifies that the pressure drop per element can be considered as constant for design purposes. Our observations showed that this assumption is not totally valid, although this might be sufficiently accurate for the design of the flow system (pump sizing). Indeed, the pressure drop induced by the second element was found to be systematically lower than for the first one (deviation of about 10% or so). We noticed that the pressure drop for the second element is very similar to the one obtained when there is only one mixing element in the tubular housing.

In order to characterize the mixing of fluid layers, we show in Fig. 5 the position of eleven massless particles injected at the inlet plane in an orderly fashion as they are transported by the flow at different instants (0.1 s, 1 s, 1.1 s, 1.45 s, 1.7 s, 3 s). The viewpoint is that of an observer located at the exit of the tube and looking inside it. In Table 2, the z-position of the elements is given for each instant. Several interesting observations can be made on the swirl induced by the elements. It can be seen first that at 0.1 s the particles are nicely aligned in the cross-section which reflects their position at the injection point. At 1 s, the particles are located right before the first mixing element and start to be influenced (onset of an angular motion) by the shape of the element. At 1.1 s, the swirling effect on the fluid layers is so strong that some intermediate particles (no. 4, 5, 7 and 8) have already changed of quadrant while particles 1, 2, 6, 10 and 11 were not affected yet. At 1.45 s, the intermediate layers marked by particles 2-5 and 7-10 have completed half a turn since the layers are now at about 180 degrees of the initial position (twisting effect). At 1.7 s, the fluid is now reaching the second element and we obtained a pattern similar to the one obtained at 1.1 s except that the layers are not in the same order. The pattern is now such that the splitting and mixing effects can be clearly identified. The initial layer sequence 1,2,...6 and 7...12 now reads as 1,4,3,5... Finally at 3 s we obtained the eventual mixing pattern. The distribution of tracers is noteworthy and illustrates clearly the role of the static elements.

	Time (s)					
	0.1	1.0	1.1	1.45	1.7	3.0
z_1	0.033	0.42	0.47	0.61	0.77	1.64
z_2	0.041	0.51	0.57	0.82	0.96	1.74
z_3	0.048	0.60	0.69	0.94	1.03	1.41
z_4	0.053	0.66	0.77	0.99	1.11	1.58
z_5	0.056	0.68	0.76	0.95	1.14	1.71
z_6	0.059	0.70	0.74	0.76	0.76	1.13
z_7	0.056	0.68	0.76	0.95	1.14	1.71
z_8	0.053	0.66	0.77	0.99	1.11	1.58
z_9	0.048	0.60	0.69	0.94	1.03	141
z_{10}	0.041	0.51	0.57	0.82	0.96	1.74
z_{11}	0.033	0.42	0.47	0.61	0.77	1.64

Table 2: Axial position of the particles

The distribution of residence time for the eleven particles along with the distance covered is displayed in Table 3.

Particle	Time (s)	Distance (m)
1	3.18	1.87
2	3.03	1.85
3	3.95	1.87
4	3.51	1.87
5	3.16	1.87
6	4.08	1.81
7	3.16	1.87
8	3.51	1.87
9	3.95	1.87
10	3.03	1.85
11	3.18	1.87

Table 3: Distribution of residence time

The following observations can be made:

- The residence time varies between 3.03 (particle 2) and 4.08 s (particle 6) which shows that the twisting of the fluid layers has strongly affected the flow pattern. Indeed, particle 6 was the fastest one at the inlet plane but has the longest residence time;

- The distance covered by the particles is very similar which proves that there is no recirculation zone in the mixer;

Another interesting feature of the flow pattern is the evolution of the velocity profile within the mixer. We show in Figs. 6 and 7 the profiles $v_z(x)$ and $v_z(y)$ at three locations along the flow axis. It can be observed that the flow evolves from an initial parabolic shape at the inlet plane to a flat profile as it approaches the mixing elements. In the mixing region, the flow is split up and rotates as it was shown in Fig. 5. After the second element, the profile tends to a plug flow as observed in (Kemblowski and Pustelnik, 1988) for another type of static mixer. It can be speculated that this effect will be more pronounced as the number of elements increases.

CONCLUDING REMARKS

The objective of the paper was to show that advanced finite element techniques could be readily used for the modelling of the very complex flow pattern inside a static mixer. The accuracy of the numerical predictions for the pressure drops was found to be very good, especially when considering the relative coarseness of the finite element mesh used. This work is now being extended to the simulation of simultaneous momentum, heat and mass transfer.

ACKNOWLEDGMENTS

Financial support provided by the Fonds FCAR, Québec, and NSERC, Canada, is gratefully acknowledged.

REFERENCES

Cybulski A. and K. Werner, Int. Chem. Eng., **26**, 171 (1986)
Gadbois M. et al., Int. J. Num. Meth. Fluids, submitted.
Genetti W.E., Chem. Eng. Comm., **14**, 47 (1982)
Kabatek J. and P. Ditl, Proc. 6th Europ. Conf. on Mixing, 545, Pavia, 1988
Kemblowski Z. and P. Pustelnik, Chem. Eng. Sci., **43**, 473 (1988)
Lecjaks Z. et al., Int. Chem. Eng., **27**, 210 (1987)
Nauman E.B., AIChE J., **25**, 246 (1979)
Nauman E.B., Can. J. Chem. Eng., **60**, 136 (1982)
Pahl M.H. and E. Mushelknautz, Int. Chem. Eng., **22**, 197 (1982)
Pelletier D. et al., Proc. 6th Int. Conf. on Num. Meth. in Laminar and Turbulent Flows, 1803, Swansea, 1989
Robichaud M. et al., Int. J. Num. Meth. Fluids, in press.
Schott N.R. et al., Chem. Eng. Progress, **71**, 54 (1975)
Shintre S.N., Proc. 6th Europ. Conf. on Mixing, 551, Pavia, 1988
Tanguy P.A., Ph.D. thesis, Laval University, 1982.

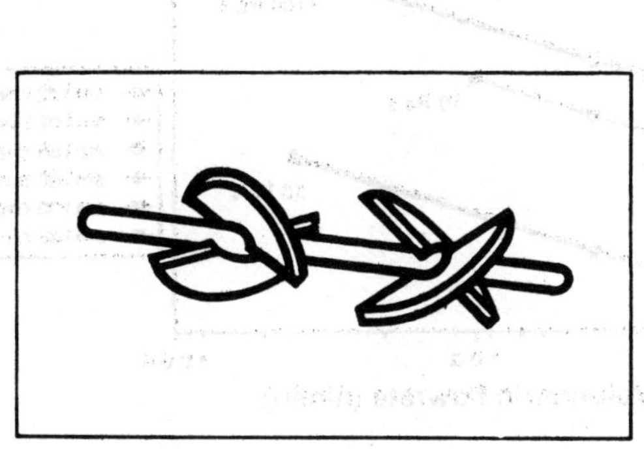

Fig. 1 : LPD Dow-Ross static mixing elements

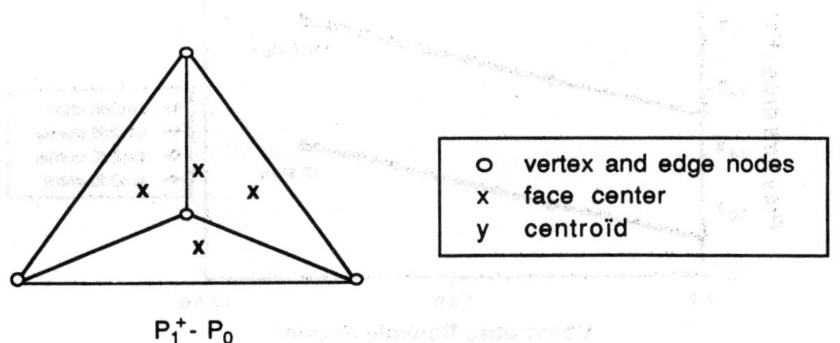

Fig. 2 : P_1^+-P_0 element

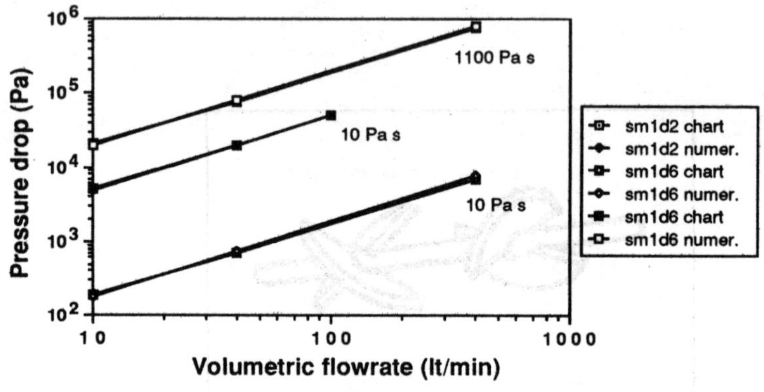

Fig. 3 : Pressure drop for one mixing element

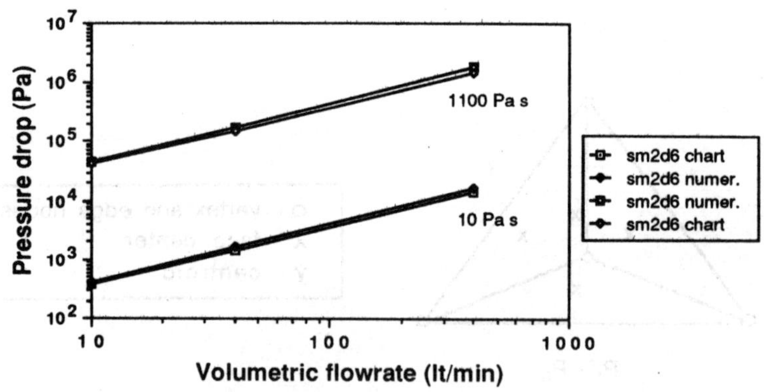

Fig. 4 : Pressure drop for two mixing elements

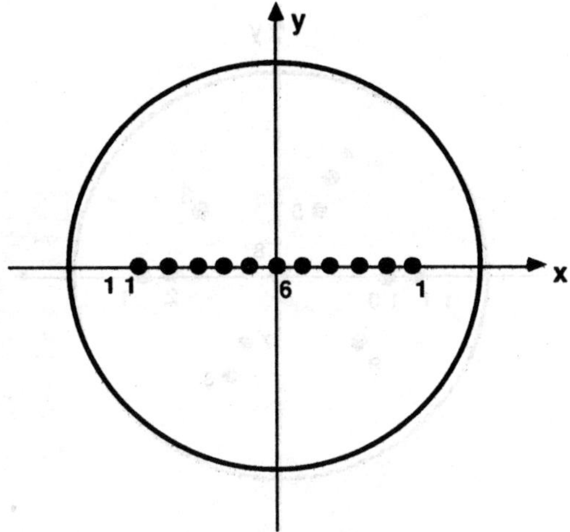

Fig. 5-a : Position of the particles in the mixer at t=0.1s

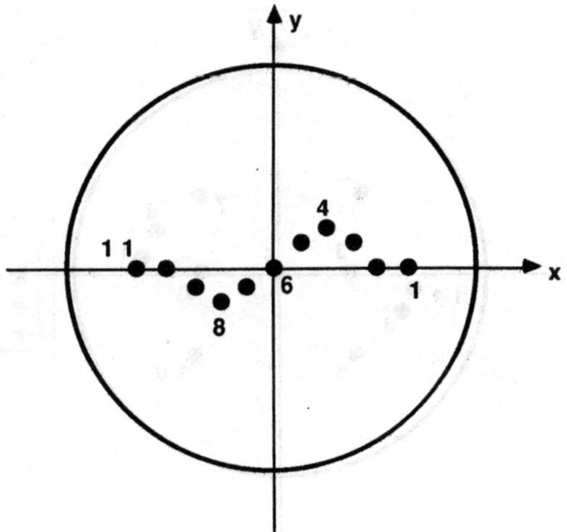

Fig. 5-b : Position of the particles in the mixer at t=1s

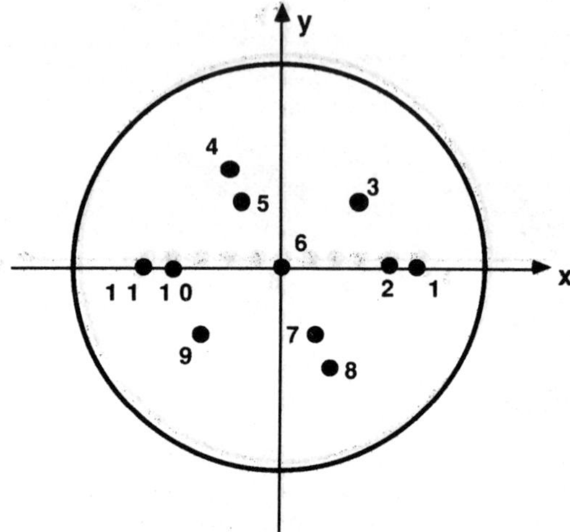

Fig. 5-c : Position of the particles in the mixer at t=1.1s

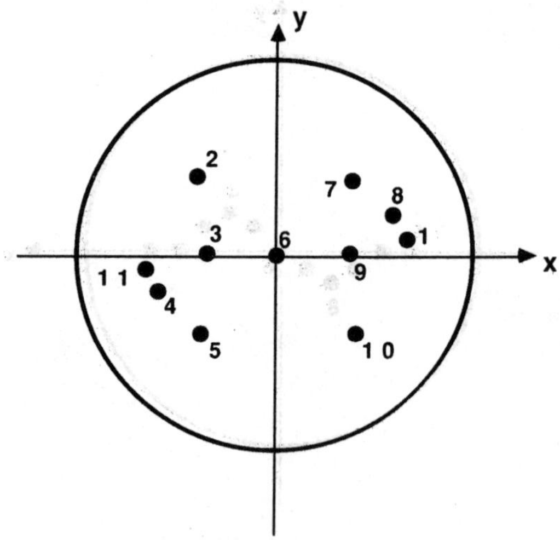

Fig. 5-d : Position of the particles in the mixer at t=1.45s

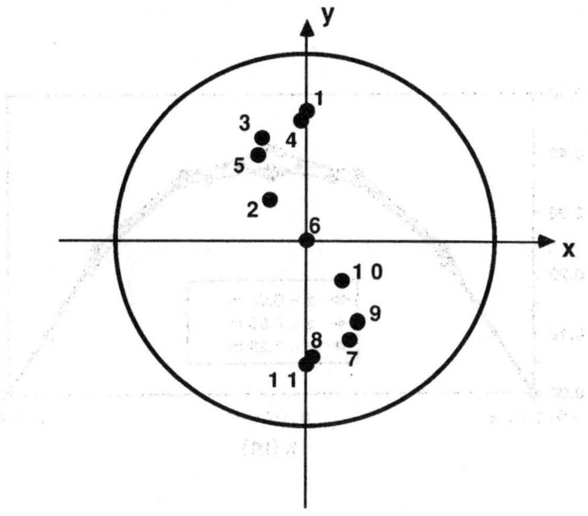

Fig. 5-e : Position of the particles in the mixer at t=1.7s

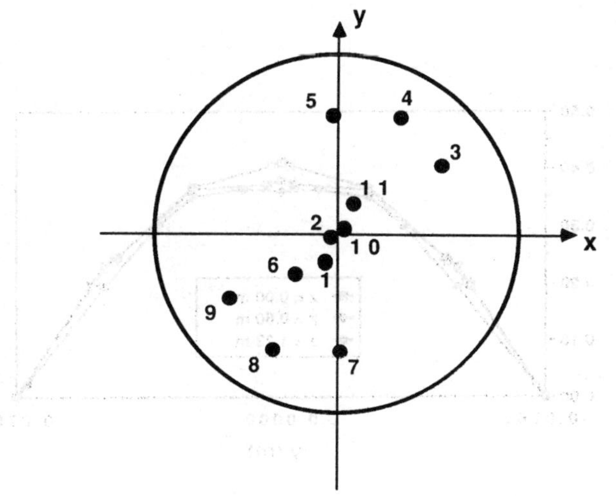

Fig. 5-f : Position of the particles in the mixer at t=3s

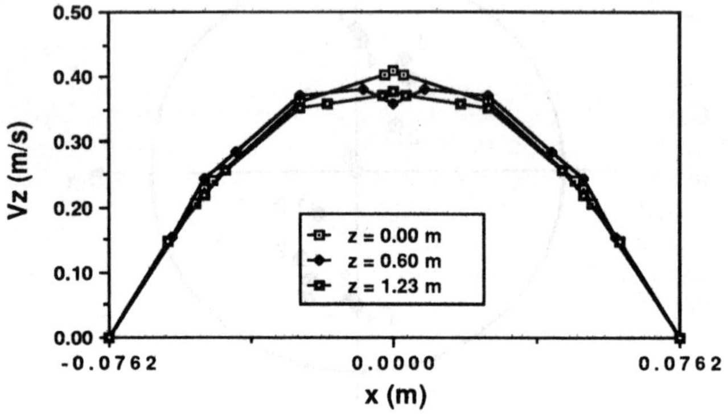

Fig. 6 : Axial velocity profile in the plane y=0

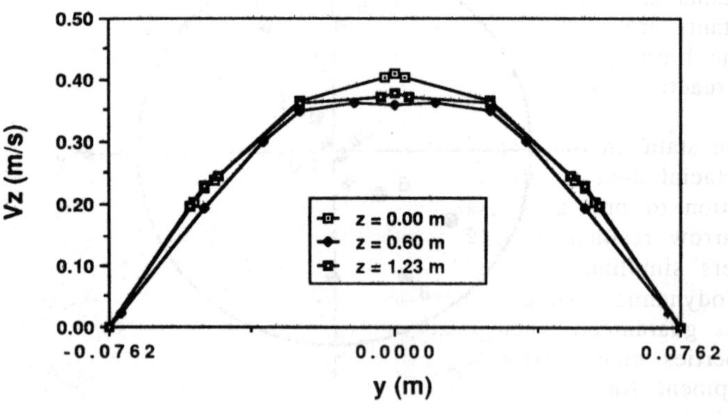

Fig. 7 : Axial velocity profile in the plane x=0

STATIC MIXERS AS GAS/LIQUID REACTORS

G. Schneider*

Gas/liquid reactions limited by mass transfer can be improved by Sulzer static mixers. The features of these as gas/liquid reactors are discussed and compared with those of conventional reactor types. Designs of reactors using static mixers as mass transfer equipment are discussed by means of typical applications.

INTRODUCTION

The process taking place in a reactor can be regarded as the combined effect of three stages: intermixing of the initial reactants, equalization of concentration by diffusion, and the actual chemical reaction. The speed of the reaction is dependent on the temperature, the concentration of the reactants and on the catalysts. In the case of gas/liquid reactions, the reactants are in different phases. The speed of the mass transfer process is the limiting factor within such systems; which in fact often determines the reaction rate.

When static mixers are used to contact gas and liquid flows, a large interfacial area is produced, resulting in optimum mass transfer. In addition to providing highly effective radial mixing static mixers feature a narrow residence time distribution. Where special designs are used, the mixers simultaneously become good heat exchange equipment. Since the hydrodynamic characteristics of static mixers are known, a reliable scale-up is guaranteed when starting out from pilot-scale results. All the properties mentioned above justify the increasing utilization of such equipment for gas/liquid reactions, in particular for reaction systems which are limited by the mass transfer effect.

* Sulzer Brothers Ltd., MRT/0655 Mixing Process Equipment, CH-8401 Winterthur, Switzerland

CONSTRUCTION AND FEATURES OF SULZER MIXERS

The Sulzer SMV mixer (Fig. 1) is used for the mixing and contacting of low viscosity fluids (gases and/or liquids). The mixer consists of several mixing elements arranged in series within a pipe. Each element is built up
from corrugated plates, arranged to form open, intersecting channels. In these channels, the fluid flow is split into sub-flows, which are repeatedly changed in direction and re-united, thereby achieving intermixing in the radial direction. This results in a uniform distribution of concentration, temperature and velocity over the whole flow cross-section. By this means, the basic requirements for a uniform controlled reaction are created. If gas and liquid are present at the same time, the gas is split up into small bubbles of 1 - 2 mm diameter. The bubble diameter is uniform and can be calculated in advance [1]. Unlike the situation in an empty pipe, the bubbles are uniformly distributed over the whole mixer volume (Fig. 2), even when the tube is oriented horizontally, provided that the flow velocity is greater than 0.3 m/s. The bubble size and thus the interfacial surface area can be influenced by the flow velocity and the geometry of the mixing elements fitted. The gas content of the liquid phase can be varied over a wide range without coalescence. The narrow residence time distribution, a consequence of the uniform velocity distribution, ensures that the maximum possible concentration gradient is available for the mass transfer process. Mass transfer is also enhanced by the continuous renewal of the interfacial surfaces of the bubbles [2]. These properties thus accelerate those reactions where rates are limited by mass transfer.

In the Sulzer SMR mixer-reactor, the mixing elements are made from tubes through which a heat transfer medium flows (Fig. 3). The mixing elements themselves therefore constitute active heat exchange surface. In addition to the properties of the SMV mixer described above, which apply equally to the SMR mixer-reactor, the latter has also a high heat transfer capacity, by virtue of a high heat transfer coefficient and a large internal heat transfer surface compared to the operating volume. A particular feature of the SMR is that the ratio of heat transfer surface to volume can be kept constant to a great extent when scale-up takes place. Using this equipment, mixing, dispersion and contacting can take place simultaneously with the transfer of heat directly at the reaction location. Isothermal conditions with endo or exothermic reactions can thus be guaranteed.

The pressure drop across the Sulzer mixer can be calculated in advance [3]. It is small and can be matched to the requirements of the process. The pressure along the mixer or reactor decreases at a constant rate, i.e. the energy dissipation is uniform over the whole mixer volume. This is also the reason why the bubbles are of uniform size and are uniformly distributed over the whole volume. Hence the much improved mass transfer values as a function of the energy input obtained by Sulzer mixers are higher in comparison with those applying for stirred vessels. In stirred vessels, there is a very high energy dissipation near to the stirrer which drops sharply as the distance from the stirrer increases (Fig. 4).

A feature common to all Sulzer mixers is the fact that the hydrodynamic behaviour has been carefully investigated and is thus known. This is the key factor enabling reliable reproduction of the results of pilot investigations on a full scale plant. Scale-ups with factors of more than 100 - 1000 have already been successfully carried out.

GAS/LIQUID MASS TRANSFER USING SULZER MIXERS

Chemical gas/liquid reactions often have reaction rates limited by the mass transfer process. Such reactions can thus only be accelerated by improvement of the physical absorption of the gas into the liquid phase. The individual parts of the process and the quantities influencing the same have to be investigated in order to be able to take appropriate measures. In practice, mass transfer can be influenced as follows:

Influencing factor/quantity:	Established by:
• physical properties of the substances in a system	nature, temperature/pressure of the system, operation of the plant
• process conditions	process engineer / operation of the plant
• hydrodynamic behaviour of the mass transfer equipment	equipment / equipment supplier

The following formula applies:

$\dot{n}$	=	k_L	x	a	x	$(c^* - c)$
Mass transfer rate	=	Mass transfer coefficient	x	Specific mass transfer area	x	Driving concentration gradient
		HTU			x	NTU
	=	Hydrodynamic characteristics of the equipment			x	Physical properties of the components in the system

The following quantities exert an influence on the individual factors:

$\dot{n}$	=	k_L	x	a	x	$(c^* - c)$
		• Diffusion • Turbulence • Mixer geometry		• Interfacial tension • Energy dissipation • Phase ratio gas/liquid • Turbulence • Mixer geo-metry		• Solubility • Pressure • Temperature • Concentration
Physical properties of the components		+		±		+
Process conditions		+		±		±
Major dimensions of the equipment		±		+		
Influence is: + marked ± moderate - slight						

The above table shows clearly that the interfacial surface area, a, has a significant influence on the mass transfer process. This interfacial surface area is in turn strongly dependent on the mixer/reactor design. A well-designed layout of the Sulzer mixer can significantly accelerate mass transfer and thus the overall accomplishment of a gas/liquid reaction, and/or significantly decrease the reactor volume required for a given throughput. From the point of view of safety, in most cases a small reactor volume is advantageous, or even mandatory.

GAS/LIQUID REACTORS WITH SULZER MIXERS

The following comments refer to coalescing gas/liquid systems. Furthermore, the characteristics of a few contacting devices of conventional design are reviewed for comparison purposes.

Stirred vessel

The stirred vessel (Fig. 5) has high local energy dissipation adjacent to the stirrer, where fine bubbles (diameter less than 1 mm) are formed, and thus a large interfacial area, a, is produced. The residence time in this zone is very short. However, outside this zone, the bubbles coalesce (diameter 3 to 5 mm) and the mass transfer surface area decreases sharply. The system is totally back mixed.

Diaphragm, perforated plate

Here also a high local energy dissipation occurs adjacent to the diaphragm or to the perforated plate (Fig. 6), thereby producing fine bubbles of short life, which coalesce immediately into larger bubbles. Equipment of this type have generally a high degree of back mixing.

Sulzer mixers in in-line configuration (Fig. 7)

Energy dissipation is uniform over the whole mixer volume. Bubbles of equal size (diameter 1 to 2 mm) are uniformly distributed over the whole reactor volume. Plug flow takes place in the equipment and the residence time is in general short (up to a few seconds). Such a mixer is used for the physical absorption of gases and for chemical reactions which take place quickly. It is utilized, for example, for the dissolution of chlorine in water, where the chlorine is subsequently to react with alkenes.

A residence time of up to 1 to 2 minutes can be achieved by arranging gaps between the mixing elements (Fig. 8). A slight fall-off in the residence time behaviour compared with the configuration described above takes place. Such a mixer is used for chemical reactions which require a longer residence time and for physical absorption tasks. Sulzer mixers are being successfully applied as part of a loop reactor utilized for the hydrogenation of aromatic substances in conjunction with a suspended catalyst.

Loop reactor with forced circulation (Fig. 9)

The liquid and gaseous reaction components are rapidly distributed into the circuit by means of Sulzer mixers. Further gas for the reaction is mixed in the SMR mixer-reactor. The heat produced by the strongly exothermic reaction is removed adjacent to its place of origin by the SMR. The reverse applies for endothermic reactions (heat transfer capacity up to 40 kW/m^3 °C). The reaction temperature can be kept constant within very narrow limits. Hence, for example, side reactions are strongly depressed, or reactions can be carried out at temperatures near to the critical limit. This means that the volume of the equipment can be kept small. Total back mixing of the reactor contents takes place and the residence time can range from a few seconds to some minutes. Since the quantity of liquid recirculated can be varied, it is possible to match the reactor characteristic optimally to the particular process requirements. For example, using such reactors, organic products can be oxidized by means of air under precisely controlled temperature conditions.

Loop reactor with natural circulation (Fig. 10)

In this reactor type also, the liquid and gaseous reactants are mixed with or dispersed into the circuit liquid by means of Sulzer mixers. The density difference between the two legs is the driving force for the circulation of the reactor contents. Total back mixing occurs also in this system. Removal of the heat of reaction is often accomplished by evaporation of the reaction product or of the solvent. Such reactors are used in the production of ammonium nitrate from nitric acid and ammonia or for the chlorination of ethylene to form EDC (ethylene dichloride), an intermediate in the production of PVC.

Bubble reactor

The bubble reactor (Fig. 11) is a loop reactor with natural circulation (air lift principle). The inner tube is completely filled with the Sulzer SMV mixer-packing. The liquid and gaseous reactants introduced at the base of this tube are rapidly mixed with or dispersed into the reactor content. A high mass transfer is thus achieved. The excess gas is separated off at the head and the liquid in the annular space between the internal tube and the reactor wall is recirculated. The bubble reactor is widely used where long residence times (minutes to hours) are required for the chemical reaction or for desorption to take place. This system is completely back mixed. Reactors having a height of 10 m and diameters of more than 1 m are being used for the oxidation of sodium sulphite to sodium sulphate (Fig. 12).

Cascade type bubble reactor

The cascade type reactor (Fig. 13) consists of several bubble reactors arranged one above the other. The reactor volume and the number of individual stages are matched to the shape of the reaction curve (Fig. 14). The gas and the liquid flow through the reactor in opposite directions (counter-current mode). Individual stages function in the co-current mode and are completely back mixed. The total reactor volume can be kept small by optimum matching. The small reactor volume and the fact that the gas and the liquid flow from stage to stage without requiring pumps are particularly advantageous in the case of reactions which take place under increased pressure. This type of reactor is used, e.g. for the oxidation of organic sulphur compounds under increased pressure and at higher temperatures.

Supply and removal of heat can take place via heat exchangers which are fitted in the annular gaps of the individual stages. A combination of mixer-packing for mass transfer and SMR mixer-reactor for heat transfer can also be fitted in the reaction space for this purpose.

Stirred vessel with static mixer

Existing equipment with insufficient mass transfer capacity or low absorption efficiencies, e.g. a stirred vessel, can be improved by the addition of an external liquid recirculation facility incorporating a static mixer (Fig. 15) with upstream gas admixing. The mixer provides a higher mass transfer, the original vessel functions as a residence time tank and as a gas separator. The performance of numerous conventional reactors have already been markedly improved by the provision of such additional equipment.

COMPARISON

As can be seen from the examples quoted, the Sulzer mixer provides high mass transfer, thereby enabling gas/liquid reactors to be compact. The comparison of the k_La values for the same energy consumption shows the superiority of the Sulzer mixers compared with conventional energy input systems.

According to Figure 4:

$$\frac{k_La \text{ of a Sulzer mixer}}{k_La \text{ of a conventional system}} = 10 \text{ to } 50$$

A further noteworthy feature of these reactors is the fact that the Sulzer mixer has no moving parts and thus requires very little maintenance. Shaft penetrations are not required, hence the problems associated with shaft sealing do not occur.

The Sulzer mixer has a favourable influence on the chemical reaction in that, where equipment with plug flow characteristics is used (in-line and cascade-type reactors), increased throughput is obtained.

General safety considerations suggest that the reactor volume is to be as small as possible and that the minimum number of seals are to be used. These two requirements are met by the use of static mixers as reaction equipment.

Sulzer static mixers have proven their effectiveness as gas/liquid reactors in many applications for low viscosity, coalescing reaction systems. The hydro-dynamic characteristics allow the reactors to be optimally matched to the particular process conditions. Optimum solutions to problems involving constructional material are also possible, since the mixer can be made from various plastics (PP, PVDF and PTFE), as well as from metallic materials, such as stainless steel, hastelloy, monel, titanium, tantalum etc. Mixing elements of small diameter (the smallest to date is 3.2 mm!) are available for test equipment. Reliable scale-up, based on experimental results, is available, due to the in depth knowledge of the hydrodynamic behaviour which has been obtained.

BIBLIOGRAPHY

[1] Streiff, F.: Anwenden statischer Mischer beim In-line Dispergieren. Maschinenmarkt 83 (1977), p. 289-295.

[2] Grosz, F., Bättig, J., Moser, F.: Gas/Liquid Mass Transfer with Static Mixing Units.
Forth European Conference on Mixing, April 27-29 1982, Paper F2.

[3] Sulzer brochure No. 23.27.06., Sulzer Mixing Process Equipment.

NOMENCLATURE

a	m^2/m^3	Specific mass transfer area
c	$kmol/m^3$	Concentration
c*	$kmol/m^3$	Concentration at equilibrium
DN	mm	Nominal diameter
g	m/s^2	Acceleration due to gravity
HTU	m	Height of a transfer unit
k_L	m/s	Mass transfer coefficient in the liquid
$\dot{n}$	$kmol/m^3 \cdot s$	Mass transfer rate
NTU	-	Number of transfer units
P	W	Power
V	m^3	Mixer volume
$\dot{V}$	m^3/s	Volume flow
η	Pa·s	Dynamic viscosity
ρ	kg/m^3	Density

Indices

G	Gas
L	Liquid
α	Inlet
ω	Outlet

Fig. 1 SMV mixer DN 350.

Fig. 2 Bubble bed in a horizontal Sulzer mixer of type SMV-16 DN 100. Air/water system. Water flow velocity 0.7 m/s, average bubble diameter 1.2 mm.

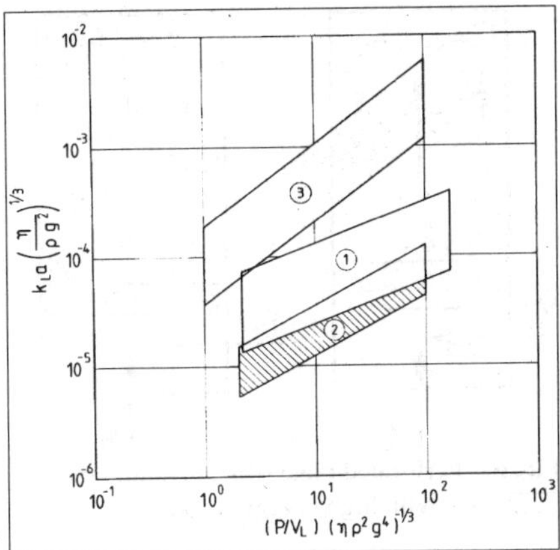

Fig. 4 Comparison of the $k_L a$ values for a stirred loop reactor (1), a stirred vessel (2) and a Sulzer SMV mixer (3) for coalescing systems.

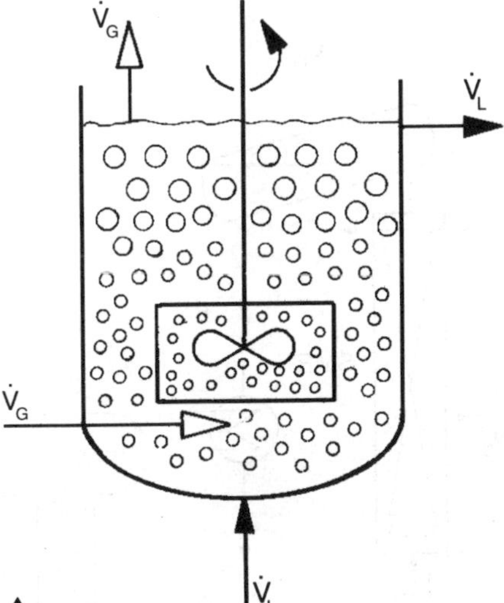

▲ **Fig. 5** Stirred vessel
▭ Zone with high energy dissipation

◀ **Fig. 3** Sulzer SMR mixer-reactor DN 80. The mixing elements are made from tubes, thus consti-tuting active heat exchange surface

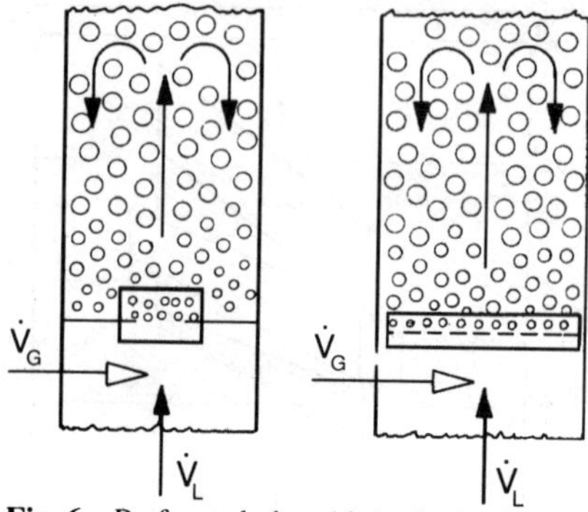

Fig. 6 Perforated plate (rhs), diaphragm (lhs).
▭ Zone with high energy intake.

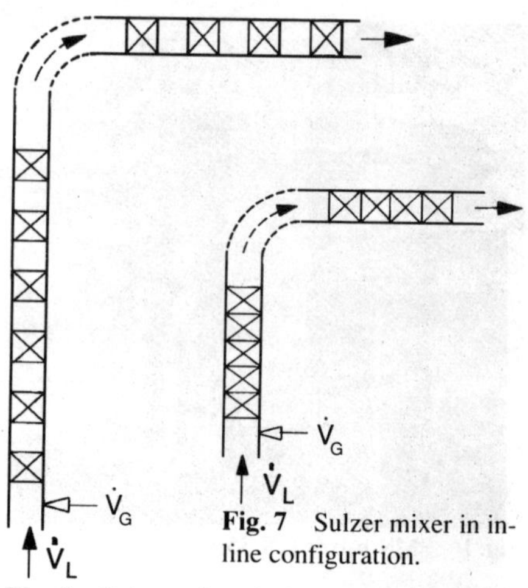

Fig. 7 Sulzer mixer in in-line configuration.

Fig. 8 Sulzer mixer in in-line configuration, mixing elements have gaps between them.

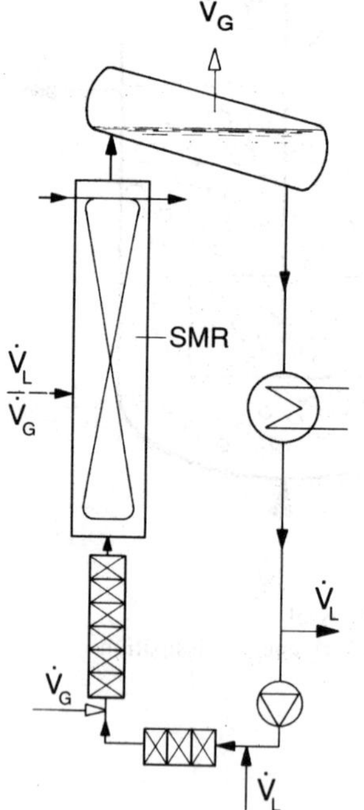

Fig. 9 Loop reactor with forced circulation.

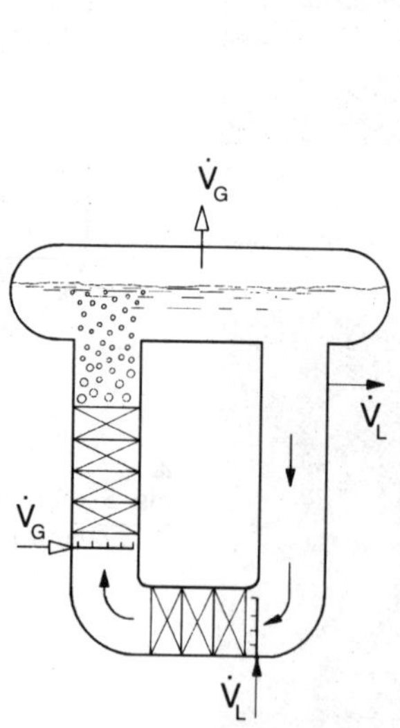

Fig. 10 Loop reactor with natural circulation.

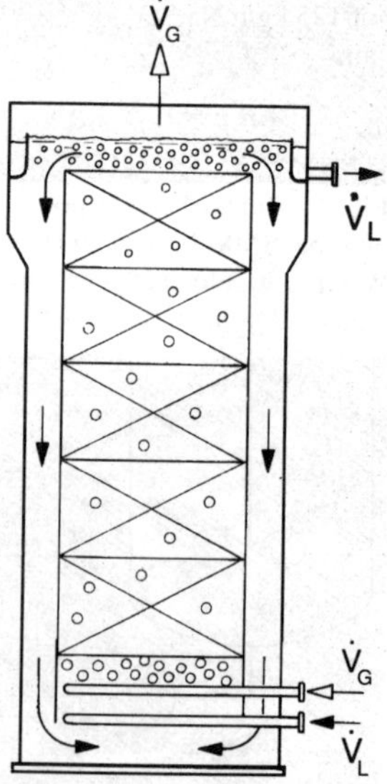

Fig. 11 bubble reactor (loop reactor using the air lift principle).

Fig. 12 Sulzer bubble reactor for the oxidation of 125 kg/h Na_2SO_3 to Na_2SO_4 using 2300 m³/h air.

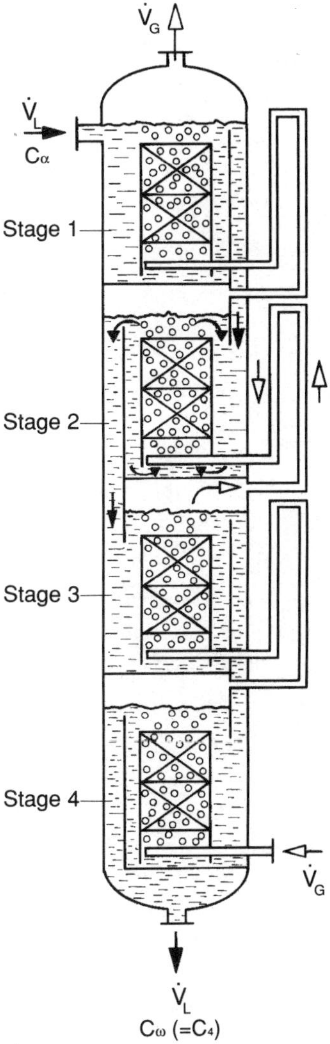

$\dot{V}_G$

$\dot{V}_L$ $C\alpha$

Stage 1

Stage 2

Stage 3

Stage 4

$\dot{V}_G$

$\dot{V}_L$ $C\omega$ (=C_4)

Fig. 13 Cascade type bubble reactor with four stages.

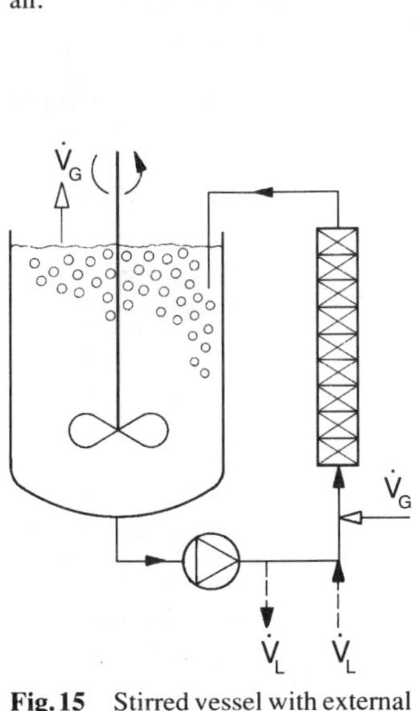

Fig. 15 Stirred vessel with external static mixer to improve mass transfer.

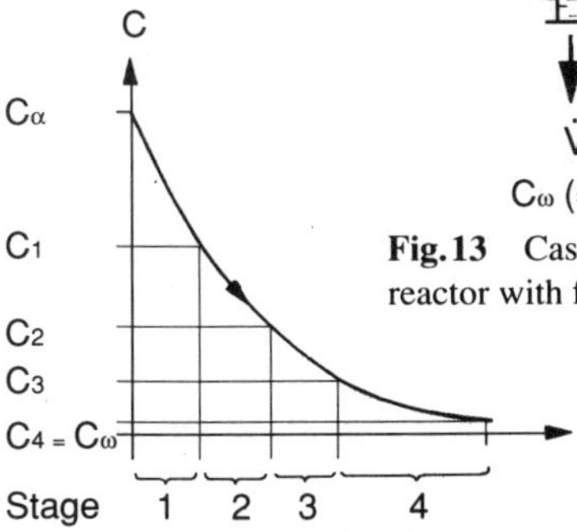

C

$C\alpha$

C_1

C_2

C_3

$C_4 = C\omega$

Stage 1 2 3 4

Fig. 14 Reaction/concentration curve for a four-stage cascade type reactor.

MIXING ENERGIES FOR SUBMERGED JET GAS-LIQUID TRANSFER

J. Varley*

In the submerged jet device described here, liquid is withdrawn from a tank and, after oxygen injection via a venturi, is recycled to the tank through a converging nozzle. A gas-liquid mixture therefore enters the tank as a submerged jet. The submerged jet mixes the contents of the tank and creates interfacial area for oxygen transfer. Submerged jet systems are typically used for waste water treatment and operate at mixing energies of 10-100 W/m^3. This is considerably lower than mixing energies of 500-1000 W/m^3 for mechanically agitated contactors, which are also often used for waste water treatment. Submerged jet systems may therefore provide a more energy efficient alternative for oxygen transfer. To explore this possibility, it is important to establish how oxygen transfer varies with mixing energy input to the system. Experiments were undertaken, on three scales of apparatus, to establish the dependence of rates of oxygen dissolution, oxygen utilisation efficiency and oxygen transfer efficiency on the mixing energy input. The mixing energy can be altered by changing either (i) velocity of gas-liquid jet, (ii) flowrate of the gas-liquid jet or (iii) volume of liquid in the tank. The experimental results show that the relationship between mixing energy and oxygen transfer rates and efficiencies depends on how the change in mixing energy is achieved. Results, similar to those presented here, can be used to determine which process parameters maximise rates of oxygen dissolution for a given mixing energy input. This is important, not only for submerged jet oxygenation systems, but also, for other gas-liquid contacting devices. Results for the submerged jet system are compared with values available in the literature for other gas-liquid contacting devices.

INTRODUCTION

Gas-liquid contacting devices are an important component of a wide variety of processes in many industries *eg* chemical and food industries. Submerged jet gas-liquid contacting devices have to date been used to a limited extent, and are mainly found in the waste water treatment industry. In the submerged jet device described here, liquid is withdrawn from a tank and, after oxygen injection via a venturi, is recycled to the tank through a converging nozzle. A gas-liquid mixture thus enters the tank as a submerged jet. (For schematic diagram of the system see Figure 1). The submerged jet mixes the contents of the tank and creates interfacial area for oxygen transfer. The efficiency of this device, as for any other gas-liquid transfer device, should be measured in terms of both its capability as a mixing device and as a gas-liquid transfer device. Mixing performance is generally measured in terms of mixing time.

* Department of Food Science and Technology, Reading University.

Gas-liquid transfer efficiency is measured in terms of utilisation efficiency (*ie* percentage of gas injected that is dissolved) and transfer efficiency (*ie* rate of gas transferred per unit power input).

Mixing time correlations derived for liquid-liquid mixing are reported in the literature for submerged jets (for a review of such correlations see Revill (1)). Using a correlation derived by Hiby and Modigell (2), Revill (1) has shown that for the same power input the mixing time is significantly higher for a submerged jet system than for a mechanically agitated device. However, in waste water treatment processes, submerged jet oxygenation systems are typically operated at lower mixing energies than other devices commonly used in this area *eg* mechanically agitated contactors. Mixing energies are typically 10-100 W/m^3 for submerged jet systems, as compared to 500-1000 W/m^3 for mechanically agitated contactors. In waste water treatment, residence times for the necessary biological activity are significantly higher than mixing times of submerged jet systems. Therefore, even if mixing times for submerged jet systems are considerably higher than for other types of gas-liquid contacting devices, they are still insignificant when compared to the bulk liquid residence time. For situations where the mixing time is insignificant as compared to the residence time, the relative energy efficiencies of different oxygen transfer devices will depend on the amount of oxygen transferred per unit mixing energy input, as opposed to the mixing time achieved at a given mixing energy.

Before any meaningful comparisons can be made between energy efficiencies of various gas-liquid transfer devices, it is therefore important to establish how gas-liquid transfer is related to mixing energy input. The work presented here attempts to determine this dependence for gas-liquid transfer in submerged jet systems.

Gas-liquid transfer rates for oxygen transfer in a submerged jet system can be determined from the following mass transfer rate equation:

$$\frac{dC}{dt} = k_L a (C_s - C) \qquad (1)$$

where C is the dissolved oxygen level, C_s is the saturation dissolved oxygen level, t is time, k_L is the mass transfer coefficient and a is the interfacial area per unit volume of liquid.

Mixing energy is an important variable for gas-liquid transfer for several reasons. Firstly, as mixing energy increases the bulk liquid velocities increase. This will lead to higher mass transfer coefficients as a result of higher rates of surface renewal at the gas-liquid interface (as expected from the generally accepted theories of Higbie (3) and Danckwerts (4)). Secondly, as a result of its effect on bulk liquid motion, mixing energy influences the path followed by the bubbles, bubble residence times, gas phase hold ups and hence gas-liquid interfacial area. Thirdly, mixing energy variations resulting from changes in jet velocity will cause variations in bubble sizes and hence in the interfacial area available for gas-liquid transfer. If however, the mixing energy is altered by varying the liquid volume bubble size will remain largely unchanged.

Mixing energy for submerged jet systems can be calculated from the following equation:

$$\text{Mixing energy} = \frac{v_j^2 \ Q_l \ \rho_l}{2 \ V_t} \qquad (2)$$

where v_j is the jet velocity, Q_l is the liquid volumetric flowrate through the nozzle, ρ_l is the liquid density and V_t is the liquid volume in the tank.

From Equation (2) it is obvious that mixing energy can be altered by varying several parameters *eg* jet velocity and liquid volume. Many correlations in the literature are available for determining bubble diameters (*eg* Calderbank (5), Pandit and Davidson (6)) and mass transfer coefficients (*eg* Linek, Vacek and Benes (7)) as a function of mixing energy (W/m^3). In the development of all these correlations, mixing energy was only altered by changing the impeller speed or the jet velocity, and not by changing liquid volume. It is therefore important to realise that such correlations are only applicable for such changes in mixing energy.

EXPERIMENTAL

Experimental apparatus

A schematic diagram of the experimental apparatus is given in Figure 1. The tank was filled with tap water. A sidestream of liquid was withdrawn from the tank by means of a variable speed centrifugal pump. The liquid was then pumped through a calibrated rotameter and into a venturi.

Nitrogen (used to deoxygenate the water) and oxygen from compressed gas cylinders were connected to a four way valve. This allowed the gas in the line to be switched simultaneously from nitrogen to oxygen. The gas flowed through the four way valve, a control valve, a series of previously calibrated rotameters and into an annular ring surrounding the venturi throat. The gas then flowed through four holes (< 1mm diameter), which were equally spaced around the venturi throat, and into the liquid flowing through the venturi. The gas-liquid mixture finally flowed through the remaining pipeline before being jetted back into the tank through a converging nozzle.

Experiments were carried out on two scales of apparatus, laboratory and pilot plant scale. A limited number of results are also presented for an industrial scale unit. These results were collected by BOC, Environmental Department, who sponsored the work presented here. The major dimensions of each scale of apparatus were as follows:

Scale	Length (m)	Width (m)	Maximum liquid depth (m)	Nozzle Diameter (m)
1	0.6	0.3	0.6	0.0050
2	1.5	0.75	1.5	0.0125
	Diameter (m)			
3	6.4		7.9	0.0200

The tank was rectangular for scale 1 and 2 and circular for scale 3 apparatus. For scale 3 apparatus, there were 6 nozzles as opposed to one for scale 1 and 2 apparatus.

Experimental procedure

The tank was filled to a measured level with tap water. The recirculation pump was switched on and the liquid flow set at the desired rate. The nitrogen line was opened and the required flowrate of nitrogen was injected into the liquid until the dissolved oxygen level had reached a standard low value (this low value was typically 1mg/l, which is about 2% of the saturation level of pure oxygen). When this value was reached, the gas in the line was switched from nitrogen to oxygen. Oxygen was subsequently injected until a standard high value of dissolved oxygen was reached in the tank (this value was typically 10 mg O_2/l, which is about 20% of the saturation level of pure oxygen). The rate of rise of dissolved oxygen was measured in the tank and in the pipeline just upstream of the nozzle using polarographic electrodes (first order response time < 7 seconds). The pressure was measured at the venturi and nozzle inlet using pressure transducers. These measurements were used to determine the oxygen transfer efficiency (see below).

Analysis of experimental measurements

Rate of oxygen dissolution and oxygen utilisation efficiency. Oxygen is dissolved in the venturi, pipeline and the tank. The oxygen utilisation efficiency was determined for (i) the overall system (*ie* tank, pipeline and venturi), (ii) the pipework (including the pipeline and the venturi) and (iii) the tank only.

The oxygen utilisation efficiency (OUE) is defined as:

$$OUE = \frac{\text{Rate of oxygen dissolution}}{\text{Rate of oxygen injection}} \times 100 \qquad (3)$$

(i) Overall system. The rate of oxygen dissolution in the overall system is given by:

$$\text{Rate of overall oxygen dissolution} = V_t \frac{dC_t}{dt} \qquad (4)$$

where C_t is the dissolved oxygen level in the tank.

(Note:- dC_t/dt was measured as the dissolved oxygen level rose from 1 to 10 mg/l, and was found to remain constant over this range).

The rate of oxygen injection to the overall system is $\rho_g Q_g$, where ρ_g is the density of the gas and Q_g is the gas volumetric flowrate. The oxygen utilisation efficiency for the overall system, $(OUE)_o$, can therefore be determined from Equations (3) and (4):

$$(OUE)_o = \frac{V_t \dfrac{dC_t}{dt}}{\rho_g Q_g} \times 100 \qquad (5)$$

(ii) Pipework. The rate of oxygen dissolution in the pipework is given by:

$$\text{Rate of oxygen dissolution in the pipework} = Q_l (C_p - C_t) \qquad (6)$$

where C_p is the dissolved oxygen level in the pipeline.

The rate of oxygen injected into the pipework at the venturi is $\rho_g Q_g$. Therefore from Equations (3) and (6) the oxygen utilisation efficiency for the pipework, $(OUE)_p$, is given by:

$$(OUE)_p = \frac{Q_l (C_p - C_t)}{\rho_g Q_g} \times 100 \qquad (7)$$

(iii) Tank. The rate of oxygen dissolution in the tank can be determined from a mass balance of the oxygen dissolved and injected in the system (see Figure 2):

$$\text{Rate of oxygen dissolution in the tank} = \text{Rate of oxygen dissolution in the overall system} - \text{Rate of oxygen dissolution in the pipework} \qquad (8)$$

From Equations (4), (6) and (8)

$$\text{Rate of oxygen dissolution in the tank} = V_t \frac{dC_t}{dt} - Q_l (C_p - C_t) \qquad (9)$$

The rate of oxygen injected into the tank is equal to the rate injected at the venturi minus the rate dissolved in the pipework *ie*,

$$\text{Rate of oxygen injection into tank} = \rho_g Q_g - Q_l (C_p - C_t) \qquad (10)$$

The oxygen utilisation efficiency for the tank, $(OUE)_t$, can then be determined from Equations (3), (9) and (10):

$$(OUE)_t = \frac{V_t \dfrac{dC_t}{dt} - Q_l (C_p - C_t)}{\rho_g Q_g - Q_l(C_p - C_t)} \tag{11}$$

Oxygen transfer efficiency. The oxygen transfer efficiency (OTE) is defined as:

$$OTE = \frac{\text{Rate of oxygen dissolution}}{\text{Power input}} \tag{12}$$

The power input was determined from the following equation:

$$P = (Q_g + Q_l) \Delta P \tag{13}$$

where P is the power input and ΔP is the pressure drop.

The pressure drop was measured (i) between the venturi inlet and the liquid surface for evaluation of oxygen transfer efficiency for the overall system, (ii) between the venturi inlet and the nozzle inlet for OTE in the pipework only and (iii) between the nozzle inlet and the liquid surface for OTE in the tank only.

Determination of power by this method ignores several factors (i) pump efficiency, (ii) power required to transfer the liquid to the venturi throat and (iii) compressibility of the gas phase. The results for oxygen transfer efficiency should therefore be considered in a qualitative and not a quantitative light.

The oxygen transfer efficiency for the overall system, the pipework, and the tank were determined from Equations (4), (6) and (9) respectively, the relevant power input determined from Equation (13), and Equation (12).

Mixing energy. Mixing energy was determined from the following equation:

$$\text{Mixing energy} = \frac{v_j^2 \; Q_l \; \rho_l}{2 \; V_t} \tag{2}$$

The mixing energy was varied by (i) altering the jet velocity which also altered the liquid volumetric flowrate and (ii) altering the liquid volume by altering the liquid depth. (Note:- since the apparatus was fixed it was impossible to alter the liquid flowrate and jet velocity independently).

RESULTS

As explained in the experimental section above rates of oxygen dissolution and oxygen efficiencies were determined experimentally for (i) the overall system, (ii) the pipework, and (iii) the tank. However, for scale 1 and 2 the results for the tank only will be discussed here. There are several reasons for limiting the discussion to dissolution in the tank. Firstly, these results can most easily be compared with results reported in the literature for other types of gas-liquid contacting device. Secondly, in general the majority of oxygen transfer takes place in the tank and therefore in terms of oxygen transfer this is the most important part of the system. Finally, a meaningful mixing energy can only be defined for the tank and not for a system consisting of a tank and pipework. However, for scale 3 apparatus results are only available for oxygen transfer in the overall system. Oxygen dissolution in the overall system will be a maximum of 20% higher than dissolution in the tank (Kite (8)).

Results showing the variation of rate of oxygen dissolution, oxygen utilisation efficiency and oxygen transfer efficiency with mixing energy are presented below.

Rates of oxygen dissolution. For scale 1 apparatus, Figure 3 shows that if all process variables except liquid depth are held constant the rate of oxygen dissolution decreases as mixing energy input increases. For scale 2 apparatus, Figure 4 shows that if all the process variables except liquid velocity are held constant the rate of oxygen dissolution increases asymptotically as mixing energy increases. Although Figure 3 and 4 are for scale 1 and 2 apparatus respectively, similar relationships exist for all three scales of apparatus.

Oxygen utilisation efficiency. Figure 5 shows that if all process variables except liquid depth are held constant oxygen utilisation efficiency decreases as the mixing energy input increases. Figure 6 shows that if all the process variables except liquid velocity are held constant the oxygen utilisation efficiency increases asymptotically as mixing energy increases.

Oxygen transfer efficiency. Figure 7 shows that if all process variables except liquid depth are held constant oxygen transfer efficiency decreases as the mixing energy input increases. Figure 8 shows that if all the process variables except liquid velocity are held constant the oxygen transfer efficiency decreases as mixing energy increases.

Note:- The rate of oxygen dissolution will obviously depend on the gas injection rate. Therefore the rate of injection or gas-liquid injection ratio (*ie*, gas volumetric flowrate/gas plus liquid volumetric flowrate) was held constant in each set of experiments, although different injection rates were used for different scales of apparatus.

DISCUSSION

Mixing energy varied by changing liquid volume

Consider the following mass transfer rate equation for transfer of oxygen from the gas to the liquid phase:

$$\frac{dC}{dt} = k_L\, a\, (C_s - C) \tag{1}$$

The decrease in rate of oxygen dissolution and oxygen utilisation efficiency with increasing mixing energy shown in Figures 3 and 5 can be explained by considering the parameters in Equation (1). The mass transfer coefficient (k_L) is likely to increase as a result of increased liquid velocities and hence higher rates of surface renewal at the gas-liquid interface. The driving force for mass transfer (C_s - C) will be largely unaltered by changes in mixing energy (providing that the liquid phase is well mixed for the entire range of mixing energies under consideration). Therefore, the observed decrease in rates of oxygen dissolution and oxygen utilisation efficiency must be a result of lower gas-liquid interfacial area at higher mixing energies. Since the size of bubbles formed at the nozzle will be little altered by changes in mixing energy affected by reducing the liquid volume, the decrease in interfacial area must be a result of shorter bubble residence times and lower gas phase hold-ups.

Changes in liquid depth make only a small change to the required power input and therefore the observed decrease in oxygen transfer efficiency with increasing mixing energy (shown in Figure 7) is mainly a result of reduction in rates of oxygen dissolution.

The results discussed above show that liquid depths should be kept high if oxygen transfer rates are to be maximised and power requirements minimised, even though this reduces the mixing energy input to the system. These results also have implications for the design of submerged jet systems for oxygen transfer. For example, if oxygen transfer rates are to be maximised liquid depths and hence tank height/diameter ratios should be high. The limit on actual height/diameter ratio will depend on several factors. These factors include: i) ease of construction of the tank, and ii) preventing excessive momentum loss when the jet hits the wall and preventing bubbles being jetted straight to the wall and then the liquid surface (this would reduce bubble residence times and hence interfacial areas available for mass transfer).

Mixing energy varied by changing jet velocity and liquid flowrate

Increasing mixing energy by holding all process variables constant except jet velocity, and hence liquid volumetric flowrate, will have the following effects on the parameters in the mass transfer rate Equation (1). Firstly, k_L is likely to increase as a result of increased liquid velocities and hence higher rates of surface renewal at the gas-liquid interface (according to the generally accepted theories (3) and (4)). Secondly, C_s - C will be unaltered. Thirdly, the interfacial area is likely to increase for several reasons: (i) as the jet velocity is increased the shear rates in the jet increase and hence the size of bubbles formed at the nozzle may be expected to decrease (as shown by *eg* (6) and Unno and Inoue (9)) and (ii) increased bubble residence times and higher gas-phase hold-ups resulting from the lower bubble rise velocities associated with smaller bubbles.

For a submerged jet system, Varley (10) showed that the diameter of bubbles formed downstream of the nozzle decreased asymptotically with increases in jet velocity. As the bubble diameter decreases the interfacial area available for oxygen transfer increases. This increase in interfacial area is likely to be the major contribution towards increased rates of oxygen transfer and oxygen utilisation efficiencies at higher mixing energies associated with higher jet velocities (as shown in Figures 4 and 6).

As the jet velocity and hence liquid flowrate is increased, the power input will increase as shown by Equation (13). Since oxygen transfer efficiency is observed to decrease as mixing energy increases (Fig 8), the increase in power required is obviously greater than the increase in oxygen dissolution rate. For a constant rate of oxygen injection, rates of oxygen dissolution increase but oxygen transfer efficiency decreases as mixing energy is increased by increasing jet velocity and liquid flowrate. Therefore, the optimum operating conditions will depend on the relative costs of oxygen and power.

Comparison with other gas-liquid contactors

Commercial data are available for oxygen transfer efficiencies for other gas-liquid contacting devices. Such data are not generally available for standard conditions or published for a range of mixing energies. Scientific papers generally discuss overall mass transfer coefficients (k_La) as opposed to oxygen or gas efficiencies for gas-liquid transfer systems. In order to compare the experimental results described here with results for other gas-liquid contactors, the results can readily be written in terms of a mass transfer coefficient for the tank. The rate of oxygen dissolution in the tank can be written as:

$$\text{Rate of oxygen dissolution in the tank} = k_La\ V_t\ (C_s - C_t) \tag{14}$$

where k_La is the overall mass transfer coefficient for the tank.

As explained above, the rate of oxygen dissolution was determined for a range of mixing energies according to Equation (9). Therefore, from this measured value and Equation (14) k_La can be evaluated for a given bulk liquid dissolved oxygen level and known volume of liquid. (The results given below were determined for a dissolved oxygen level of 0 mg/l). Either the overall mass transfer coefficient (k_La) or the 'duty' (k_La V) can compared with values quoted in the literature for other gas-liquid transfer devices. Table 1 shows mass transfer coefficients for submerged jet systems together with those given by Middleton (11).

TABLE 1 - Comparison of gas-liquid transfer devices

Device	ε_l (%)	k_La (s^{-1})	V (m^3)	k_La V (m^3 s^{-1})	Power per unit volume (kW m^{-3})
Agitated tank (baffled)	0.9	0.02-0.2	0.002-100	10^{-4}-20	0.5-10
Bubble column	0.95	0.05-0.01	002-300	10^{-5}-3	0.01-1
Packed tower	0.05	0.005-0.02	0.005-300	10^{-5}-6	0.01-0.2
Plate tower	0.15	0.01-0.05	0.005-300	10^{-5}-15	0.01-0.2
Static mixer (bubble flow)	0.5	0.1-2	Up to 10	1-20	10-500
Submerged jet	>0.99	0.0004-0.003	0.035-470	10^{-5}-0.04	0.001-0.5

When comparing the various devices in Table 1 above, it should be noted that the liquid phase volume fraction is much greater for the submerged jet device than for all other devices. Experimental results are therefore needed at much higher gas injection ratios if accurate and detailed comparisons are to be made. The maximum duty found for the submerged jet system was 0.04 m^3 s^{-1} at a mixing energy input of 16 W m^{-3} on scale 3 apparatus. This duty is comparable to that given for agitated tanks in Table 1 but the power input is considerably lower.

The overall mass transfer coefficient has been measured for a wide range of gas-liquid transfer devices. Experimental data is often correlated in the following form:

$$k_La = x\ (P/V)^y\ (v_g)^z \qquad (15)$$

where (P/V) is the power input per unit volume (mixing energy), v_g is the superficial gas velocity and x, y, z are constants. Values of y are typically 0.4-0.95 for agitated vessels (*eg* (11) and (7)). (Mixing energy was varied by altering the agitator speed and not the liquid volume).

For experimental results similar to those presented here, y was found to vary between 0.30 and 0.35 for submerged jet systems. These values of y apply to changes in mixing energy affected by changes in either the liquid depth or the jet velocity and hence liquid flowrate.

CONCLUSIONS

For meaningful comparisons to be made between energy efficiencies of various gas-liquid transfer devices, it is important to establish how gas-liquid transfer is related to mixing energy input.

Rates of oxygen dissolution, oxygen utilisation and transfer efficiencies were measured for a range of mixing energies, on three scales of apparatus. Rate of oxygen dissolution, oxygen utilisation efficiency and transfer efficiency decrease as mixing energy is increased by decreasing the liquid volume. Therefore, if oxygen transfer rates are to be maximised and power requirements minimised, liquid depths and tank height/diameter ratios should be kept high. Rate of oxygen dissolution and oxygen utilisation efficiency increase and oxygen transfer efficiency decreases as mixing energy is increased by increasing jet velocity (and hence liquid flowrate). The optimum operating conditions therefore depend on the relative costs of oxygen and power. The dependence of oxygen transfer rates and efficiencies on mixing energies was explained by considering the influence of mixing energy on the parameters in the mass transfer rate equation, particularly the gas-liquid interfacial area.

Overall mass transfer coefficients (k_La) for submerged jet systems were compared to data available in the literature for other gas-liquid contacting devices. Although the results for the submerged jet system were for a narrow range of gas phase hold ups, compared to those available for other devices, the results indicate that comparable rates of oxygen transfer can be obtained at lower mixing energy inputs.

SYMBOLS USED

a = interfacial area per unit volume of liquid (m^{-1})

C = concentration of dissolved oxygen (kg/m^3)

ε_l = liquid phase hold up (%)

k_L = mass transfer coefficient (m/s)

k_La = overall mass transfer coefficient (s^{-1})

OTE = oxygen transfer efficiency (kg/kW h)

OUE = oxygen utilisation efficiency (%)

P = Power (W)

ΔP = Pressure drop (N/m^2)

Q = volumetric flowrate (m^3/s)

t = time (s)

V_t = volume of liquid in the tank (m^3)

v_g = superficial gas velocity (m/s)

v_j = jet velocity (m/s)

Greek symbols

ρ = density (kg/m^3)

Subscripts

g = gas

l = liquid

o = overall

p = pipework

t = tank

s = saturation

REFERENCES

1. Revill, B.K., 1985, Jet Mixing. In 'Mixing in the Process Industries'. Edited by Harnby, N., Edwards, M.F., and Nienow, A.W., Butterworths, England.

2. Hiby, J.W., and Modigell, M., 1978, Paper presented at 6th CHISA Conf., Prague.

3. Higbie, R., 1935, Trans. Am. Inst. Chem. Engrs. 35, 365.

4. Danckwerts, P.V., 1951, Ind. Eng. Chem. 43, 365.

5. Calderbank, P.H., 1958, Trans. I. Chem. E. 37, 173.

6. Pandit, A.B., and Davidson, J.F., 1986, 'Bubble Break-Up in Turbulent Liquid', International Conference on Bioreactor Fluid dynamics, Cambridge, England, 109.

7. Linek, V., Vacek, V., and Benes, P., 1987, Chem. Eng. J. 34, 11.

8. Kite, O.A., 1990, Personal discussions about unpublished work, BOC.

9. Unno, H., and Inoue, I., 1980, Chem. Eng. Sci. 35, 1571.

10. Varley, J., 1990, 'Submerged Jet Oxygenation', PhD Thesis, Cambridge, England.

11. Middleton, J.C., 1985, Gas-liquid dispersion and mixing. In 'Mixing in the Process Industries'. Edited by Harnby, N., Edwards, M.F., and Nienow, A.W., Butterworths, England.

Figure 1 Schematic diagram of a submerged jet oxygenation system

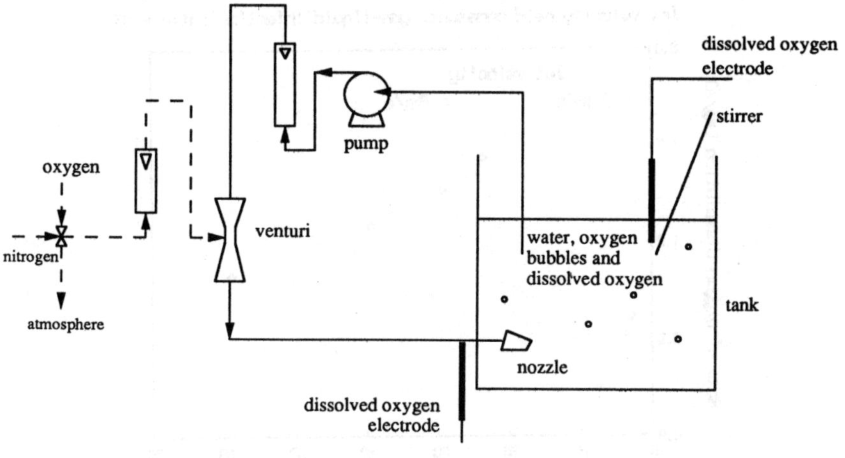

Figure 2 Oxygen dissolution in a submerged jet oxygenation system

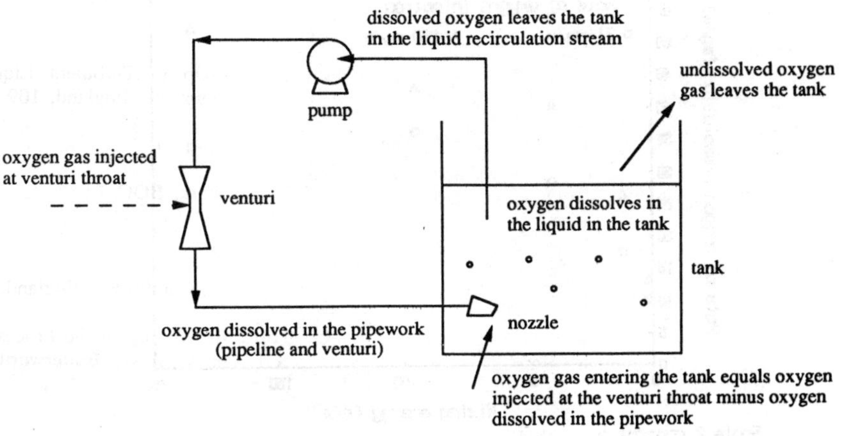

Figure 3 Rate of oxygen dissolution as a function of mixing energy

Jet velocity held constant, gas-liquid injection ratio = 6%

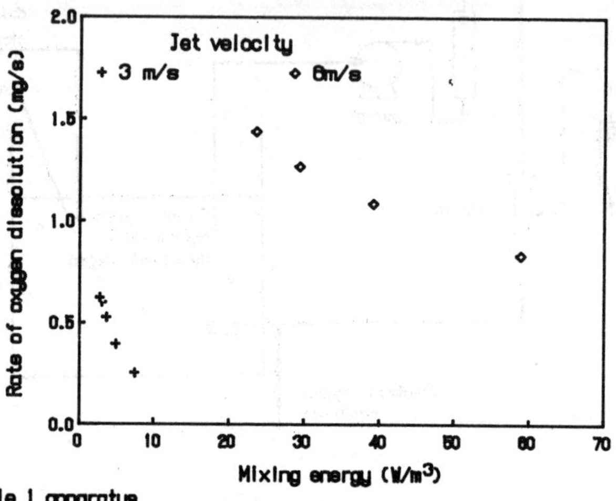

Scale 1 apparatus

Figure 4 Rate of oxygen dissolution as a function of mixing energy

Liquid depth held constant (1.0m)

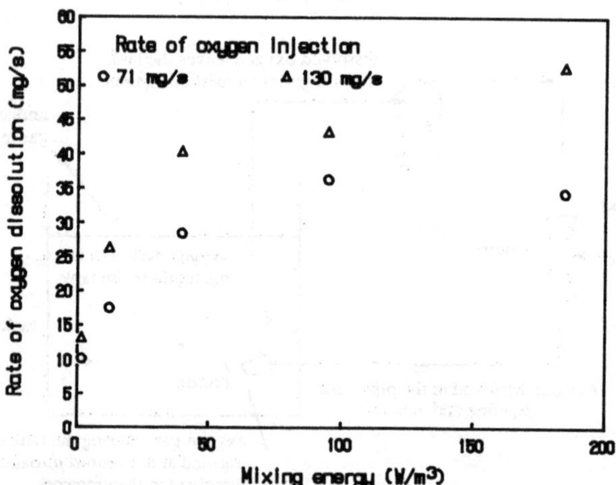

Scale 2 apparatus

Figure 5 Oxygen utilisation efficiency as a function of mixing energy

Jet velocity held constant (12m/s)

Gas-liquid injection ratio = 6% for scale 1 and 2 and 8% for scale 3

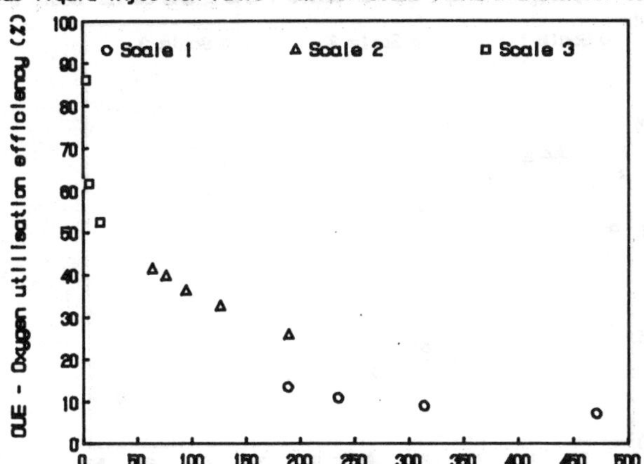

For scale 1 and 2, OUE for tank only; for scale 3, OUE is for overall system

Figure 6 Oxygen utilisation efficiency as a function of mixing energy

Liquid depth held constant

Gas-liquid injection ratio = 3% for scale 1 and 2 and 8% for scale 3

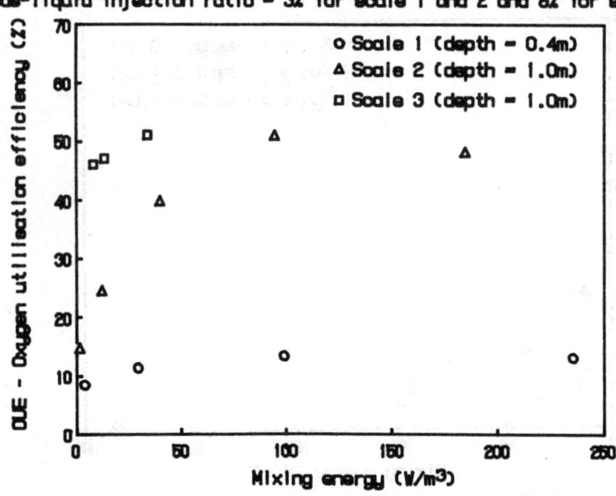

For scale 1 and 2, OUE for tank only; for scale 3, OUE is for overall system.

Figure 7 Oxygen transfer efficiency as a function of mixing energy

Jet velocity held constant (12m/s)

Gas-liquid injection ratio = 6% for scale 1 and 2 and 8% for scale 3

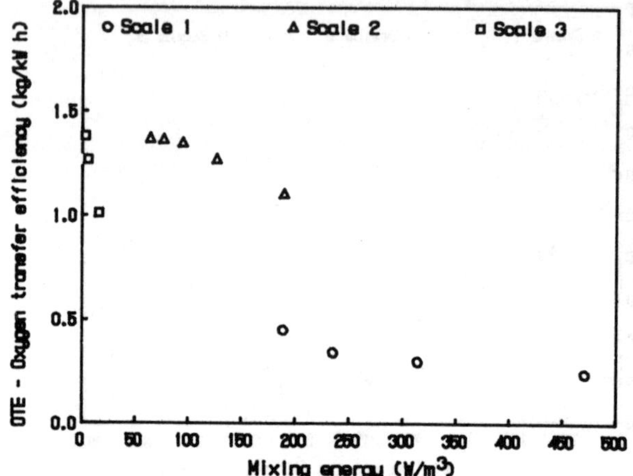

For scale 1 and 2 OTE is for tank only; for scale 3 OTE is for overall system

Figure 8 Oxygen transfer efficiency as a function of mixing energy

Liquid depth held constant

Gas-liquid injection ratio = 6% for scale 1 and 2 and 8% for scale 3

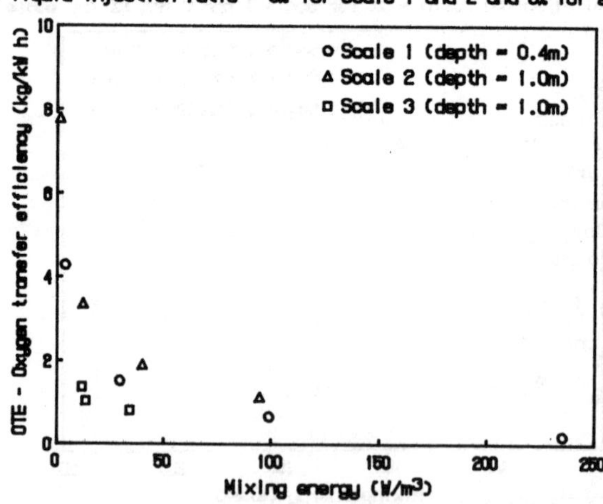

For scale 1 and 2 OTE is for tank only; for scale 3,OTE is for overall system

A FUNDAMENTAL STUDY OF GAS-INDUCING IMPELLER DESIGN

G.M. Evans*, C.D. Rielly*, J.F. Davidson* and K.J. Carpenter†

Gas-inducing impellers allow almost complete utilisation of a solute gas by recycling gas from the head space, without the need for an external loop and compressor. The impellers considered here have a hollow shaft and blades and use pressure reduction, created by liquid acceleration over the blades, to induce a flow of gas from the free surface. Theory is developed to predict the critical impeller speed for the start of gas induction and also for the flow rate of gas induced at higher impeller speeds. The model uses two dimensionless constants which depend on the slip between the fluid and the blade and also on the blade shape and orifice position; these constants are measured experimentally prior to the start of gas induction. Good agreement is achieved between the model predictions and experimental values of the critical impeller speed and induced gas flow rates. Simple, cylindrical section blades are considered here, but the methodology described may be applied to complex blade shapes and could be used for optimising gas-inducing impeller design.

INTRODUCTION

The fraction of gas absorbed in a single pass through a gas-liquid stirred vessel is small and almost complete utilisation of the solute requires recycling and recompression of the head-space gas. In many cases, *e.g.* in oxygenations and hydrogenations or with toxic gases, continuous recompression through an external loop is hazardous and undesirable. A self-inducing impeller uses acceleration of the liquid phase over the blades to locally reduce the pressure behind the blades and induce a flow of gas, from the head-space, through a hollow shaft. Self-inducing impellers have found applications in froth flotation and waste-water treatment processes, but have not, as yet, been widely used in the chemical industry.

The design of self-inducing impellers considered here consists of a hollow shaft connected to hollow blades (see Figure 1). A hole in the shaft above the free liquid surface allows gas to flow through the shaft and to leave through orifices on the impeller blades. At a critical impeller speed, N_C, the reduction in pressure at the outlet orifice on each blade is sufficient to overcome the static head of liquid and a flow of gas is induced. Increasing the impeller speed above N_C leads to an increase in the induced gas flow rate.

Braginsky (1964) applied Bernoulli's equation between the free surface and the orifice on the blade to calculate the critical speed, N_C, at which a gas flow is first induced,

$$d\left(\frac{v^2}{2}\right) + d\left(\frac{p}{\rho}\right) = 0 \tag{1}$$

* Department of Chemical Engineering, Pembroke Street, Cambridge
† ICI Fine Chemicals Manufacturing Organisation, Blackley, Manchester

where $v = 2\pi NR$, and R is the radial distance of the orifice from the shaft axis. If the depth of submersion of the orifice is h, the induction speed from eq.(1) is

$$N_C = \sqrt{\frac{gh}{2k(\pi R)^2}} \tag{2}$$

where k is a constant which allows for head losses within the aerator (Braginsky, 1964). Joshi & Sharma (1977) determined experimentally that k ranged from 0.9-1.3 for pipe and flattened cylindrical blades, and similarly, Zundelevich (1979) obtained a value for k of 1.15 for a turbo-aerator. Sawant & Joshi (1979) extended this approach to include the effects of liquid physical properties on N_C, and proposed

$$N_C = \sqrt{\frac{0.21\,gh}{4\,R^2}\left(\frac{\mu}{\mu_w}\right)^{0.11}} \tag{3}$$

which shows only a weak dependence on viscosity and no dependence on surface tension or density.

Eq.(2) may be rearranged to predict the pressure drop across the outlet orifice, as a function of impeller speed, and thus may be used to obtain an estimate of the induced gas flow rate (see Theory section). Zundelevich (1979) demonstrated experimentally that this approach over-predicted the induced gas flow rate, which was found to be proportional to the speed of the impeller raised to a power less than one. A number of empirical correlations have also been proposed (Martin, 1972; Sawant *et al.*, 1981; Raidoo *et al.*, 1987) to predict gas induction flow rates for different impeller systems, however, no theoretical study has as yet shown why expressions derived from eq.(2) cannot be applied successfully to impeller speeds higher than the critical induction speed.

The aim of this paper is to derive expressions for the pressure reduction in the vicinity of the outlet orifice: (i) prior to gas induction, in a single phase flow; and, (ii) after induction, in a gas-liquid flow; and thus to predict the critical speed for gas induction and the gas flow rate induced for $N > N_C$. The paper sets out a methodology for the design of gas-inducing impellers by comparing theory with experimental results for a simple impeller shape — cylindrical impeller blades are used here, since the flow around the front part of these blades may be represented by potential flow theory and thus some of the model constants might be derived.

THEORY

The Critical Impeller Speed for Gas-Induction

In the system illustrated in Figure 1 the liquid inside the vessel rotates as a result of the drag exerted by the impeller blade. For a cylindrical blade, the radial flow component is small compared to the tangential component. The absolute tangential velocity of the liquid ahead of the outlet orifice, v_L, is assumed to be a fraction K of the blade speed at the orifice position,

$$v_L = 2\pi KNR \tag{4}$$

where N is the impeller rotational speed and R is the radius of the outlet orifice. K is called the blade slip factor, and is a function of the blade design and vessel geometry ($0 < K < 1$). The magnitude of the pressure reduction depends on the *relative* velocity, U, of the outlet orifice and the liquid ahead of the blade

$$U = (1-K)\,2\pi NR \tag{5}$$

The relative liquid velocity over the outlet orifice, $u_L(\theta)$, depends on its angular position on the surface of the blade, such that

$$u_L(\theta) = (C_p(\theta))^{1/2}(1-K)\,2\pi NR \tag{6}$$

where $C_p(\theta)$ is a pressure coefficient (Abbot & von Doenhoff, 1959), defined as

$$C_p(\theta) = \frac{(P_0 + \rho_L gh) - P_1(\theta)}{\frac{1}{2}\rho_L U^2} \tag{7}$$

$P_1(\theta)$ is the absolute pressure on the surface of the blade at an angular position θ measured from the rear stagnation point and P_0 is the absolute head space pressure. The pressure coefficient is a function of the blade shape and orifice position, and for inviscid flow, $C_p(\theta)$ may be predicted from potential theory; e.g. for uniform flow around a 2-D cylinder $C_p(\theta)$ is given as

$$C_p(\theta) = 4\sin^2\theta \tag{8}$$

Bernoulli's equation (1) can be applied to the streamlined flow around a cylindrical blade to determine the surface pressure $P_1(\theta)$ at different angular positions

$$P_1(\theta) = P_0 + \rho_L gh - \frac{1}{2}\rho_L C_p(\theta)\{\,2\pi NR\,(1-K)\}^2 \tag{9}$$

where h is the depth of the impeller below the liquid free surface. Gas is induced from the head-space when $P_1(\theta) = P_0$, *i.e.* from eq.(9), when

$$N_C = \sqrt{\frac{gh}{2C_p(\theta)\,\{(1-K)\,\pi R\}^2}} \tag{10}$$

Liquid Height Inside the Impeller Prior to Induction

The model constant $C_p(\theta)$ can be predicted from potential flow theory for simple blade shapes provided the fluid is inviscid and the flow field around the blade is known, *e.g.* eq.(8). However, for applications involving complex blade designs, and unknown flow fields the constants K and $C_p(\theta)$ must be measured experimentally or predicted from numerical solutions of the Navier-Stokes equations. In the following section experiments are described to estimate these constants from measurements of the height of liquid inside the hollow shaft, before induction has started.

Prior to induction, the change in liquid level, Δh, inside the shaft, for the system shown in Figure 1 and Figure 2(b), can be obtained by equating the surface pressure outside the orifice,

$P_1(\theta)$, with the radial pressure component, ΔP_r, plus the hydrostatic pressure inside the impeller. The radial pressure difference due to the forced vortex rotation of the liquid inside the horizontal impeller blade is given by

$$\Delta P_r = \int_0^R \rho_L (2\pi N)^2 r \, dr = 2\rho_L (\pi N R)^2 \tag{11}$$

and so,

$$P_1(\theta) = 2\rho_L (\pi N R)^2 + \rho_L g(h - \Delta h) + P_0 \tag{12}$$

Using eqs.(9) and (12) the change in liquid height inside the impeller shaft is

$$\Delta h = \frac{2(\pi N R)^2 \{1 + C_p(\theta)(1-K)^2\}}{g} \tag{13}$$

Note that gas induction does not start when $\Delta h = h$ because the radial pressure difference must still be overcome before gas can reach the outlet orifice. The radial distance Δr (see Figure 2(b)), which the liquid is pushed outwards within the blade is obtained by equating ΔP_r for the liquid inside the blade with the pressure outside the orifice $P_1(\theta)$, *i.e.* when $\Delta h = h$ in eq.(12):

$$P_1(\theta) - P_0 = \Delta P_r = \int_{\Delta r}^R \rho_L (2\pi N)^2 r \, dr \tag{14}$$

Integrating and substituting for $P_1(\theta) - P_0$ from eq.(9) gives

$$\frac{\Delta r}{R} = \sqrt{1 + C_p(\theta)\,(1-K)^2 - \frac{gh}{2(\pi N R)^2}} \tag{15}$$

Gas induction starts only when $\Delta r = R$, in which case eq.(15) reduces to eq.(10).

An alternative design of self-inducing impeller introduces the gas to the blade through a riser tube, as shown in Figure 2(a). The liquid inside the riser tube is at a constant radial distance from the rotating shaft, and therefore experiences no radial pressure gradient. The change in liquid height inside the riser tube is obtained by equating the pressure inside the tube, $P_0 + \rho_L g(h - \Delta h)$, with the surface pressure at the gas orifice $P_1(\theta)$ from eq.(9), yielding

$$\Delta h = \frac{2C_p(\theta)\,\{\pi N R\,(1-K)\}^2}{g} \tag{16}$$

Note that for the riser tube feed, gas is induced when $\Delta h = h$, and eq.(16) reduces to eq.(10), indicating that there should be no difference in the critical speed for gas induction for the alternative feed positions. Equations (13), (15) and (16) for the dependence of Δh and Δr on impeller speed are not used directly for predicting the critical induction speed, but they do provide a method for characterising impeller shapes and orifice positions (*i.e.* finding K and $C_p(\theta)$) from data collected *prior to induction.*

Gas Induction Rate

Once gas induction has started, the volumetric gas flow rate *per orifice*, Q_G, can be calculated by assuming that the dominant pressure losses are due to flow through the outlet orifice, *i.e.* wall friction inside the hollow shaft and blade, and pressure losses at the inlet orifice, are negligible.

$$Q_G = C_o A_o \sqrt{\frac{2(P_0 - P_1(\theta))}{\rho_G}} \qquad (17)$$

In eq.(17) A_o is the cross-sectional area of the orifice and C_o is a discharge coefficient. The pressure drop across the orifice, $(P_0 - P_1(\theta))$ is given by eq.(9), and so

$$Q_G = C_o A_o \sqrt{\frac{\rho_L}{\rho_G}\left\{C_p(\theta)\,[2\pi RN(1-K)]^2 - 2gh\right\}} \qquad (18)$$

Equation (18) only applies when there is liquid at the orifice outlet. If a gas cavity forms at the orifice outlet then the gas flowrate is determined by the pressure inside the ventilated cavity,and the rate at which the bubbles are carried into the bulk fluid. Under these conditions eq.(18) is not applicable.

EXPERIMENTAL

The experimental study was carried out in a 0.290 m diameter, fully baffled perspex tank, shown in Figure 1. A cylindrical section impeller was located 0.100 m from the base of the tank, and the total liquid height was fixed at 0.400 m, except for a series of experiments in which the effect of the blade submersion depth was examined. For the latter experiments the liquid height was varied between 0.150 and 0.400 m. The impeller had an overall diameter of 0.130 m and consisted of two 0.0126 m diameter perspex cylindrical blades, spaced 180° degrees apart. A 0.005 m diameter hole was drilled along the axis of each blade to allow gas to pass down the hollow shaft and exit through a 0.0035 m diameter orifice on the surface of each blade. The orifice locations along the blade were set at 0.025, 0.040 and 0.055 m from the axis of the shaft, and at any one time only one orifice opening was used, the remaining two being sealed. The cylindrical blade could be rotated to vary the angular position of the orifice relative to the rear stagnation point.

A hollow shaft allowed gas to be drawn from the head-space above the liquid surface through the sealed bearing, and the induced gas flow rate was measured using a soap bubble meter. A frequency counter was used to record the rotational speed of the shaft. The seal and bearing at the base of the impeller hub provided extra support for the shaft at high speed and also gave a means by which the liquid head inside the shaft could be measured, while the impeller was rotating. Figure 1 shows the U-tube manometer formed between the shaft and a vertical glass tube located outside of the vessel. With the shaft rotating the liquid heights in both legs were equal, and the change in liquid height inside the shaft could be accurately recorded from the manometer leg outside the vessel. Thus, the surface pressure at the orifice, $P_1(\theta)$ could be calculated from eq.(12) for the shaft feed location.

An alternative method of introducing the gas was through a riser tube attached to the blade, directly above the outlet orifice; for the majority of the experiments the gas orifice was located at $\theta = 90°$, on the bottom edge of the blade. Figure 2(a) shows that the gas phase above the free surface was able to travel down the hollow riser tube and exit through the gas orifice. The riser tube was used to study the effect of the radial pressure gradient acting on the liquid inside the horizontal blade, for comparison with the case where the gas was introduced through a hollow shaft and blade. To ensure similarity of the velocity field generated by the impeller for the two cases, a solid riser tube was attached to the blade for the shaft feed experiments. Consequently, the liquid velocity profiles inside the vessel were unchanged irrespective of whether shaft, or riser tube feeds were used.

Two series of experiments were performed in this study. The first consisted of using the cylindrical blade with no riser tube attachments, as shown in Figure 1. The position of the gas-liquid interface inside the shaft or impeller was measured as a function of impeller speed for different radial and angular positions of the outlet orifice, and for various submersion depths, prior to gas induction. When the gas-liquid interface was located inside the rotating shaft the height was obtained from the liquid level in the manometer leg outside the vessel, and when the interface was inside the horizontal blade the radial position was obtained by freezing the motion of the blade using a stroboscope. The radial distance of the interface from the axis of the shaft was read from a scale on the surface of the blade.

In the second series of experiments the change in vertical height, and radial position, of the gas-liquid interface was measured as a function of impeller speed, for both shaft and riser tube feeds. A stroboscope was used to freeze the motion of the impeller, and the liquid height in the riser was read from a scale on the riser tube. In addition, induced gas volumetric flow rates at different impeller speeds were obtained for both the shaft and riser tube feeds.

From the experimental results it was not possible to evaluate separately the model constants, K and $C_p(\theta)$. It was possible, however, to estimate the dimensionless function $C_p(\theta)(1-K)^2$ from measurements made prior to gas induction, which was used to predict the critical impeller speed, N_C, and the gas flow rate induced, for $N > N_C$. The comparison between theory and experiment is discussed in the following section.

RESULTS AND DISCUSSION

Predicting Δr From Δh Measurements

In Figure 3 the change in liquid height has been plotted against the impeller speed for an orifice angular position of 90° at a radial distance of 0.055 m (no riser tubes attached). The solid line is a parabolic least squares fit and shows that, in agreement with eq.(13), the change in height is proportional to the square of the impeller speed. Consequently, the function $C_p(\theta)(1-K)^2$ in eq.(13) was calculated to be equal to 1.35, and from this value predictions of the radial position of the miniscus, Δr, were calculated from eq.(15). Both sets of values have been plotted in Figure 3, which shows good agreement between the theoretical prediction for Δr, and the experimental data.

Effect of Angular Position of the Gas Orifice

The influence of the coefficient $C_p(\theta)$ on the pressure at the orifice was studied by changing the angular position of the gas orifice; three positions at 0°, 45°, 90°, and 135° were investigated. In these experiments it was assumed that the blade slip factor remained constant over the range of θ studied. The ratio $C_p(\theta) / C_p(90°)$ was calculated from the Δh against impeller speed measurements using eq.(13). The theoretical pressure coefficient ratio, for a cylinder in uniform flow, eq.(8), is also shown in Figure 4. The agreement between the experimental values for $C_p(\theta) / C_p(90°)$ and those predicted from potential flow was poor. At $\theta < 90°$ (orifice on the rear of the blade) the experimental points deviated from the curve, because of flow separation; Achenbach (1975) notes that flow separation takes place at about 90°-100° from the rear stagnation point, at a cylinder Reynolds number of order 10^4-10^5. The average Reynolds number for the cylindrical blade used in these studies, based on the relative liquid velocity, was about 2.5×10^4. The deviation for $\theta > 90°$ may be due to circulation around the blade (Schewe, 1983).

Effect of Radial Position of the Gas Orifice

The function $C_p(90°)(1 - K)^2$ for three radial positions of the orifice were obtained from the Δh measurements given in Figure 5. It can be seen from the graph that for all three radial orifice positions the change in liquid height inside the hollow shaft was proportional to the square of the impeller speed. Consequently, $C_p(90°)(1 - K)^2$ for each radial position was calculated from a least squares fit to eq.(13), and the results are plotted in Figure 6. where it can be seen that $C_p(90°)(1 - K)^2$ decreases with increasing radial position. Assuming that $C_p(90°)$ remains approximately constant, then the results show that K increases with increasing radial distance. Woo *et al.* (1989) showed that for a cylinder in a shear field (*e.g.* a cylindrical impeller blade rotating about a central shaft), a secondary flow is generated parallel to the cylinder axis. This secondary flow in the radial direction may affect the value of K.

Effect of Submersion Depth of the Gas Orifice

For the arrangement shown in Figure 1, the effect of the depth of submersion of the blade on the critical gas induction speed was examined by keeping the blade clearance from the base of the vessel constant at 0.100 m, and changing the liquid height in the vessel. Predictions of N_C from eq.(10) for different submersion depths were obtained by assuming a constant value of $C_p(90°)(1 - K)^2 = 1.35$ obtained from Δh measurements made prior to gas induction at a liquid depth of 0.4 m. Figure 7 shows a comparison of these predictions with experimental results for the minimum speed at which gas bubbles were observed at the orifice. It can be seen from the graph that agreement is within 10%, with eq.(10) consistently underpredicting N_C. The reason for this trend is that the slip between the blade and the fluid is dependent on the geometry of the vessel; although the results do suggest that the variation in K is not significant. At the lowest submersion depth ($h = 0.05$ m) the proximity of the free surface has its largest effect on the relative velocity between the fluid and the blade.

Comparison Between Shaft and Riser Feeds

From the analyses of the pressure acting on the liquid within the impeller, eqs.(13) and (16), there should be a difference in the Δh measurements at the same impeller speed, for the shaft and riser feeds (see Figure 2). This is the result of the radial pressure gradient acting on the liquid inside the rotating blade when the gas is introduced via the hollow shaft. If the gas is introduced through a riser tube, attached directly above the orifice, there is no radial pressure gradient to be overcome.

In Figure 8 the change in vertical height of the gas-liquid interface is plotted against the impeller speed for both the shaft and riser feeds, for an orifice radial position of 0.055 m and submersion depth of 0.280 m. The graph shows that in each case the change in liquid height is proportional to the square of the impeller speed; the solid lines are parabolic least squares fits to the data. For the same impeller speed the change in height is less for the shaft feed system than for the riser feed, due to the radial pressure gradient in the liquid inside the blade: the radial pressure drop, when added to the hydrostatic head of the liquid inside the shaft, is equivalent to the pressure reduction at the orifice. The function $C_p(90°)(1-K)^2$ for both the shaft and riser feeds were calculated from the Δh measurements using eqs.(13) and (16), respectively. The values for $C_p(90°)(1-K)^2$ were found to be 1.39 for the shaft feed, and 1.44 for the riser feed; a difference of only 3.5%.

Prior to induction the Δh measurements for the shaft and riser feed setups were different at the same impeller speed, however, both eqs.(15) and (16) reduce to the same eq.(10) for the critical gas induction speed and it was expected that N_C would be equal for both cases. The point of induction is determined only by the liquid pressure outside the gas orifice, which is identical for both the shaft and riser tube feeds, there being no difference in the liquid flow patterns generated. The riser tube feed Δh *vs* N measurements were extrapolated to yield a critical impeller speed of 5.65 rps for gas induction. For the shaft feed the Δh measurements were extrapolated to the submersion depth at a rotational speed of 4.44 rps. Gas induction did not start at this point, since there was still a radial pressure drop in the horizontal blade to be overcome. Setting Δr equal to R in eq.(15), and using $C_p(90°)(1-K)^2 = 1.39$, a critical impeller speed of 5.75 rps was obtained. The results from the shaft and riser feed extrapolations are in good agreement and also compare favourably with a critical impeller speed of 5.7 rps obtained by visually observing the lowest speed at which bubbles were ejected from the gas orifice.

Prediction of the Rate of Gas Induction

It was shown in the previous section that the critical impeller speed at which gas induction is first initiated is independent of the internal path taken by the feed gas. The rate of gas induction should also be independent of the method of introducing the gas, provided the blade profile remains unchanged. Figure 9 shows that the measured induced gas flow rate for a single orifice, at a given impeller speed, is similar for both the shaft and riser tube feeds: the exiting gas stream from the orifice appeared to be a constant flow of discrete bubbles and no gas cavities clinging to the blades were observed over the range of gas flowrates examined. Figure 9 also shows the theoretical curve for the induced gas flow rate, predicted by eq.(18) using an average value of $C_p(90°)(1 - K)^2 = 1.42$, obtained from the Δh measurements *prior* to gas induction. The orifice discharge coefficient C_o was taken to have a constant value of 0.78, corresponding to an orifice Reynolds numbers of about 500 and an area contraction ratio of 0.5 (Ower & Pankhurst, 1977).

Figure 9 shows that eq.(18) grossly overpredicts the induced gas flow rate; one reason for this may be that the liquid density is used in these calculations, rather than that of the two-phase mixture in the region close to the orifice. The presence of gas outside the orifice locally lowers the mean density of the fluid, $\overline{\rho}$, close to the orifice, thereby decreasing the reduction in pressure associated with the fluid flow. Yokosawa *et al.* (1986) and Watanabe *et al.* (1990) showed that the presence of low gas void fractions in a cross-flow over a cylinder can lead to a decrease in the drag and pressure coefficients. Replacing the liquid density, with the mean density of the two-phase mixture in the kinetic energy term of eq.(9) yields

$$P_1(\theta) = P_0 + \rho_L g h - \frac{1}{2}\overline{\rho} C_p(\theta) \{ 2\pi N R(1-K)\}^2 \tag{19}$$

The static head term in eq.(19) remains the same as in eq.(9), since the liquid head above the orifice is unchanged, even though the height of the gas-liquid mixture increases with increasing gas throughput. The two-phase mixture density at the orifice can be calculated from the densities of the gas and liquid phases, and the gas void fraction, ε, in the fluid stream near the orifice.

$$\overline{\rho} = (1 - \varepsilon)\, \rho_L + \varepsilon \rho_G \tag{20}$$

The void fraction in the fluid passing over the orifice is not easy to estimate. To investigate the effect of the mean density of the two-phase mixture on the induced gas flow rate, the gas velocity in the mixture is taken to be of order Q_G / A_o and the liquid velocity to be of order $u_L(\theta)$. Then, for a homogeneous two-phase flow (no relative velocity between the phases), the gas void fraction is approximately

$$\varepsilon \approx \frac{\dfrac{Q_G}{A_o}}{\dfrac{Q_G}{A_o} + u_L(\theta)} \tag{21}$$

Eqs.(19) and (20) can be combined to obtain an expression for the induced gas flow rate,

$$Q_G = C_o A_o \sqrt{\left[\frac{\rho_L (1-\varepsilon)}{\rho_G} + \varepsilon\right] C_p(\theta)\,[\,2\pi NR\,(1-K)]^2 - \frac{2\rho_L g h}{\rho_G}} \tag{22}$$

which was solved numerically to obtain the volumetric flow rate of gas per orifice, for known values of $C_p(\theta)$ and K, and using eq.(21) as an approximation to the gas void fraction.

The induced gas flow rates predicted by solving eq.(22) are also shown in Figure 9 and are in reasonable agreement with the experimental data for impeller speeds up to approximately 8.5 rps. At this impeller speed the gas void fraction estimated from eq.(21) was around 40% and significant coalescence of bubbles may have occurred, so that the homogeneous flow model would be unlikely to apply. The estimate of the void fraction is very approximate in these calculations, and only serves to illustrate that the two-phase nature of the fluid passing the blades can have a significant effect on the rate of gas induction. The prediction of the surface pressure at the orifice after induction requires further detailed study to elucidate the effects of the local void fraction in the two-phase flow over the impeller.

CONCLUSIONS

The experimental results presented in this study support the theoretical model developed to predict both the onset and rate of gas induction of a self-inducing impeller. The model is based on a pressure coefficient for the blade shape and a blade slip factor to allow for the relative velocity between the blade and the fluid. Experiments were described to determine the factor $C_p(\theta)(1-K)^2$ from the change in position of the gas-liquid interface inside the rotating shaft or blade, prior to gas induction. Predictions of the critical speed for gas induction were in good agreement with experiment. It was also shown that the critical impeller speed and the flow rate of gas induced, were independent of the feed method to the orifice. This result was predicted from the theory and showed that the pressure force driving the gas flow was only a function of the liquid velocity over the orifice.

It is important to note that the pressure coefficient and slip factor provide a method by which the blade and vessel design, and the resulting liquid flow patterns, inter-relate to determine the overall gas-induction characteristics of a self-inducing impeller. They are dimensionless quantities and, therefore, the values determined on a small scale should be the same for full-scale equipment. The value $C_p(\theta)(1-K)^2$ is the important quantity in the theory, but the purpose of splitting this product into two factors is to illustrate that both blade and vessel geometry, and the position of the orifice may be varied individually to give the optimum design. The results presented here are for a simple impeller shape, but the methodology set out for determining $C_p(\theta)$ and K also applies to more complex designs. Experiments using more realistic designs of gas-inducing impellers and on a larger scale tank are the subject of future work.

Acknowledgements: This work was carried out with support from an SERC / ICI Fine Chemicals collaborative award, which GME gratefully acknowledges.

REFERENCES

ABBOT, I.H. & VON DOENHOFF, A.E. (1959) "Theory of Wing Sections", p.42, Dover Publications, New York.

ACHENBACH, E. (1975) "Total and Local Heat Transfer from a Smooth Circular Cylinder in Cross-Flow at High Reynolds Number", *Int. J. Heat Mass Transfer*, **18**, 1387-1396

BRAGINSKY L.N. (1964) "The Study of Aeration Characteristics of Impellers for Autoclave Leaching Processes", Sci. Research Report, R. and D. Institute, NIIKhIMMASH, Leningrad, USSR.

JOSHI, J.B. & SHARMA, M.M. (1977)"Mass Transfer & Hydrodynamic Characteristics of Gas-Inducing Type of Agitated Contactors", *Can. J. Chem. Eng.*, **55**, 683-695.

MARTIN, G.O. (1972) "Gas-Inducing Agitator", *Ind. Eng. Chem. Process Des. Develop.*, **11**, 397-404.

OWER, E. & PANKHURST, R.C. (1977) "The Measurement of Air Flow", p.165, Pergamon Press, London.

RAIDOO, A.D., RAGHAV RAO, SAWANT, S.B. & JOSHI, J.B. (1987) "Improvements in Gas-Inducing Impeller Design", *Chem. Eng. Commun.*, **54**, 241-264.

SAWANT, S.B. & JOSHI, J.B. (1979) "Critical Impeller Speed for the Onset of Gas Induction in Gas-Inducing types of Agitated Contactor", *The Chem. Eng. J.*, **18**, 87-91.

SAWANT, S.B., JOSHI, J.B., PANGARKAR, V.G., & MHASKAR, R.D. (1981) "Mass Transfer & Hydrodynamic Characteristics of Denver Type of Flotation Cells", *The Chem. Eng. J.*, **21**, 11-19.

SCHEWE, G. (1983) "On the Force FluctuationsActing on a Circular Cylinder in Crossflow from Subcritical up to Transcritical Reynolds Numbers", *Journ. Fluid Mech.*, **133**, 265-285

WATANABE, H, IHARA, A. & ONUMA, S. (1990) "Effect of a Few Small Air Bubbles on the Performance of Circular Cylinder at Critical Flow Range in Water", ASME *Journ. Fluids Eng.*, **112**, 67-72

WOO, H. G. C., CERMAK, J. E. & PETERKA, J. A. (1989) "Secondary Flows and Vortex Formation Around a Circular Cylinder in Constant Shear Flow", *Journ. Fluid Mech.*, **204**, 523-542

YOKOSAWA, M., KOZAWA, Y., INOUE, A. & AOKI, S. (1986) "Studies of Two-Phase Cross Flow Part II: Transition Reynolds Number and Drag Coefficient", *Int. J. Multiphase Flow*, **12**, 169-184.

ZUNDELEVICH, Y. (1979) "Power Consumption and Gas Capacity of Self-Inducing Turbo Aerators", *A.I.Ch.E.J.*, **25**, 763-773.

NOTATION

A_o	orifice cross-sectional area	m^2
C	impeller clearance	m
C_o	orifice discharge coefficient	
C_p	pressure coefficient	
g	gravitational acceleration	$m^2 s^{-1}$
h	submersion depth	m
K	blade slip factor	
k	head loss coefficient in eq.(2)	
N	impeller speed	rps
N_C	critical impeller speed for gas induction	rps
P_0	pressure in the head space	$N m^{-2}$
P_1	surface pressure outside the orifice	$N m^{-2}$
Q_G	gas volumetric gas flow rate per orifice	$m^3 s^{-1}$
R	radial position of orifice	m
U	relative velocity of fluid upstream of blade	$m s^{-1}$
u_L	relative velocity of blade and fluid over the orifice	$m s^{-1}$
v	fluid absolute tangential velocity	$m s^{-1}$
Δh	change in liquid height of the gas-liquid interface	m
ΔP_r	radial pressure drop	$N m^{-2}$
Δr	change in liquid radius of the gas-liquid interface	m
μ	viscosity	Pa s
θ	angular position from the horizontal (measured from the rear of the blade)	

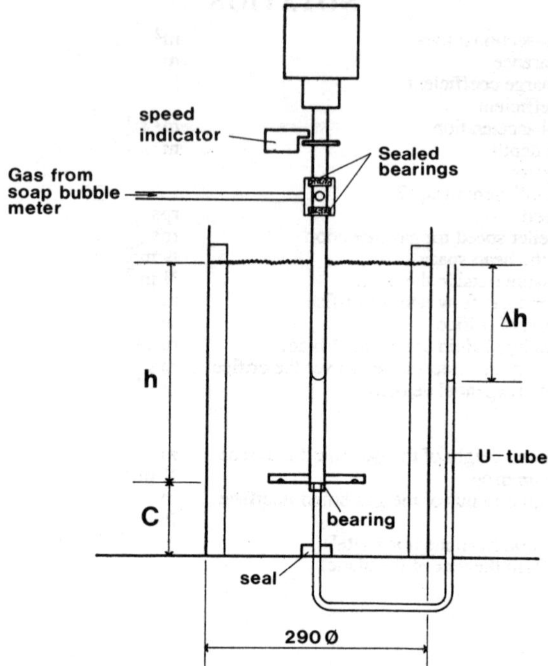

Figure 1 Experimental apparatus with hollow-bladed, gas-inducing impeller.

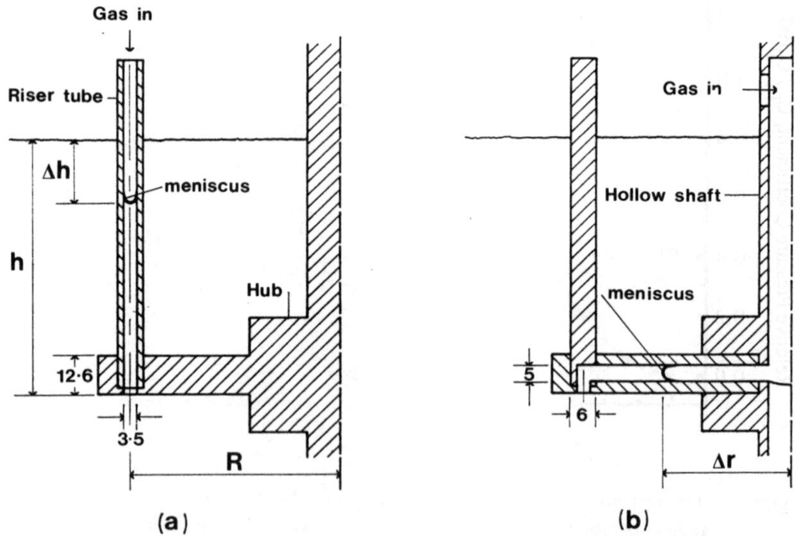

Figure 2 Alternative feed positions for a gas-inducing impeller (all dimensions in mm)
(a) Riser feed; (b) Shaft feed.

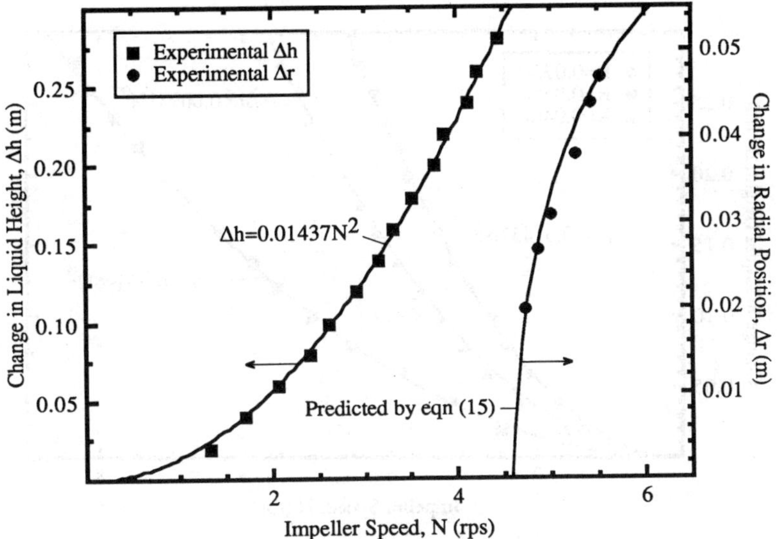

Figure 3 The change in liquid height and radial position of the gas-liquid interface in the hollow shaft or blade for an orifice radial position of 0.055 m and angular position of $\theta = 90°$ (no riser attachments). $C_p(\theta)(1-K)^2 = 1.35$

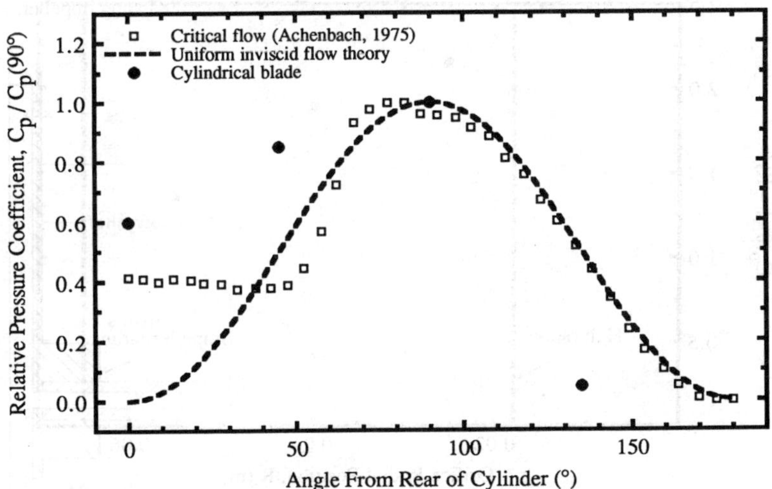

Figure 4 The variation of pressure coefficient ratio with angle from the rear for an orifice radial position of 0.055m (no riser tube attachments). The dashed line represents the potential flow solution, eq.(8).

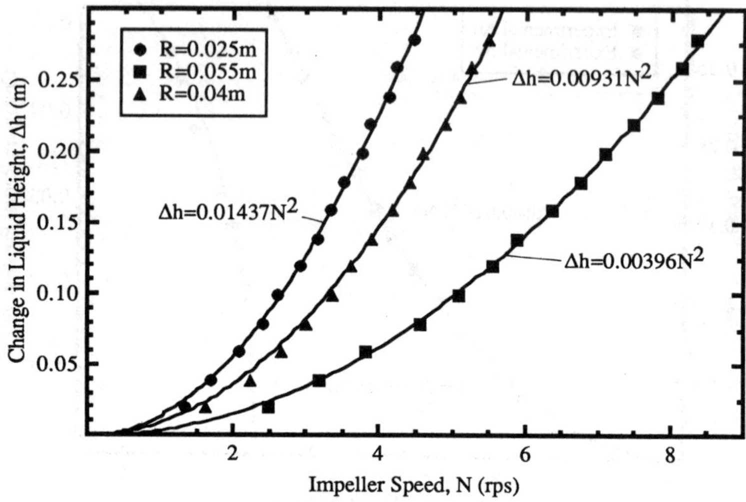

Figure 5 The change in liquid height in the hollow shaft for various orifice radial positions and angular position of $\theta = 90°$ (no riser attachments).

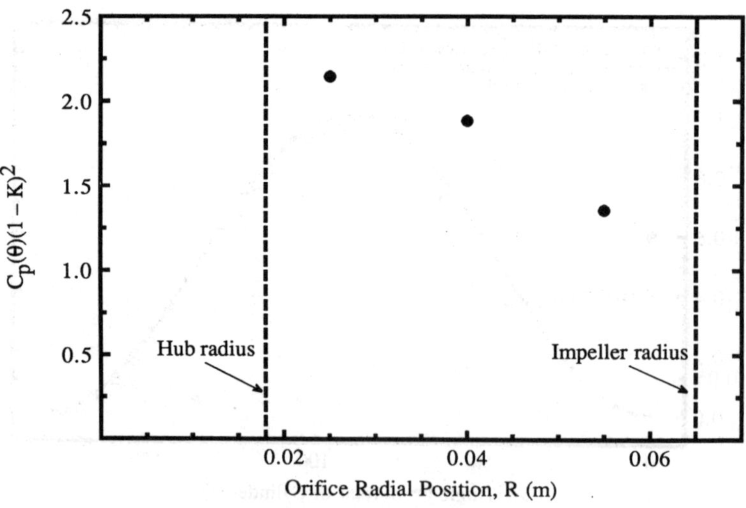

Figure 6 The effect of orifice radial position on the $C_p(\theta)(1 - K)^2$, for an orifice angular position of $\theta = 90°$ (no riser attachments).

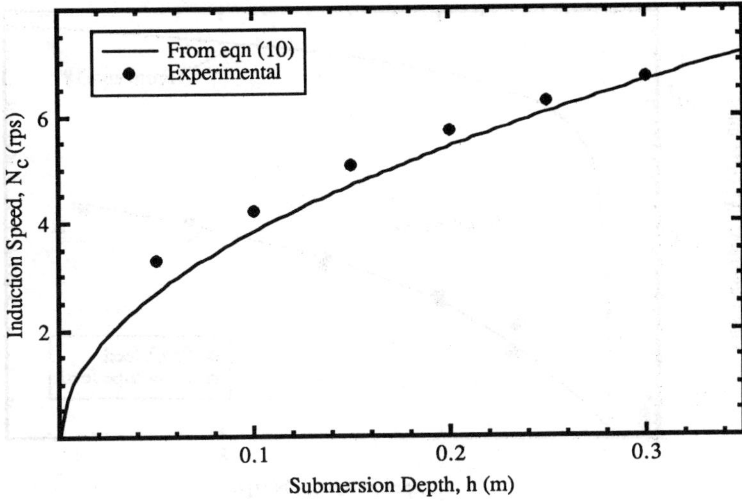

Figure 7 Comparison of theory and experiment for the effect of submersion depth on critical induction speed, for an orifice radial position of 0.055 m and angular position of $\theta = 90°$ (no riser attachments). $C_p(\theta)(1-K)^2 = 1.35$.

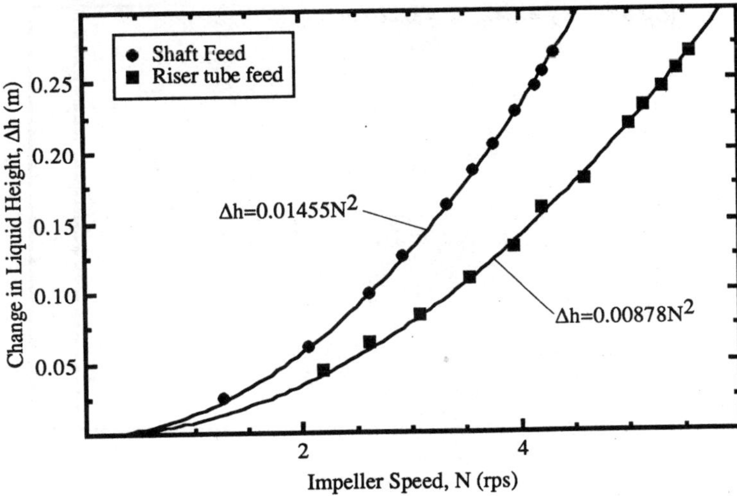

Figure 8 Comparison of the change in liquid height for the shaft and riser tube feeds. The orifice radial position is 0.055 m and angular position is $\theta = 90°$ ($h = 0.280$ m).

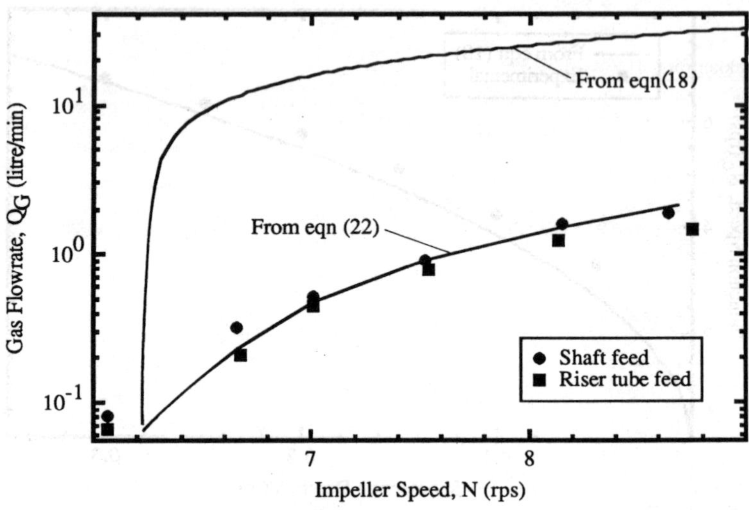

Figure 9 Comparison of the induced gas flow rate for the shaft and riser tube feeds. The orifice radial position is 0.055 m and angular position is $\theta = 90°$. The solid lines shows the theoretical predictions from eq.(18) and from eq.(22). ($C_p(\theta)(1 - K)^2 = 1.42$, $h = 0.280$ m).

THE USE OF PROFILED AXIAL FLOW IMPELLERS IN GAS-LIQUID REACTORS

A. Bakker and H.E.A. van den Akker*

In recent years a variety of profiled axial flow impellers has been developed. The aim of the present research work is to investigate the performance of these impellers and to contribute to the understanding of the gas dispersion mechanism. Three impeller types have been compared: a standard inclined blade impeller and the profiled A315 and Leeuwrik impeller. Important parameters are pumping capacity, turbulence intensity, gas dispersion performance, power consumption and mass transfer. It is concluded that in most of these respects the profiled impellers offer advantages over traditional impellers.

INTRODUCTION

Gas-liquid stirred vessels are often equipped with simple impellers which are cheap to make, like disc turbines or inclined blade impellers. Because of the industrial importance a lot of research has been done to describe the hydrodynamic properties of these impellers, see for instance Warmoeskerken et al. (1, 2), Chapman et al. (3) and Nienow et al. (4).

However, during the last few years several axially pumping, hydrodynamically profiled impellers have become available on the market. Examples of such impellers are the Lightnin A310 and A315, the Ekato Intermig and Interprop and the Prochem hydrofoil. In general, the manufacturers claim that at a given level of power input these impellers are capable of dispersing more gas and yield higher mass transfer coefficients than conventional impellers. For example, it is reported that mass transfer could be enhanced by up to 30% on average by replacing a disc turbine by a Lightnin A315 system (Oldshue (5)).

The aim of the present research work is to investigate the performance of a number of profiled impellers and to contribute to the understanding of the gas dispersion mechanism when such impellers are used. Therefore various impellers are compared on the basis of pumping efficiency, turbulence intensity, power consumption, gas handling capacity and mass transfer performance. Impellers used are a Lightnin A315, the Dutch Leeuwrik impeller and a traditional inclined flat blade turbine.

* Kramers Laboratorium voor Fysische Technologie, Delft University of Technology, Prins Bernhardlaan 6, 2628 BW, Delft, The Netherlands

THEORETICAL BACKGROUNDS

The spatial distribution of the gas bubbles in a stirred vessel is largely determined by the overall flow pattern and the strength of the liquid flow. Two classes of gas-liquid dispersions can be distinguished.

In the first class a disk turbine is used as impeller. The gas issuing from the sparger rises into the impeller and is dispersed from there. It has been shown that the turbulence induced in the vessel is extremely non uniform with very high dissipation levels in the immediate vicinity of the impeller and very much lower levels elsewhere (Laufhütte and Mersmann (6)). Therefore the spatial bubble size distribution is also non-uniform. Bubble break-up occurs mainly in the impeller outflow. The resulting bubbles are small. Due to coalescence much larger bubble sizes are found in the liquid bulk (Greaves and Barigou (7)).

The second class of stirred dispersions is obtained by using a downwards pumping impeller. As the rising gas-flow is counteracted by the downward liquid flow, the gas does not enter the impeller directly but is distributed over the contents of the vessel by the strong circulatory flow. Only part of the gas is recirculated and redispersed by the impeller. As a consequence bubble size may, apart from the initial size, to a large extent depend on processes of breakup and coalescence in the liquid bulk. Therefore, turbulence intensity should be distributed as uniformly as possible. At high gassing rates, however, the downward liquid flow will not be sufficient to overcome and deflect the rising gas flow. The gas will rise directly into the impeller and the homogeneity of the dispersion becomes worse. This so-called indirect loading to direct loading transition has been described by (2). It will be clear that it is essential to postpone this transition as long as possible. The strong liquid flow which is required to do so, renders pumping efficiency an important parameter.

Defining the pumping efficiency

The pumping efficiency η can be defined as the ratio of the gain in enthalpy of the fluid, $\Delta p.Q_l$, to the power input P by the impeller. The pressure rise Δp over the impeller as determined by, among other things, rotational speed N and impeller geometry is converted into velocity head of a circulatory flow:

$$\Delta p = K_w \frac{1}{2} \rho_l \bar{v}^2 \sim \rho_l Q_l^2/T^4 \tag{1}$$

The proportionality constant K_w depends on the vessel geometry only and is not known in general. Now, by using the dimensionless power number Po and the pumping number N_q:

$$Po = P/(\rho_l N^3 D^5); \qquad N_q = Q_l/(N D^3); \tag{2}$$

the following relation can be found:

$$\eta = \Delta p\, Q_l/P \sim (D/T)^4 N_q^3/Po \tag{3}$$

Since the absolute value of η can not be calculated from eq. (3), the impellers will be compared on basis of the pumping efficiency relative to the inclined blade impeller (45↓6): $\eta' = \eta/\eta(45{\downarrow}6)$

The impellers used

Three impellers are compared during the present research: a Lightnin A315, the Dutch Leeuwrik impeller and a traditional inclined flat blade turbine.

Gas-liquid dispersion using downwards pumping impellers with flat inclined blades has been discussed by several authors. The influence of sparger geometry and impeller position in two-phase and three-phase systems has been investigated (3, 4). Mass transfer and power consumption proved to be strongly related to the process of cavity formation (2). Furthermore, it has been reported that with respect to mass transfer efficiency an inclined blade impeller performs at least as well as a Rushton turbine (2).

The design of the Leeuwrik impeller was carried out by the Netherlands Organization for Applied Scientific Research TNO (Vermeulen and Pohlmann (8)). This impeller is meant to create a large liquid flow as well as a strong trailing vortex. The large flow should increase the homogeneity of the dispersion whereas the vortex should decrease the size of the recirculated bubbles. The plan form of the blades is based on the shape of a delta wing. Wings of this type yield a large lift and produce strong trailing vortices (fig. 1a). These delta wings can be mounted on a hub (fig. 1b). Profiling the blades and adjusting the pitch gives the shape of the Leeuwrik impeller (fig. 1c). Van den Akker and Bakker (9) visualized the trailing vortex by means of cavitation techniques, see fig. 2.

The A315 is a relatively new impeller design, commercially available from Mixing Equipment Co (Rochester U.S.A.). This four-bladed impeller is characterized by the large blades and rounded edges (figure 3). According to Oldshue et al. (5, 10) this impeller is capable of dispersing 86% more air than a disc turbine at the same power input. At the same power input, gassing rate and torque, but different D/T ratio, it was found that gas-liquid mass transfer could be enhanced by 30% on average by using an A315 system instead of a conventional disc turbine system.

EXPERIMENTAL

The experimental equipment is shown in figure 4. First, a vessel was used for laser-doppler measurements of velocities in a liquid-only mode of operation. This vessel was provided with a draft tube being 0.19m in diameter. The liquid level above the draft tube was 0.20m. Laser-doppler measurements were made at 0.20m and 0.60m below the impeller. The pumping capacities of the impellers were obtained from the measured velocity profiles.

Second, a flat-bottomed vessel of diameter T=0.44m was used for the gas dispersion experiments. The vessel was equipped with four baffles (width W=0.1T). The liquid height H was equal to the tank diameter T throughout the experiments. A ring sparger of diameter d_s=0.076m, provided with 31 holes and mounted at a separation distance of S=0.7D was used for most gas-liquid experiments. An additional separation distance S=0.3D was tested with the A315. Impeller-bottom clearance equalled C=0.4T in most cases.

Mass transfer experiments were done using a conventional dynamic method. The mass transfer coefficient $k_l a$ was calculated from the increase in oxygen concentration after a step change of inlet gas from nitrogen to air. Distilled water was used as continuous phase during these experiments.

In both vessels power demand was calculated from the impeller rotational speed and the torque on the shaft. The torque was determined with a Vibro-torque transducer mounted in the shaft. Impeller dimensions can be found in table 1.

SINGLE-PHASE FLOW

The draft-tube vessel was used for measuring the pumping capacity of the various impellers. The average velocity ($\bar{v}_{ax}$) and the fluctuating component (v'_{ax}=RMS($v_{ax}-\bar{v}_{ax}$)) of the axial velocity have been measured at different radial distances in the draft tube at an impeller speed N=10Hz and at a distance of 0.20m below the impellers. The results for the three different impellers have been plotted in figure 5a,b. Figure 5c shows the relative turbulence intensity $v'_{ax}/\bar{v}_{ax}$.

The outflow of the Leeuwrik impeller (LS6) is dominated by the trailing vortex. The average velocity profile is rather flat near the centre of the draft-tube but shows a steep increase near the blade tip. The vortex also causes a high turbulence intensity. The velocity profiles of the 45↓6 and the A315 look rather different when compared with that of the LS6. With these impellers the average velocity profile flattens near the blade tip. The differences in shape of the turbulence profiles are remarkable. The turbulence intensity is constant near the centre for the LS6 but increases near the blade tip. For the 45↓6 and the A315 the turbulence intensity first decreases and only shows a weak increase towards the tip. On the average, the relative turbulence intensity is lowest for the A315. This will lead to lower shear stresses, which can be an advantage in shear-sensitive fermentation broths.

The results for the pumping efficiency are given in table 1. The A315 clearly has the highest pumping efficiency. These results qualitatively agree with the findings of Weetman and Oldshue (11). In fact, they find a smaller difference between the A315 and the 45↓6 but this is possibly due to a difference in vessel geometry and especially the fact that radial flow is suppressed in the draft tube. It is also possible that the larger diameter of the A315 is advantageous in draft-tube systems. The LS6 does have a larger efficiency than the 45↓6 but the difference is rather small compared to the difference with the A315. Probably the LS6 loses too much energy in the vortex near the blade tip: a substantial part of the power supplied to the impeller is directly converted into turbulent kinetic energy rather than into velocity head of a mean circulatory flow. The high value of Po of the LS6 is striking too.

GAS-LIQUID DISPERSION

Power consumption

The 0.44m standard vessel was used for two-phase experiments. Power consumption and mass-transfer coefficients were measured as a function of gassing rate and impeller speed for the different impellers.

TABLE 1 - Impeller diameter, power number, pumping capacity and pumping effficiency in the two vessels

Impeller	Draft tube vessel				Standard vessel	
	D(m)	Po	N_q	η'	D(m)	Po
45↓6	0.167	1.27	0.66	1.00	0.176	1.75
LS6	0.168	2.55	0.93	1.42	0.168	2.48
A315	0.178	0.76	0.66	2.12	0.178	0.76

The increase in radial flow produced by the 45↓6 when placed in the standard vessel, which is suppressed in the draft-tube, leads to an increase of almost 40% in the power number (table 1). The power numbers of the LS6 and the A315 do not differ significantly in the draft tube and the standard vessel, which means that these impellers maintain a strong axial flow.

The gassed power curves from the LS6 and the 45↓6 have been plotted in fig. 6b. The ratio between gassed and ungassed power consumption is denoted by P_g/P_u and Fl denotes the gas-flow number.

When using a LS6 the gassed power shows a steep increase at low impeller speeds. It was determined visually that this increase coincides with a transition from indirect loading and good dispersion to direct loading and bad gas dispersion. At the same time, the flow pattern switches from axial flow to radial flow. This transition in flow pattern leads to a sharp increase in power number. The gas-flow number at which this transition occurs increases with impeller speed. At high impeller speeds (N>6Hz) such a flow transition was not observed, in the investigated gassing range.

The power curves from the 45↓6 agree well with those obtained by (2). The lower value of P_g/P_u at Fl>0.05 for N=6Hz compared to N=4Hz is caused by the larger cavity size at higher impeller speeds.

Two sparger positions have been tested for the A315: S=0.3D and S=0.7D. The power curves for these two positions are plotted in figure 7. The power drop in the curves coincides with an indirect-direct loading transition. This occurs earlier for the smaller separation distance. Therefore, it can be concluded that a large separation distance is beneficial.

For this large separation distance the power curves have been studied more closely (fig. 8). At low impeller speeds the power consumption increases with gassing rate. This agrees fairly well with the findings of Lally (12). The flow pattern near the impeller has been studied photographically. It is found that the process of cavity formation is analogous to that for inclined blade impellers. At low gassing rates, Fl<0.01, vortex cavities are formed. At increasing gassing rates the cavity size increases and growing cavities are formed. This coincides with the increase in P_g/P_u at low speeds (N=4Hz and N=5Hz). At still higher gassing rates (Fl≈0.04, depending on impeller speed) large cavities are formed, power consumption decreases and gas dispersion becomes worse. Consequently, in order to achieve good gas dispersion this impeller should be operated in the vortex/growing cavity regime.

In general the A315 is capable of dispersing more air than the 45↓6. For example at an impeller speed of N=6Hz the indirect loading to direct loading transition occurs at Fl≈0.045 for the A315 and at Fl≈0.03 for the 45↓6. Because of the lower power number of the A315 power consumption is also lower.

Mass transfer

The mass-transfer coefficient was determined as a function of gas flow rate and impeller speed for all three impeller types. Figure 6a shows the results for the LS6 and the 45↓6. Figure 9 shows the results for the A315.

Study of the mass transfer curves for the 45↓6 makes clear that stirrer hydrodynamics does not only influence the power curves but, as a consequence, also effects mass transfer. The dips in the curves correspond to the drop in power consumption at the formation of large cavities. The opposite is not always true: for the LS6 the increase in power consumption at the axial-radial flow transition at N=4Hz does not lead to an increase in $k_l a$. Here the larger power dissipation does not lead to a better gas-dispersion. The mass transfer curves for the A315 are in general flatter and do not show such dips.

Figure 10 shows $k_l a$ as a function of specific power consumption at constant superficial gas velocity (v_{sg} = 0.01 m/s). Both the A315 and the LS6 yield a mass transfer coefficient which is larger than when using the 45↓6. An explanation could be that the larger pumping efficiency of the A315 and the LS6 results in a larger gas-holdup and hence in a larger mass transfer. When compared at equal impeller-bottom clearance (C/T=0.4) the difference between the A315 and the LS6 is only small. For the A315 $k_l a$ increases when clearance is decreased to C/T=0.3, possibly due to an increase in gas holdup. Because of it's stability and good mass transfer performance it seems advantageous to use an A315 rather than a 45↓6 or a LS6.

Some remarks with respect to interpreting this result should be made. Firstly, the geometries used are not optimal. Mixing Equipment Co. advises to use a larger ring sparger for the A315. Secondly, our comparison is made for impellers at equal D/T ratio. The D/T ratio is important when retrofitting existing installations. In such a case impeller speed, power consumption and torque are given parameters. This imposes restrictions on the D/T ratio of the new impeller. A comparison at different D/T ratios might give slightly different results.

CONCLUSIONS

Several aspects of the behaviour of three axial flow impellers have been studied. It may be concluded that modern, profiled impellers like the A315 and the Leeuwrik impeller can offer advantages over traditional inclined blade impellers. The A315 gave the best overall performance from the impellers tested.

The profiled A315 and Leeuwrik impeller do have larger pumping efficiencies than the inclined blade impeller. This can lead to a better gas dispersion and higher mass transfer coefficients. The A315 creates lower turbulence intensities and shear stresses than the other two impellers, which can be advantageous in shear sensitive fermentation broths.

It is expected that impeller performance can be improved by optimizing the reactor geometry. However, the influence of liquid properties and geometrical parameters like impeller-bottom clearance, D/T ratio and sparger has not yet been studied. Further research in this area is useful.

For a better understanding of the processes which occur in the vessel detailed knowledge of the internal structure of the gas-liquid dispersion is necessary. This will be the subject of future study.

These investigations are supported by the Netherlands Technology Foundation (STW, DTN 44.0566). The A315 impeller was kindly put at our disposal by Mixing Equipment Co.

NOMENCLATURE

C = impeller-bottom clearance (m)

d_s = sparger diameter (m)

D = impeller diameter(m)

Fl = gas flow number (-)

$k_l a$ = mass transfer coefficient (s^{-1})

K_w = friction coefficient (-)

N_q = pumping number (-)

Q_g = gas flow ($m^3 s^{-1}$)

Q_l = liquid flow through the draft tube ($m^3 s^{-1}$)

P = power consumption (W)

Po = impeller power number (-)

S = impeller-sparger separation (m)

T = vessel diameter (m)

v_{sg} = superficial gas velocity (ms^{-1})

$\bar{v}_{ax}$ = local mean axial liquid velocity (ms^{-1})

v'_{ax} = local velocity fluctuation (ms^{-1})

$\bar{v}$ = average axial liquid velocity in the draft tube vessel (ms^{-1})

Δp = pressure rise at impeller (Pa)

η = pumping efficiency (-)

ρ_l = liquid density (kgm^{-3})

REFERENCES

1. Warmoeskerken M.M.C.G., Smith J.M., 1982, Proceedings 4th Eur.Conf.Mixing, BHRA Fluid Engineering, Cranfield U.K., 237-246

2. Warmoeskerken M.M.C.G., Speur J., Smith J.M., 1984, Chem.Eng.Com. 25 11-29

3. Chapman C.M., Nienow A.W., Cooke M., Middleton J.C., 1983, Chem.Eng.Res.Des. 61, 71-95 and 167-185

4. Nienow A.W., Konno M., Bujalski W., 1986, Chem.Eng.Res.Des. 64, 35-42

5. Oldshue J.Y., 1989, Chemical Engineering Progress 5, 33-42

6. Laufhütte H.D., Mersmann A.B., 1985, Proceedings 5th. Eur.Conf.Mixing, BHRA Fluid Engineering, Cranfield U.K., 331-340

7. Greaves M., Barigou M., 1988, Proceedings 6th Eur.Conf.Mixing, BHRA Fluid Engineering, Cranfield U.K., 313-320

8. Vermeulen P.E.J., Pohlmann J.W., 1980, "Ontwikkeling van de Leeuwrikschroef", Internal report MT-TNO, Ref.nr. 80-015586, (Dutch)

9. Van den Akker H.E.A., Bakker A., 1989, MIXING XII, Trout Lodge U.S.A.

10. Oldshue J.Y., Post T.A., Weetman R.J., Coyle C.K., 1988, Proceedings 6th Eur.Conf.Mixing, BHRA Fluid Engineering, Cranfield U.K., 345-350

11. Weetman R.J., Oldshue J.Y., 1988, Proceedings 6th Eur.Conf.Mixing, BHRA Fluid Engineering, Cranfield U.K., 43-50

12. Lally K.S., 1987, Mixing Equipment Technical Report #A315-2

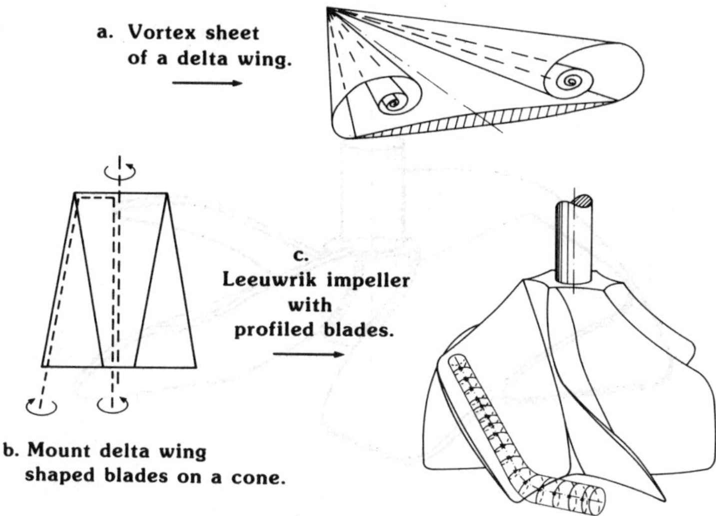

Figure 1 a,b,c The design of the Leeuwrik impeller.

Figure 2 The trailing vortex behind the blade of a three-bladed Leeuwrik impeller as visualized in a cavitation tunnel. The vortex is filled with vapour.

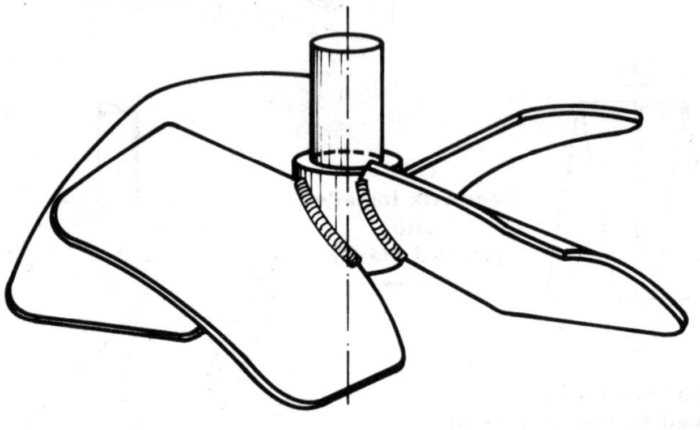

Figure 3 The A315 impeller.

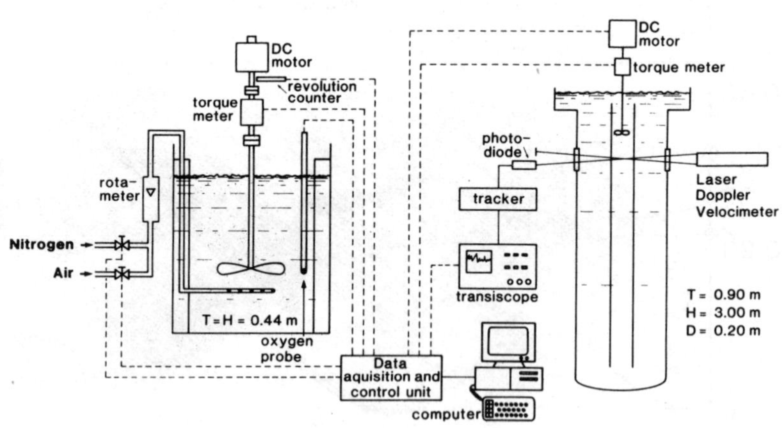

Figure 4 The experimental equipment.

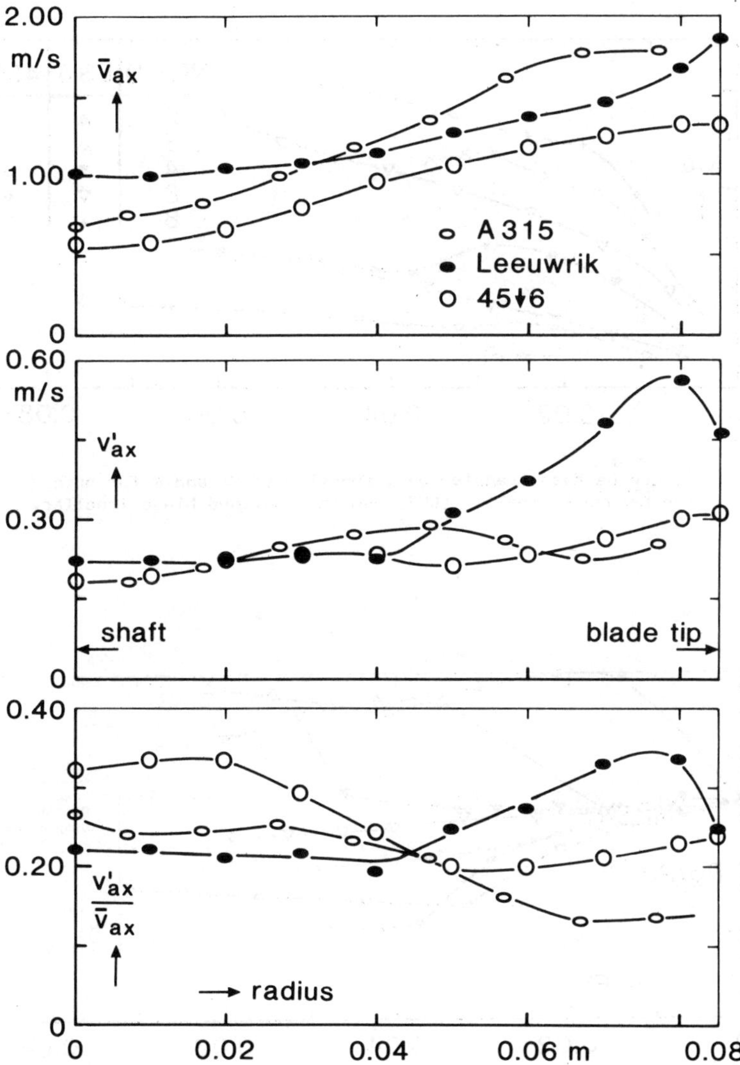

Figure 5 a,b,c Axial velocity and turbulence intensity below the impellers as measured in the draft-tube vessel.

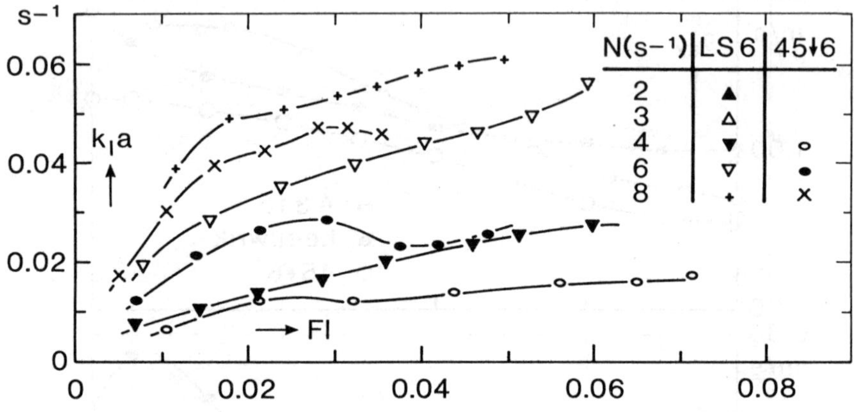

Figure 6a Mass transfer as a function of Fl and N for both the Leeuwrik impeller (LS6) and the inclined blade impeller.

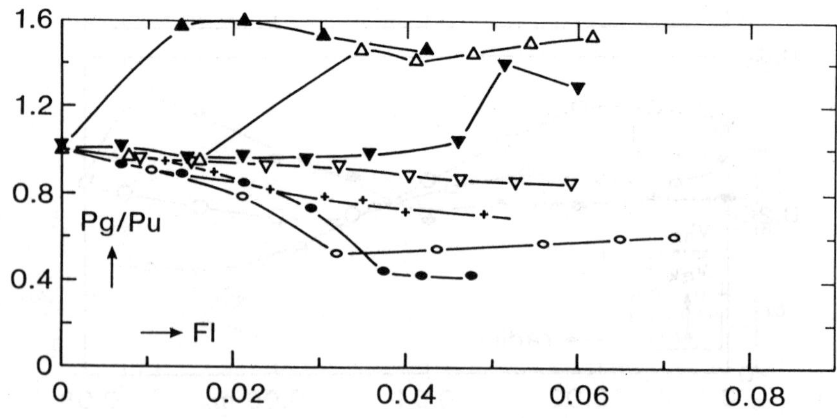

Figure 6b P_g/P_u as a function of Fl and N. Legend as in figure 6a.

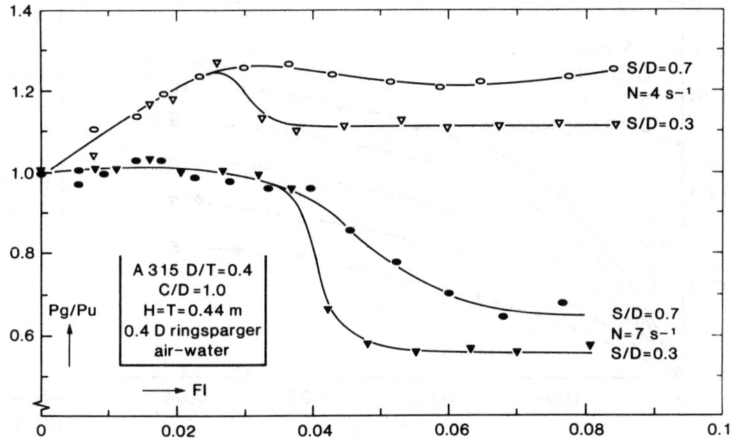

Figure 7 The power curves for the A315 impeller at two different impeller-sparger separation distances.

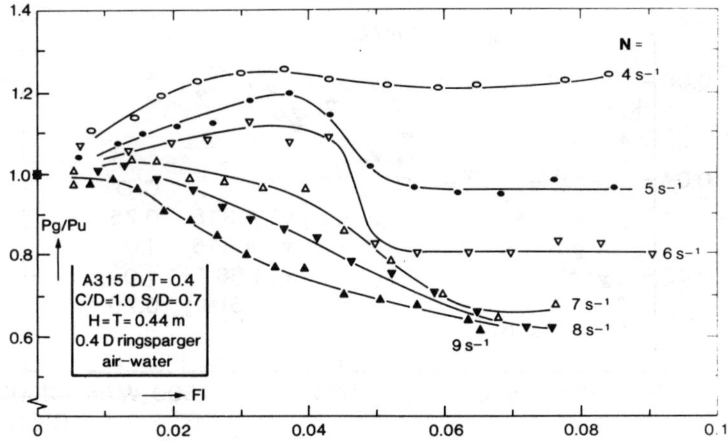

Figure 8 The power curves for the A315 impeller at the large impeller sparger separation distance.

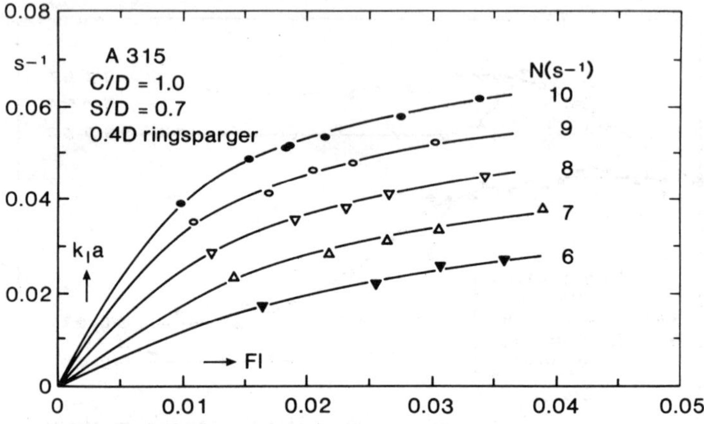

Figure 9 $K_l a$ as a function of gas-flow number and impeller speed for the A315 impeller.

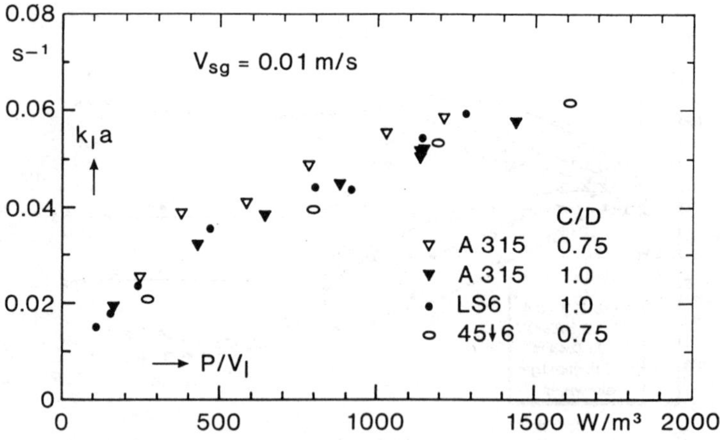

Figure 10 The mass transfer coefficient as a function of power consumption at a superficial gas velocity v_{sg}=0.01 m/s.

THE EFFECT OF PHYSICAL PROPERTY RANGES ON CORRELATIONS FOR MINIMUM IMPELLER SPEEDS TO DISPERSE IMMISCIBLE LIQUID MIXTURES

P.C.Lines* and K.J.Carpenter*

A series of experiments has been carried out for a range of immiscible liquid pairs at 20 litre scale using a 3-blade retreat curve impeller in a partially baffled vessel to measure the speed (Njd) at which no undispersed layer of either phase persists. Regression analysis of the results shows some coefficients on variables to be different to those currently in the published literature. The results presented here demonstrate that two separate equations: one for light-phase dispersed in the heavy-phase and one for heavy-phase dispersed in the light-phase give better statistical fits to the data.

INTRODUCTION

The impeller speed (Njd) at which one immiscible liquid is just dispersed in another has been investigated by several authors (1-7) and a variety of different correlations produced. Most equations are developed from dimensional analysis and some assume a basic relationship between dimensionless numbers. However Carpenter (8) has pointed out that the assumption of basic relationships between dimensionless groups can force the size and even the sign of the exponent on a variable. Using the modified Suratman number, for example, inherently assumes that the droplet size correlates to the impeller diameter. The reason for doing small-scale experiments should be to see if underlaying mechanisms appear without prior assumptions about relationships between dimensionless numbers. In one of his papers, Skelland (4), produces two different correlations from the same data; one based upon dimensional analysis and the other based upon empirical correlation. He remarks that the "equation obtained by the application of standard statistical techniques .....depends upon the form assumed for correlations". It therefore seems prudent to make the minimum number of prior assumptions and in this analysis the basic starting point is only that the correlation between Njd and the mixture properties takes an exponential form.

* ICI Fine Chemicals Manufacturing Organisation

Hexagon House, Blackley, Manchester M9 3DA

Most correlations in the published literature have been developed from tests with 4 or 5 liquid systems, although Pavlushenko (6) examined 14, and Skelland (4) 11 systems. In every case the range of at least one variable is limited. This is shown in Table 1 and can be seen to be especially true of the volume fraction. Our liquid mixtures have been chosen to cover as large a range as possible for each variable. This aims to widen the applicability of the correlation and to cover the range of properties encountered in fine chemicals manufacture. The 21 systems investigated here are designed to give a wide range of physical properties and also to minimise the correlation coefficients between the variables. It is known (9) that the solution of least squares equations may be dubious because of strong simple correlations between some or all of the independent variables. Since regression analysis with stepwise variable selection is applied here to the natural logarithms of the variables, using least squares equations, it is important to minimise simple correlations between variables.

APPARATUS AND PROCEDURE

Materials

The fluid mixtures used during this investigation are listed in Table 2, together with their physical properties. All the materials are of "technical" grade and no purification was undertaken, although the calcium chloride solutions had to be filtered free from small amounts of undissolved solids. Fluid properties are taken from an ICI data bank which allows temperature variations to be taken into account. For the few materials whose properties are not available from the literature, viscosities and surface tensions have been measured on samples pre-saturated with the other phase of a liquid pair. Shell Edelex 13, for example, is a refined medium viscosity oil whose properties can vary from sample to sample. Interfacial tensions are always calculated from the surface tensions of pure liquids, rather than measured, since they do not correlate strongly with Njd. It has also been shown (7) that literature values of interfacial tension correlate with Njd better than measured values; see later discussions.

Apparatus

Figure 1 shows a sketch of the vessel which is set up to model a typical glass-lined steel reactor used by the Fine Chemical Manufacturing Organisation of ICI. The dimensions and positions of the impeller and baffles are fixed for all the experiments discussed here. They conform to German National Standards for the plant-scale equipment. The vessel is a 305 mm internal diameter glass cylinder with a dished base: allowing good visual observations. Two "beavertail" baffles (37 mm wide) are used to minimise vortex formation. Mixing is effected by an impeller (184 mm diameter) on the vertically centred shaft, spaced at C = 0.1T and rotated by a variable speed motor drive. The shaft and impeller are made from stainless steel; the baffles from polypropylene. A loose fitting cover is fitted to the vessel which also has a bottom centre drain. This facilitates simple and rapid changes to the liquid volume fractions. An electric motor provides an infinitely variable output speed of 10-1000 rpm, and the speed is measured by a shaft-mounted digital tachometer. The motor and instruments all conform to Zone 1 electrical classification. No attempt was made to control the liquid temperature which rarely changes by more than two degrees during a set of experiments.

Operational Procedure

In all the trials reported here, the liquid height is equal to the vessel diameter. After setting up the baffle and agitator, 18 litres of heavy phase are charged, followed by 2 litres of light phase. If this is the first test on a new mixture, it is vigorously agitated to ensure mutual saturation. During each experiment the impeller speed is gradually increased in small steps until visual observation shows that dispersion is just achieved. The last pockets of clear liquid persist on the surface around the agitator shaft and require a substantial increase in speed to disperse. This excludes the small amount of heavy phase material which remains undisturbed in the bottom run-off pipe. Since this is only about 20 mls of liquid, and represents at most about 2% by volume of the heavy phase, it is neglected. In these trials the just-dispersed speed (Njd) is reached when a small pocket of undispersed liquid; whose diameter is about one tenth of the vessel diameter, remains undispersed on the surface around the shaft. Once this condition has been achieved the impeller speed is increased further, and then decreased in small steps until Njd is again reached. This is repeated until a common or average value is obtained. The direction of approach does not seem to impart any consistent hysteresis on the value of Njd.

The temperature, phase fraction and Njd are recorded before stopping the impeller motor and allowing the phases to separate. During settling, the time to settle and the appearance of the droplets are noted. This is used to identify the continuous (and dispersed) phase according to the methods described by Quinn and Sigloh (10) and by Selker and Sleicher (11). After the phases have separated, 2 litres of the heavy phase is drained from and 2 litres of the light phase charged to the vessel. Thus the light phase volume fraction is increased from 0.1 to 0.9 in steps of 0.1; although in some experiments increments of 0.5 in the volume fraction are used. At initially dispersed-phase fractions of about 0.7, dispersed to continuous phase inversion occurs and extra care is taken to allow long times between speed changes.

RESULTS AND DISCUSSION

With the impeller sited in the dished bottom at a low clearance, the heavier liquid consistently forms the continuous phase until its volume fraction falls to 0.2-0.3 of the total liquid volume. The dispersion of the lighter liquid in the heavier liquid commences in the heavy, continuous, phase and gradually rises through the lighter phase as the impeller speed increases. The dispersion first reaches the surface in front of the baffle(s) where it is deflected upwards. Undispersed material disappears last from small dead-spaces on vortices behind baffle(s) and especially from the vortex of undispersed light phase around the impeller shaft. Overall flow patterns in the fluid are essentially circumferential and the central core remains the most difficult portion to disperse. Even when the heavy phase volume fractions falls to 0.2, the impeller is still in the heavy phase and an inversion from heavy to light phase continuous will occur as the impeller is started from rest. The inversion is consistently observed with each combination of liquids whether the change is from organic to aqueous continuous or vice versa. When inversion has occurred however, the dispersion characteristics change. The heavy liquid becomes very finely dispersed in the light liquid and of this light phase, the volume around the shaft is the last portion to contain the dispersion. The heavy liquid also forms a region of undispersed material directly below the impeller which resembles an "inverted vortex" and stretches from the vessel bottom to the impeller shaft. This region of undispersed heavy liquid gradually narrows as the speed increases and finally disappears. These observations led us to consider treating such results separately during analysis.

Each liquid is taken from a single stock and all stocks are of normal technical quality. The only apparent problem with quality is caused by the presence of undissolved material in calcium chloride solutions. This impurity is filtered from the solutions and accounted for in the calculation for the solution strength. Tap water is used for all the aqueous phases, except for side-by-side experiments on two systems where tap water and de-ionised water were used. In each system there are no observable differences between the results for tap and de-ionised water. Impurities in the materials are always less than 5% and usually less than about 2% by weight. This means that the literature values for physical properties will be accurate, with the possible exception of the surface tension, which is known to be sensitive to impurity levels. In this analysis the interfacial tension is calculated from the difference between the literature values of the surface tensions for each liquid pair. This simplistic approach is justified on two counts: firstly van Heuven (7) shows that Njd correlates better with literature values for interfacial tension than with measured values, and secondly most correlations, including the two developed here, show a weak dependency on interfacial tension (see Table 3). It has also been demonstrated (5) that a correlation developed from tests on "pure" materials can be used to predict Njd with equal confidence for the same liquid pairs contaminated with various surface active agents. It can be argued that although interfacial tension is important in correlating droplet size, it does not strongly influence the prediction of Njd. Thus the literature values for interfacial tension will predict Njd to acceptable accuracies despite the presence of impurities in the solvents.

The experiments discussed here were undertaken in two separate campaigns. A list of 14 mixtures was originally chosen to give a wide range of physical properties. This was subsequently expanded to 21 mixtures: the additional liquid pairs being designed to reduce the correlation coefficients between variables. For example Figures 2 and 3 display the reduction in correlation coefficient between interfacial tension and continuous viscosity brought about by the inclusion of the new mixtures. These results represent the analysis of one value for each variable. In practice each liquid combination is used several times, and the number of times a particular mixture appears can cause the correlation coefficient to be weighted. The correlation coefficient for the 21 mixtures, shown in Figure 3 represents one value per mixture whereas Figure 4 shows the coefficient for the same 21 mixtures, weighted by several experiments on each mixture. Coefficients can range between plus and minus 1, with zero as the perfect result. In practice a figure of less than 0.05 represents no correlation and less than 0.5 represents a level which is unlikely to affect regression analysis. In the "Statgraphics" (12) programme used to analyse the data, a value of plus or minus 1 will cause the regression analysis to fail.

The method of analysing data employed by most previous authors is to statistically fit assumed dimensionless groups to the results to obtain a relationship of the kind.....

$$G_1 = K.\ G_1^{a}\ .\ G_2^{b}\ \text{-----}\ G_n \qquad \text{------}1$$

.....where G represents dimensionless groups, e.g. Re, Fr, Ar, Su. This method however, relies on identifying all the groups initially, and allowing sufficient variation in every physical property which comprises each dimensionless group. Thus, for example van Heuven and Beek (7) predict ....

$$N \propto \rho_c - \phi_d . \Delta\rho^{-0.53}$$

and ....

$$N \propto \mu_c^{0.08}$$

.....as a result of their choice of dimensionless groups although both ρ_c and μ_c are not varied throughout their study. Similarly Skelland and Seskaria (2) deduce the relationship .....

$$N \quad \left(\frac{\mu_c}{\mu_d}\right)^{0.11} \times \left(\frac{\Delta\rho}{\rho_c}\right)^{0.25} \qquad \text{-------}2$$

although

$$0.45 < \frac{\Delta\rho}{\rho_c} < 0.57 \qquad \text{-------}3$$

$$0.74 < \frac{\mu_c}{\mu_d} < 1.13 \qquad \text{-------}4$$

......with an average deviation of the results from the equation of 10.17%.

Godfrey and Reeve (1) suggest that most of the correlations developed so far can be approximated by ......

$$Re = K.\ Su_i^{x}\ .\ Ar^{y} \qquad \text{--------5}$$

......which they fit to their data to give ......

$$N \propto \sigma_i^{0.14}\ .\ \Delta\rho^{0.33}\ .\ \rho_c^{-0.53}\ .\ \mu_c^{0.06} \qquad \text{--------6}$$

......despite the fact that in their experiments, changes in the heavy continuous phase seem to indicate a positive power relationship to ρ_c. Table 3 displays the exponents for each variable in the authors' correlations when all the group relationships are rearranged to show individual variable relationships. In this table, complete agreement is only reached on the lack of effect of the dispersed phase density. There is a good agreement on the level of contribution to the correlation from the density difference and the dispersed phase viscosity, but elsewhere there is little or no common ground. It is believed that the use of a different type of agitator and baffle is unlikely to change the general form of the correlation. Several authors (eg.2,4,6) have used both axial- and radial-flow impellers and have obtained correlations which have the same variables and exponents: the differences showing only in the constant.

It was decided that a statistical fit of a dimensioned equation to the data including the variation of each individual property was the best approach. This would avoid the introduction of variables into the correlation which had not been fully investigated. In the past dimensionless numbers have proved useful in understanding scale-independent processes. For liquid-liquid dispersion however, the various dimensionless groups give no consistently enhanced understanding of the fundamental processes. If simple dimensionless relationships really do exist, then it may be possible to deduce them from the fit to physical properties rather than by pre-imposition.

The raw data is loaded into Statgraphics files and examined for "outliers" or unusual values of Njd. This is conveniently done by plotting a frequency histogram of Njd against experiment number. Figure 5 displays such a graph showing the predicted (solid line) and observed (squares) values, together with the 95% intervals for predictions (bars). Outlaying values can be readily identified and checked against original records. The data can then be correlated against a suitable relationship which could be of the general form:

$$Njd = k.\ \rho_c^{a}.\ \Delta\rho^{b}.\ \mu_c^{c}\ .\ \mu_d^{d}\ .\ \sigma_i^{e}\ .\ (g + h\ \phi_d)^{f} \qquad \text{------7}$$

The coefficients in equation 7 can be evaluated by a multiple regression analysis of the data but the constants g and h must be determined independently before the main analysis. This is achieved by plotting a normalised Njd against the volume fraction of the dispersed phase. These graphs display two distinct types of behaviour; one attributable to organic in aqueous dispersions and the other to aqueous in organic dispersions (both light in heavy and heavy in light dispersions). For organic in aqueous dispersions, the normalised value of Njd increases steadily with volume fraction although there is some scatter in the slopes; especially at low dispersed-phase volume fractions. For most aqueous-in-organic dispersions the normalised value of Njd rises to a maximum, before falling to a lower value at high dispersed-phase volume fractions. The maximum value is different for different mixtures. The water-chlorobenzene mixture does not display a maximum value at all. Thus at least one mixture in the aqueous-in-organic dispersion group shows the trends of the organic-in-aqueous dispersions group and the reverse is also true. Neither group correlates well with the relationship for the dispersed-phase volume fraction proposed in equation 7. It was therefore decided to combine aqueous-in-organic with organic-in-aqueous dispersions, but to make the distinction between light-phase dispersed and heavy-phase dispersed. The form of the correlation is assumed to be ......

$$Njd = K.\ \rho_c^{a} . \Delta\rho^{b} . \mu_c^{c} . \mu_d^{d} . \sigma_i^{e} . \phi_d^{f} \qquad \text{------ 8}$$

..........although a check is always made to see if the dispersed-phase density gives a better fit than the continuous-phase density. Model fitting is tried initially using regression analysis on the logarithms of the variables with none of the variables excluded. This is used to give an initial idea of the fit of the model and to display which variables do not correlate significantly. significance levels greater than 0.05 show a 95% certainty that the variable is not significant. The process is repeated using forward stepwise variable selection, including all and then excluding some variables, until a good fit is obtained with a low standard residual error. When <u>all</u> the data is tested, the correlation becomes ....

$$Njd = 0.075\ \rho_c^{0.154} . \Delta\rho^{0.33} . \mu_c^{-0.12} . \sigma_i^{0.03} . \phi_d^{0.10} \qquad \text{------ 9}$$

..........with a 91.9% fit and standard residual error of 0.07. This confirms the general levels of influence of density difference and interfacial tension, obtained by other authors but contradicts the dependencies of Njd on continuous phase viscosity and density. A similar form of correlation is obtained for mixtures of light-phase dispersed in heavy-phase........

$$Njd = 0.039\ \rho_c^{0.22} . \Delta\rho^{0.32} . \mu_c^{-0.133} . \phi_d^{0.08} \qquad \text{------ 10}$$

..........but with a 93.9% fit and a standard residual error of 0.058. The negative power for the continuous phase density which has been obtained in all previous work, (see Table 3) can be explained by the use of dimensionless group regression with a limited range of heavy continuous-phase liquids. In postulating the dimensionless groups to be fitted to the data, the fit is already constrained by the fixed relationships between the physical properties comprising each group, so that for example, if ....

$$Re = K.\ (Su_i)^{x}\ (Ar)^{y} \qquad \text{------- 11}$$

.........then ....

$$N \propto \mu_c^{(1 - 2\ (x + y))}\ .\ \rho_c^{(x + y - 1)}\ \Delta\rho^y\ .\ \sigma^x \qquad \text{------- 12}$$

Since both the modified Suratman and Archimedes groups contain ρ/μ^2 they are likely to be large or very large, and x and y are likely to be small so as to predict reasonable values for Re. Thus (x + y) is unlikely to be greater than 1 and ρ_c will have a negative exponent. The same type of arguments can be used to explain the similar results from most other authors.

A different form of correlation is obtained for mixtures of heavy-phase dispersed in light phase ....

$$Njd = 0.567\ \Delta\rho^{0.37}\ .\ \mu_d^{0.07}\ .\ \phi_d^{0.12} \qquad \text{------ 13}$$

..........with a 93.7% fit and a standard residual error of 0.064. This is based upon all the liquid mixtures and property ranges outlined in Table 1 and 2 but covers only a small range of dispersed-phase fractions from 0.4 to 0.1. The same data can also be fitted to an equation of the form........

$$Njd = 15.14\ \rho_c^{-0.38}\ .\Delta\rho^{0.35}\ .\ \mu_c^{0.04}\ .\ \mu_d^{0.06}\ .\ \sigma_i^{0.08}\ .\ \phi_d^{0.14} \qquad \text{------ 14}$$

..........which is similar to equations developed by previous authors. However, the continuous phased density and viscosity, as well as the interfacial tension are not significant and have been omitted in equation 13.

CONCLUSIONS

The form of the equations presented here are different from most previous literature and this is not due to the geometry but because ......

(1) a wider range of physical properties has been tested.

(2) a range of liquid mixtures has been chosen to minimise correlation coefficients between the variables.

and (3) no prior assumptions are made about the groupings of variables.

It is believed that this gives greater confidence in the use of the equations derived here for design purposes. Good fits are obtained for the equations despite using a wider range of physical properties than before. The experiments have only been carried out at up to 20 litres and are therefore only recommended for direct use at that scale. It appears that the just dispersed condition depends more upon the bulk flow characteristics than upon the turbulence parameters and properties which are believed to determine drop size.

ACKNOWLEDGEMENTS

The author wishes to thank MG Peacock, DM McKay and ICI Fine Chemical Manufacturing Organisation for permission to publish this work.

NOMENCLATURE

Ar	Archimedes Number $(g.\rho_c.\Delta\rho.D^3)/\mu^2_c$	
D	impeller diameter	(m)
Fr	Froude Number N^2D/g	
G	dimensionless group	
g	acceleration due to gravity	(m/s^2)
N	impeller speed	(1/s)
Re	Reynolds Number $ND^2\rho/\mu$	
Su_i	Modified Suratman Number $(\rho_c.\sigma_i.D)/\mu^2_c$	
ϕ	volume fraction	
μ	viscosity	(Ns/m^2)
ρ	density	(kg/m^3)
σ	interfacial tension	(N/m)

Subscripts

c	continuous-phase
d	dispersed-phase
i	interfacial
jd	just dispersed
m	mean

REFERENCES

1. Godfrey, J.C.; Reeve, R.N.; Grilc, V.; Kardelj, E.. I. Chem. E. Symp. Series No. 89 (Fluid Mixing II), 8, 107, (1984).

2. Skelland, A.H.P.; Seskaria, R..I. and E. C. Proc. Des. Dev., 17, 56,1978.

3. Esch, D.D.; D'Angelo, P.J.; Pike, R.W. Can. J. Chem. Eng., 49, 872,1971.

4. Skelland, A.H.P.; Ramsay, G.G. Ind. Eng. Chem. Res. 1987, 26, 77 - 81.

5. Skelland A.H.P.; Moeti, L.T. Ind. Eng. Chem. Res. 1989, 28, 127 - 130.

6. Pavlushenko, I.S.; Ianishevskii, A.V. J. Appl. Chem. USSR, 1958,31,No. 9, 1334 - 1340.

7. Van Heuven, J.W.; Beek, W.J. Proc. Int. Solvent Extraction Conf. (ISEC),1971, Paper 51, 70 - 81.

8. Carpenter, K.J. Paper 25, 5th European Conference on Mixing, Wurzburg 10 - 12 June 1985.

9. Davies, O.L.; Goldsmith, P.L. (Eds.) "Statistical Methods in Research and Production", ICI/Longman, 1976.

10. Quinn, J.A.; Sigloh, D.B. Can. J. Chem. Eng. 1963, 41, 15 - 18.

11. Selker, A.H.; Sleicher, C.A. Can. J. Chem. Eng. 1965, 43, 298.

12. "Statgraphics" Statistical Graphics Corporation 1988.

Table 1

Ranges of physical properties investigated by various authors

Reference	ϕ_d	μ_c Ns/m^2	μ_d Ns/m^2	ρ_c kg/m^3	ρ_d kg/m^3	$\Delta\rho$ kg/m^3	σ_i N/m
Godfrey & Reeve(1)	0.15 0.70	0.0003 0.0084	0.0003 0.0084	659 1585		167 587	0.0084 0.05
Skelland & Seskaria(2)	0.5 -	0.00046 0.0143	0.00046 0.0143	894 1041	894 1041	41 106	0.0063 0.0437
Esch & D'Angelo(3)	0.355 0.65	0.00086 0.037	0.00039 0.215	998 1840	679 875	123 965	0.0281 0.054(*)
Pavlushenko Ianishevskii(6)	0.1 0.7	0.0002 0.0194	0.000215 0.0253	828 1595	828 1595	20 595	0.009 0.0486
van Heuven & Beek (7)	0.004 0.35	0.001	0.00033 0.00111	998	659 903	95 339	0.0085 0.0495
this work	0.1 0.9	0.00031 0.0173	0.00031 0.0173	655 1619	655 1619	37 682	0.019 0.064

* also σ_c between 0.055 and 0.073 N/m.

Table 2

Liquid mixtures used to measure Njd

Ref No.	Description	Density Diff (kg/m^3)	Interfacial Tension (N/m)
1	Perchloroethylene/40% $CaCl_2$	224	0.0524
2	35% $CaCl_2$/Hexane	682	0.0643
3	40% $CaCl_2$/Methyl Isobutyl ketone	594	0.0619
4	10% NaCl/Chlorobenzene	37	0.0414
5	20% NaCl/Hexane	492	0.0605
6	Perchloroethylene/10% NaCl	549	0.0413
7	40% $CaCl_2$/Chlorobenzene	288	0.0525
8	Perchloroethylene/70% Ethylene Glycol	533	0.0190
9	25% $CaCl_2$/Chlorobenzene	121	0.0447
10	25% $CaCl_2$/Toluene	361	0.0495
11	Water/Methyl Isobutyl ketone	196	0.0482
12	Perchloroethylene/25% $CaCl_2$	391	0.0446
13	40% $CaCl_2$/Toluene	528	0.0573
14	Water/Toluene	130	0.0436
15	20% NaCl/Methyl Isobutyl ketone	346	0.0555
16	Perchloroethylene/Water	622	0.0387
17	20% NaCl/Toluene	280	0.0504
18	Chlorobenzene/Water	110	0.0388
19	60% Ethylene Glycol/Toluene	210	0.0258
20	Water/Shell Edelex 13 oil	132	0.0398
21	40% $CaCl_2$/Shell Edelex 13 oil	530	0.0535

Table 3

Exponents on the variables in equations for Njd derived by various authors

Reference	ϕ_d	μ_c	μ_d	ρ_c	ρ_d	Δ_o	σ_i
Godfrey & Reeve(1)	0	0.05	0	-0.53	0	0.33	0.14
Skelland & Seskaria(2)	0	0.111	-0.111	0	0	0.25	0.30
Esch & D'Angelo(3)	0	0.22	0	-0.61(!)	0	0	(+)0.39
Skelland & Ramsay (4)	0.1	0.08	0.07	-0.56(*)	0	0.43	0.01
Pavlushenko & Ianishevskii(6)	0	0.06	0.04	-0.33	0	0.08	0.15
van Heuven & Beek(7)	0.9	0.08	0	-0.54	0	0.38	0.08
this work (all data)	0.1	-0.12	0	0.154	0	0.33	0.03
this work (light disp'd)	0.08	-0.133	0	0.22	0	0.32	0
this work (heavy disp'd)	0.12	0	0.07	0	0	0.37	0

! average value based upon continuous phase volume fraction
\+ actually surface tension of the continuous phase
* average value based upon phase volume fraction

Figure 1: Basic vessel configuration.

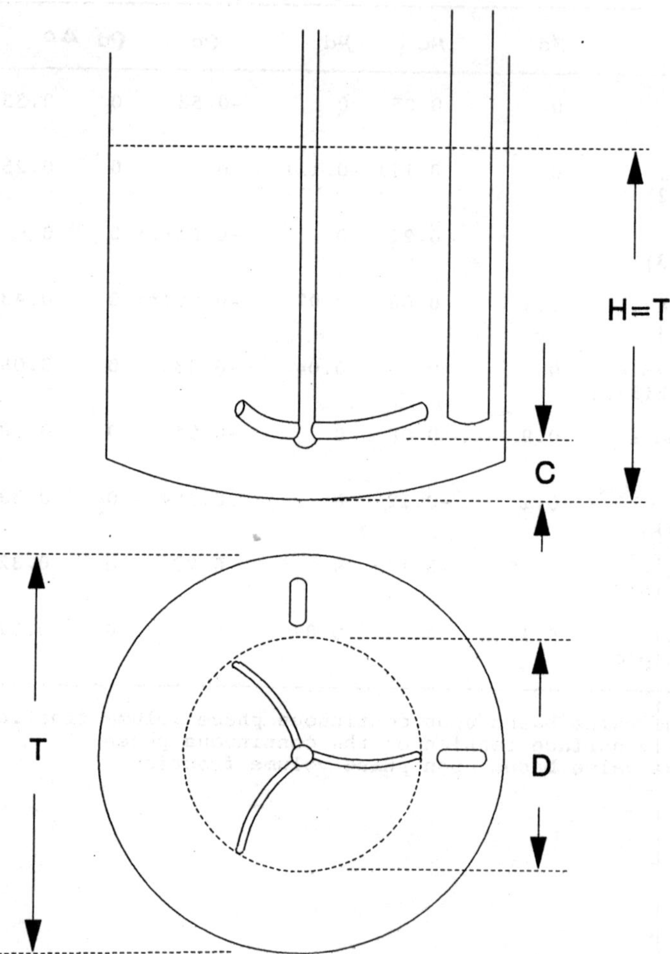

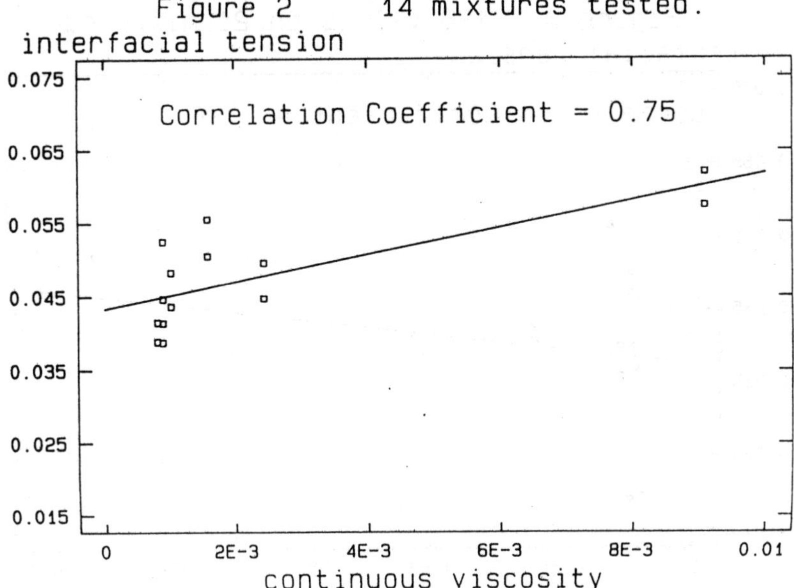
Figure 2 14 mixtures tested.
interfacial tension
Correlation Coefficient = 0.75
0.075
0.065
0.055
0.045
0.035
0.025
0.015
0
2E-3
4E-3
6E-3
8E-3
0.01
continuous viscosity

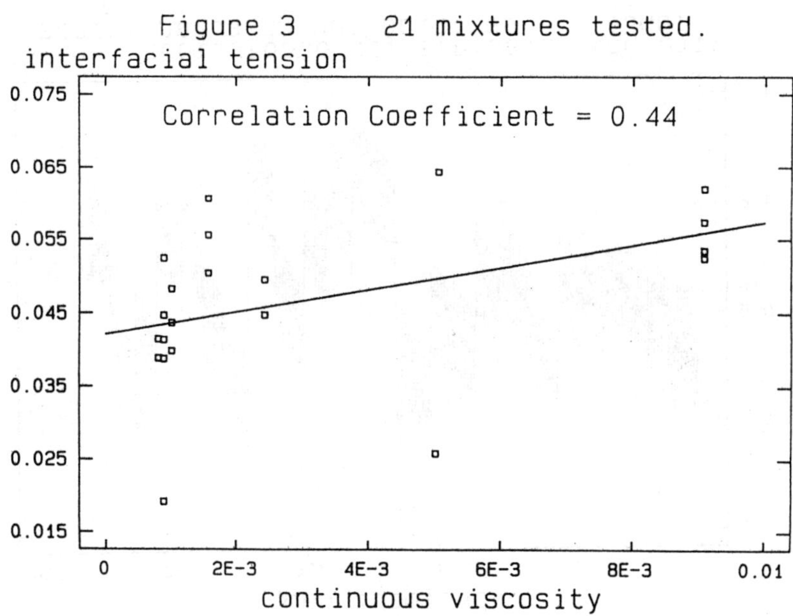
Figure 3 21 mixtures tested.
interfacial tension
Correlation Coefficient = 0.44
0.075
0.065
0.055
0.045
0.035
0.025
0.015
0
2E-3
4E-3
6E-3
8E-3
0.01
continuous viscosity

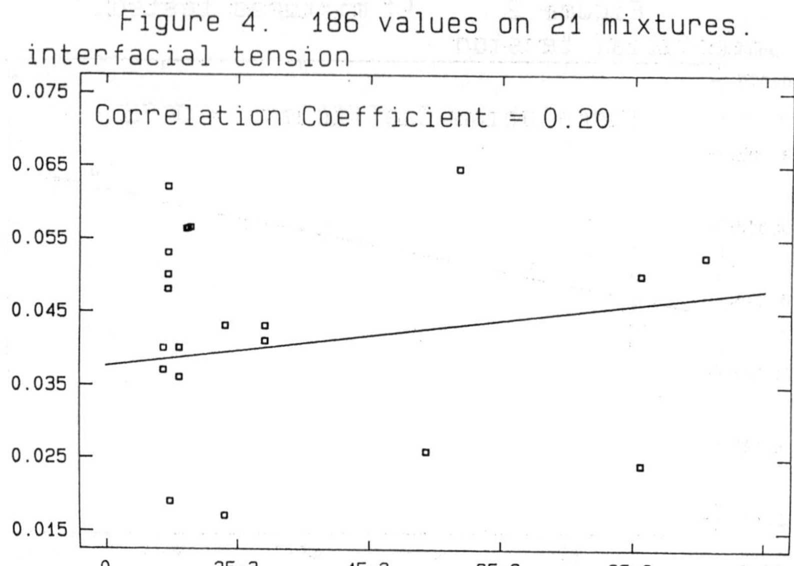
Figure 4. 186 values on 21 mixtures.
interfacial tension
0.075
0.065
0.055
0.045
0.035
0.025
0.015
Correlation Coefficient = 0.20
0
2E-3
4E-3
6E-3
8E-3
0.01
continuous viscosity

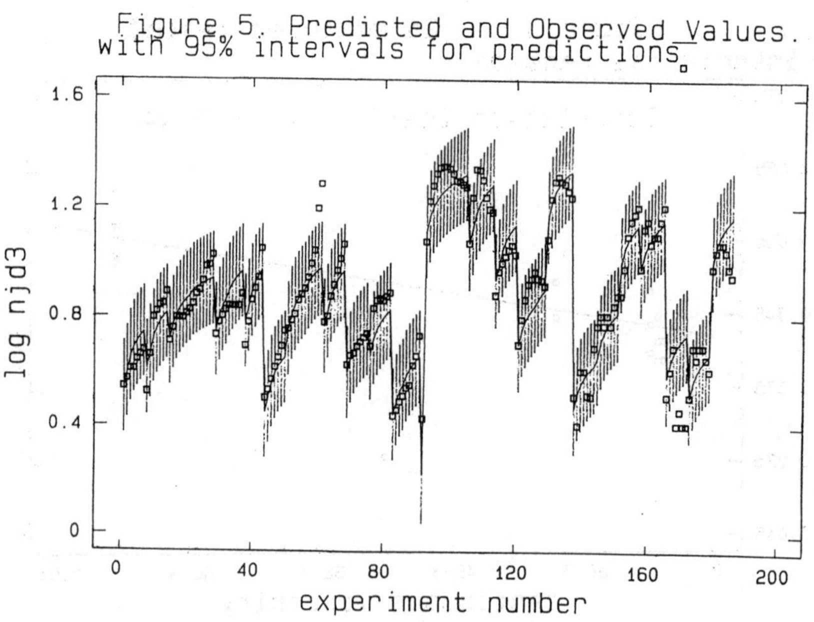
Figure 5. Predicted and Observed Values.
with 95% intervals for predictions
1.6
1.2
0.8
0.4
0
log njd3
0
40
80
120
160
200
experiment number

APPLICATION OF A LIGHT TRANSMITTANCE TECHNIQUE FOR INTERFACIAL AREA MEASUREMENT IN LIQUID-LIQUID DISPERSIONS

L. Portingell* and S. Ruszkowski*

A light transmittance technique for direct measurement of interfacial area has been applied by BHR Group Ltd. The light transmittance probe has been calibrated by taking photographs of a dispersion simultaneously with light transmittance readings. To validate the light transmittance technique some of the measurements made by Calabrese et al on silicone oil/water dispersions were replicated. An investigation was also carried out on coalescence effects at low dispersed phase concentrations.

INTRODUCTION

Many processes carried out in the chemical industry involve the contacting of immiscible liquids. Mass transfer between the phases is an important design parameter and knowledge of the effect of agitation on droplet diameter, hence interfacial area, is necessary.

A problem with using much of the experimental data available in the literature is the lack of agreement between different workers. Godfrey and Reeve (5) published a review which showed that different published correlations differed by up to a factor of ten in their prediction of power per unit mass required to maintain complete dispersion of droplets during scale-up by a geometrical factor of ten. These discrepancies can be attributed to a number of reasons:

- different experimental techniques used by different workers
- the inherent difficulties of working with two phase liquid liquid systems, which can be very sensitive to trace contaminants
- the practice of publishing correlations but little or no data makes it difficult to assess whether lack of agreement is due to differences in experimental results, or an artefact of the correlation

The only way many of the questions of scale-up will be resolved is by carrying out experiments in large scale vessels and at process conditions typically used in industry. Before embarking on such a programme we felt it necessary at BHR Group Ltd to validate an experimental technique which could be widely applied, and test the inherent reproducibility of the experiments.

* BHR Group Ltd, Cranfield, Bedford, MK43 0AJ, UK

The interfacial area of agitated dispersions has, in the past, most often been inferred from photographic measurements of the droplet diameters (eg Calabrese et al (2), Chen and Middleman (3)). The photographic technique is, however, very time consuming. It has most often been applied by photographing the dispersion from outside the vessel, and this limits its usefulness to small scale transparent vessels and measurements of droplet diameters adjacent to the vessel wall only at high dispersed phase concentrations. These are obvious disadvantages for investigations at the high dispersed phase concentrations which are of practical interest, and for investigations of scale up in realistically sized equipment.

Another technique, which overcomes the disadvantages of photography which are listed above, is that of sampling. Samples of the dispersion are removed from the vessel, and droplet diameters are measured externally. The main difficulties with this technique are of avoiding errors due to sampling, and assuming the droplets in the sample do not coalesce.

An alternative technique is light transmittance. Applications to the measurement of interfacial area of dispersions were described by Langlois (1). The light transmittance technique offers a rapid means of measuring interfacial areas of dispersions, which can be used at concentrations of up to 40%, in opaque vessels and without sampling problems.

The main disadvantage is that it is an intrusive technique which causes a hydrodynamic distrubance at the point of measurement. Care must also be taken to ensure the probe is clean each time it is used and not contaminating the system in any way.

This paper describes the development of a light transmittance technique for measuring the interfacial area of dispersions in stirred vessels. This includes calibration, experiments used for validating the technique, and results obtained from measurements of organic/aqueous dispersions in a 0.17 m diameter glass vessel.

DEVELOPMENT OF PROBE

Probe Details

The light transmittance probe used was a commercially available instrument - a Brinkmann colorimeter. The instrument consists of a control unit from which light passes via a fibre optic bundle into a stainless steel probe tip. The proportion of light which returns to the control unit is proportional to the frontal area of the dispersed phase droplets.

Light is introduced into the dispersion through a fibre optic cable, and the light transmitted through the dispersion is received by another fibre optic, via a mirror which folds the light path through 180° (Figure 1).

The probe tip was fitted with two discs of soft hydrophilic contact lens material on its glass and mirrored surfaces to prevent the sticking of organic phase droplets. The control unit was fitted with a filter to reduce the intensity of the returning light. The control unit was connected to a computer which used a sampling program to average the transmittance readings.

Previous work on liquid/liquid dispersions using a light transmission probe was done by Langlois et al (1). Langlois proposed that the interfacial area of dispersion is related to the ratio of incident and reflected light by the equation:

$$a = \frac{1}{\beta}\left[\frac{I_o}{I} - 1\right] \qquad (1)$$

If the local volume fraction of dispersed phase is known, or assumed to be constant throughout the vessel the Sauter mean diameter can then be calculated using:

$$D_{32} = \frac{6\phi}{a} \qquad (2)$$

Langlois carried out calibrations for a number of liquid pairs, and found that β is a function only of the refractive index ratio, m, of the dispersed and continuous phase (Fig 3).

Equation (1) differs from the form usually used to relate transmittance to a concentration in colorimetry:

$$\log\,(I_o/I) = kc\lambda \qquad (3)$$

Equation (3), however, relates to a system where light absorption, rather than light scattering, is taking place. Calderbank (7) used a design of light transmittance probe which caused liquid droplets to behave as though they were absorbing light, rather than scattering, and used an expression of the form of equation (3) to relate transmittance to droplet surface area.

It is worth noting that equation (1) can be derived from equation (3) by expansion and neglecting higher order terms. As a further check we fitted Langlois' calibration data to equation (3), but found that this gave a worse fit than equation (1).

Glass Bead Experiments

To check the reproducibility of the probe measurements and the validity of equation (1), a series of experiments involving glass bead/water dispersion was carried out.

A 0.105 m diameter glass vessel fitted with baffles and a lid was filled with distilled water. 0.25 g of glass beads sieved to between 180 and 200 microns were added. The probe was set to read 100% transmittance ($I/I_o = 1.0$) in distilled water and then placed into the bead/water dispersion. The probe was reset in distilled water after a set of 30 readings; 150 readings were averaged overall. This was repeated using three probe tips of path lengths 1.0 cm, 1.6 cm and 2.0 cm. The weight of the beads in the dispersion was increased to 2.0 g at o.25 g intervals.

Interfacial area is related to transmittance by

$$a = \frac{1}{\beta\lambda} \left[\frac{I_o}{I} - 1 \right] \tag{4}$$

It was assumed that the effect of path length, λ , would be incorporated simply by analogy with equation (3). This assumption was subsequently verified by measurements for the range of path lengths of paractical interest for this work.

The specific interfacial area of the suspension is directly related to the mass of glass beads in the suspension:

$$a = \frac{6M}{\rho_s V d_p} \tag{5}$$

Substitution in equation (4) gives

$$M = \text{constant} \times \frac{1}{\lambda} \left[\frac{I_o}{I} - 1 \right] \tag{6}$$

Hence a plot of M against $((I_o/I)-1)/\lambda$ should give the same straight line for all the probes. It is not possible to carry out a check of the calibration constant, β , using this technique. The apparent refractive index of the glass beads is different from the refractive index of glass because of the inclusion of microscopic air bubbles.

Graphs of $((I_o/I)-1)/\lambda$ vs weight of glass beads are plotted for the three different path lengths in Figure 4. All three graphs fit well to straight lines, indicating that the dependence of a on I_o/I is correct. The graphs also fit the same straight line, which validates the inclusion of the path length factor in the equation. The experiment was not continued above 1.5 g of glass beads, since it was no longer possible to maintain a homogeneous suspension.

This experiment was also used as a standard check for the probe as the results were reproducible from day to day. The volume fraction of glass beads in the suspension never exceeded 0.07%. The experiment demonstrates the effectiveness of the light transmittance technique even at very low dispersed phase concentrations, given the right conditions.

Photographs

To check the β values suggested by Langlois (Fig. 2), and hence calibrate the probe, photographs were taken of various dispersions.

These were taken using a Single Lens Reflex camera attached to a rigid endoscope placed within the dispersion (Fig. 5). Photographs of different organic/aqueous dispersions were then analysed on a Joyce-Loebl Magiscan Bubble Analyser. Photographs of glass beads of a known diameter were used to calibrate the Analyser. The camera was fitted with an extension tube to provide sufficient magnification of the droplets. This at the same time gave a very narrow depth of field which eliminated errors due to parallax.

Photographs were taken of three silicone oils and of isododecane. The silicone oils used were three grades of Dow Corning 200 oils 1 cS, 5 cS and 20 cS. At least 300 droplets were photographed for each calibration.

Measurements using the light transmittance probe were made simultaneously with the photographs. Sauter mean diameters were calculated from the light transmittance measurements using Langlois' recommended β values. The calculation of Sauter mean diameter from a measurement of interfacial area (and vice versa) requires a knowledge of the local volume fraction of dispersed phase. It was assumed that there was no significant variation of droplet diameter throughout the vessel, and hence the average dispersed phase volume fraction could be used in calculations of Sauter mean diameter. This was subsequently confirmed with measurements using the light transmittance probe.

The photographic diameters were compared with those calculated from the light transmittance formulae and new β values were calculated (Table 1).

The corrected β values, calculated from photographs, differ somewhat from Langlois' reported values. Factors which could contribute to these differences include variation of the refractive index from its nominal value and effects due to variations in geometry between the BHR Group Ltd probe and Langlois' probe.

The dimensions of the probe used by Langlois were much larger than those used by BHR Group Ltd. Since the process we are dealing with is light scattering, rather than light absorption, it is clear that for a given angle of scattering the transmittance will be a function of <u>both</u> path length and probe diameter. The linear relationship assumed for the effect of path length (equation 4) is only likely to hold for relatively small changes in the path length. This factor may contribute to the lack of exact agreement between the glass bead tests with probes of different path length (Figure 4).

Photographs taken at two dispersed phase concentrations of 0.15% and 0.3% gave identical Sauter mean diameters. This agreed with light transmittance results which also gave identical Sauter mean diameters.

Photographic results of the same dispersion at two different speeds showed larger droplets at the lower speed. The ratio of droplet diameters of high and low speeds was the same for both the photographic and light transmittance techniques.

<u>Effective Range of Instrument</u>

The theoretical graph of light transmittance reading vs interfacial area (Fig. 6) shows that for transmittance readings larger than approximately 20%, there is a large change in reading with interfacial area. Below this value the transmittance is insensitive to large changes in interfacial area. Experimental scatter increases as the readings decrease which means that readings below 20% are inconclusive.

The graph also shows the effect of using different dispersed phases on the range of the instrument. For dispersed phases with lower refractive indices, higher dispersed phase concentrations can be reached before the cut off point than for materials with higher refractive indices.

Figure 7 shows the experimental results using isododecane as the dispersed phase. The curve shown in Figure 7 shows a more complicated behaviour than the theoretical curves in Figure 6. The Brinkmann reading flattens off more quickly for the measured curve than for the theoretical one. The transmittance is a function of interfacial area, rather than of concentration directly. The local interfacial area measured by the probe is itself a function of concentration, since at high concentration droplet diameter increases, due to increased coalescence, and the distribution of droplet sizes throughout the vessel becomes non-uniform. The flattening off of the curve in Figure 7 is thus due to both a flattening off of the light transmittance probe response, and to an increase in the droplet size.

MEASUREMENTS OF DISPERSION

Experimental

Equipment All interfacial area and Sauter mean diameter measurements were carried out in a 0.17 m diameter, flat bottomed glass vessel. The vessel had a 4 litre capacity and was fitted with four stainless steel baffles. The baffles had a width of 0.018m (T/10) and a wall clearance of 0.006m (T/30). The vessel was also fitted with a stainless steel lid, however all experiments were carried out with an air/dispersion interface present. All dimensions are given in Table 2.

A Rushton disc turbine diameter 0.102m were used for the experiments. The impeller and the impeller shaft were made from stainless steel. Details of impeller dimensions are given in Table 2.

Torque measurements were carried out using a Coesfeld Rheosyst 1000 "Duo" unit. This senses the instantaneous load on the stirrer. This particular unit consists of two electronic force transducers with torque ranges of 0 to 100 mNm and 0 to 1 Nm.

Glass and stainless steel were the only materials allowed to come into contact with the dispersion. This was to minimise the possibility of contaminating the dispersion, and sticking of the dispersed phase to any lyophilic materials (eg rubber gaskets).

Conditions The vessel was filled with four litres of distilled water and the motor switched on. The lighter phase was added slowly whilst the impeller rotated. The standard amount used was 6 ml (0.15% by volume).

The dispersed phases used were Dow Corning silicone oils of various viscosity grades, and isododecane. Table 3 shows the physical properties of all materials used. The impeller clearance was set at T/2. The speed range was 130 to 160 rpm The minimum dispersion speed was 130 rpm, and above 160 rpm air began to be entrained into the dispersion.

The experiments to check the effect of impeller speed were carried out with a T/3 Rushton turbine, at a clearance of T/3. The range of speed was 400 to 435 rpm. The T/3 impeller gave a wider range of speeds than the T/2 impeller, over which the dispersed phase was completely dispersed and surface aeration did not take place.

Procedures A rigorous cleaning procedure was used to ensure the freedom from contamination of all surfaces which came into contact with the dispersion. Our practical experience is that liquid-liquid systems can be very sensitive to traces of contamination. The procedure was adopted to ensure day-to-day reproducibility of the measurements.

The cleaning procedure used involved soaking the vessel and internals in hot water and surfactant. The vessel was then rinsed with tap water followed by distilled water. The vessel and internals were then soaked in concentrated chromic acid and finally rinsed thoroughly with distilled water. Physical properties of all materials for the tests were measured to check for possible contamination. The pH and conductivity of the distilled water were monitored daily using a pH probe and a conductivity probe. The interfacial tension of the two phases was measured using the pendant drop technique.

Sixty minutes were allowed for dispersions to reach equilibrium before any probe readings were taken. The probe was preset to read 100% transmittance in distilled water and then placed approximately 25mm below the dispersion surface. Five samples in five seconds were taken and the process repeated ten times.

RESULTS AND DISCUSSION

Reproducibility of Results

Experiments were carried out using a glass vessel and analytical cleaning procedure to determine whether reproducible results could be achieved using silicone oil. The experiments were carried out by cleaning the vessel, preparing the dispersion, measuring the droplet diameter, then disposing of the dispersion before preparing the next dispersion using fresh materials and a freshly cleaned vessel

Four viscosity grades of silicone oil were found to give reproducible results at a concentration of 0.15% V/V. Further tests were carried out using 1cS silicone oil with concentrations ranging from 0.1% V/V to 0.55% V/V. Typically the standard deviation of the Sauter mean droplet diameter was in the range 6 to 9% for any given set of experimental conditions.

Test for Coalescence at Low Concentrations

Previous work reported in the literature has often been limited to very low dispersed phase concentration. No doubt this is partly for practical experimental reasons, but in some cases workers were interested only in the measurement of droplet break-up phenomena and were attempting to limit their work to "non-coalescing" systems (2). Before undertaking any extensive experimental investigations, we investigated the assumption of non-coalescence.

An experiment was carried out on a 0.15% V/V dispersion of 1cS silicone oil in distilled water, light transmittance readings were taken at an impeller speed of 160 rpm. The speed was then dropped to 130 rpm and finally returned to 160 rpm. An hour was allocated at each speed for the dispersion to reach equilibrium.

If the system had really been non-coalescing, then the dispersed phase would have reached an equilibrium droplet diameter at 160 rpm, and this diameter would not have changed when the speed was reduced to 130 rpm.
The results shown in Table 4 show that the droplet diameter increases with decreasing impeller speed and returns to its original value as the speed is increased again. Hence coalescence occurs even at very low dispersed phase concentrations.

Effect of Dispersed Phase Concentration

Experiments were carried out using silicone oil dispersions up to dispersed phase concentration of 0.55% V/V and using isododecane up to 11% V/V. There was no significant effect on Sauter mean diameter of **dispersed phase** concentration for silicone oil up to the maximum tested concentration of 0.55%. The results of Calderbank (quoted by (3)), indicate that the upper limit for which the dispersed phase droplet size is independent of concentrations might be of the order of 1%.

For isododecane above a dispersed phase concentration of 2.1% V/V a linear dependence can be seen (Fig 8) This agrees with previous literature which states a linear dependence.

Even though coalescence occurs in systems with very low dispersed phase concentrations, the systems are "non-coalescing" in the sense that Sauter mean diameter is independent of dispersed phase concentration.

The effect of dispersed phase concentration on droplet diameter is often reported (e.g. 2, 7) as being correlated by:

$$D_{32} \propto (1 + \gamma \emptyset) \quad (7)$$

Godfrey et al (6) reports that at higher dispersed phase concentrations (above approximately 20%) the relationship is no longer linear, and is better described:

$$D_{32} \propto (1 + \gamma \emptyset^{b}) \quad (8)$$

with b being approximately 0.8.

Most values for γ are approximately 3, although Calderbank reports γ in the range 3 to 9. Our data give a γ of approximately 15 - much higher than other results. This might be an effect of the low impeller speed (to avoid surface aeration) at which the experiments were done.

Speed Correlation

The effect of speed on Sauter mean diameter is widely reported in the literature (eg 2, 3) as:

$$D_{32} \propto N^{-1.2} \quad (9)$$

The effect of agitator speed on Sauter mean diameter was measured for a 20cS silicone oil/water dispersion. The droplet diameters were calculated over the range 400 rpm to 435 rpm for a T/3 Rushton turbine in the 0.17m vessel. The speed range was limited to this because surface aeration occurred above a speed of 435 rpm whilst the oil was not fully drawn down below a speed of 400 rpm. The impeller clearance was T/3 for these experiments.

Analysis of the results via a muliplicative regression of D_{32} on speed indicated that:

$$D_{32} \propto N^{-1.26}$$

A graph of the results is shown in Figure 9. The speed range for this work was very narrow, and ideally this effect of speed should be explored for a much wider range. Nevertheless the data followed this relationship very well (Fig. 9).

Power Measurements

The power consumption of the impellers was measured using a Coesfeld Rheosyst viscometer. This is used to drive the impeller and gives a readout of torque on the agitator shaft. The power consumption and power number are calculated from:

$$P = 2\pi N\tau \quad (10)$$

$$Po = \frac{P}{\rho N^3 D^5} \qquad (11)$$

The result for average power number is 4.3 ± 0.2, significantly lower than the nominal value of 5.0 used for the power number of a Rushton turbine. There is good agreement with Nienow's power number measurements which were 4.5 ± 0.2, 4.7 ± 0.2 in vessels of .22 and .29 in diameter.

The results show that great care is needed when comparing data from small vessels with data from large vessels. The use of a nominal value of 5.0 for the power number of a Rushton turbine could lead to serious errors in a comparison or correlation based on power input.

The change in power number implies that there is some change in the flow patterns or the way in which the trailing vortices behind the impeller blades are generated. This could in turn mean that there is a change in the droplet coalescence and break-up processes, compared to a larger vessel. There is some doubt whether the results from a small scale system can be related directly to a geometrically similar large scale system. This doubt will only be removed after work in large scale vessels is carried out.

Replication of the Work of Calabrese

One of the objectives of this study was to check how closely the droplet diameter measurements made independently in another laboratory could be reproduced.

The work reported by Calabrese (2) was chosen for this study. This is some of the most recent work available on liquid-liquid systems, it has been carried out with great attention to detail, and much of the experimental detail is available in the literature.

Calabrese's study was carried out using five different grades of silicone oil, all of which exhibit the same interfacial tension with water. His objectives were to determine the relative contributrion of interfacial tension and dispersed phase viscosity to drop stability and to develop a partial correlation for mean droplet size distributions that remain valid in the limit of an inviscid dispersed phase, The measurements were made in a range of vessel sizes from T = 0.14 to T = 0.39 m. We could not readily obtain a vessel of a size used by Calabrese, so a vessel of 0.17 m diameter was used.

Photographs were taken of a 0.15% V/V dispersion of 5 cS silicone oil at an impeller speed of 160 rpm. The Sauter mean diameter was 170 microns.

The correlation given by Calabrese predicts a Sauter mean diameter of 232 microns. However, this correlation does not incorporate the effect of working a different power per unit mass due to changes in vessel geometry which were necessitated by the use of the 0.17 m vessel. Godfrey (6) found that the effect of changes in vessel geometry was well correlated by:

$$D_{32} \propto \varepsilon^{-0.4} \qquad (12)$$

Correction of Calabrese's predicted diameter (assuming that impeller power numbers were the same for both this work and Calabrese's, since Calabrese does not report measurement of power number) gives a final corrected diameter of 189 microns.

This agreed remarkably well with the measured value, the difference being only 10% of the predicted value. Calabrese reports a standard deviation for his correlation , of measured points about the predicted value, of 23%.

CONCLUSIONS

1. A light transmittance technique is effective for the measurement of interfacial area in liquid-liquid dispersions.
2. Coalescence occurs even at very low dispersed phase concentrations. However, low concentration systems are non-coalescing in the sense that droplet diameter is independent of dispersed phase concentration.
3. Good reproducibility of droplet diameter measurements can be achieved, provided great care is taken in maintaining the cleanliness of the system. Where differences are reported between similar systems, these are likely to be due to changes in the system itself, for example different purity of materials, or the effects of contaminants or materials of construction.
4. Care must be taken when comparing results on the basis of power input, since power number for Rushton turbines is a function of scale and sensitive to geometrical variations.

SYMBOLS

a = Specific interfacial area (m^2/m^3)
b = Exponent in equation 8 (-)
c = Concentration (mol/l)
D = Impeller diameter (m)
d_p = Diameter of glass beads (m)
D_{32} = Sauter mean diameter (m)
H = Height of liquid (m)
I = Transmitted light intensity (W/m^2)
Io = Incident light intensity (W/m^2)
k = Constant in equation 3 (-)
M = Mass of glass beads (kg)
m = Refractive index ratio = n_d/n_c (-)
N = Impeller speed (1/sec)
n_c = Refractive index of continuous phase (distilled water) (-)
n_d = Refractive index of dispersed phase (-)
Po = Power number (= $P/\rho N^3 D^5$) (-)
P = Power input (W)
T = Tank diameter (m)
V = Vessel volume (m^3)
We = Weber number (= $\frac{N^2 D^3 \rho}{\sigma}$) (-)
β = Constant (-)
γ = Constant in equation 7,8 (-)
ε = Power per unit mass (w/kg)
ρ = Density of liquid continuous phase (kg/m^3)
ρ_s = Density of solids (kg/m^3)
λ = Path length of light transmittance probe (m)
$\emptyset$ = Dispersed phase volume fraction (-)
σ = Interfacial tension (N/m)
τ = Torque (Nm)

REFERENCES

1. Langlois, G. E., Gullberg, J. E. et al, 1954, The Rev Sci Ins 25, 360.

2. Calabrese, R. V. et al, 1986, AIChEJ 31, 657.

3. Chen, H. T. and Middleman, S., 1967, AIChEJ 13, 989.

4. Bujalski, W., Nienow, A. W. et al, 1987, Chem Eng Sci 42, 317.

5. Godfrey, J. C., Reeve, R. N., 1984, Inst Chem Engs Symp Series 89, 107.

6. Godfrey, J. C., Obi, F. I. N., Reeve, R. N., 1987, Inst Chem Engs Symp Series 108, 123

7. Calderbank, P. H., 1958, Trans I ChemE 36, 443.

TABLE 1 - Calibration Results

LIQUID	$m = n_d/n_c$	D_{32} PROBE (MICRONS)	D_{32} PHOTO (MICRONS)	1 / β (m) (LANGLOIS)	1 / β (m) (BHR Group Ltd)
1 cS 200 OIL	1.04	164	125	2904	3810
5 cS 200 OIL	1.050	179	170	2416	2544
20 cS 200 OIL	1.052	148	157	2283	2152
ISODO-DECANE	1.067	161	144	1670	1867

TABLE 2 - Impeller and Vessel Dimensions

	RUSHTON TURBINE
Diameter	0.102
Blade angle to horizontal	90°
Projected blade width	0.020 (0.2D)
Blade length	0.025 (0.25D)
Number of blades	6
Disk diameter	0.071 (0.7D)
Blade thickness	0.0016 (0.015D)
Disk thickness	0.0016
	VESSEL
Diameter	0.17 (T)
H:T ratio	1
Baffle-width	0.018 (T/10)
Baffle-wall clearance	0.006 (T/30)
Baffle-base clearance	0.009 (T/20)
Impeller blade - baffle spacing	0.01

Note: all dimensions in m.

TABLE 3 - Physical Properties of Materials Used

Distilled water:

Surface tension	68.4 mN/m
pH	5.4
Conductivity	4 μ S/cm
Temperature	20°C

LIQUID	DENSITY g/ml	APPROXIMATE VISCOSITY cS	INTERFACIAL TENSION WITH DISTILLED WATER mN/m
1cS 200 OIL	0.829	1	34.55
5cS 200 OIL	0.922	5	36.83
20cS 200 OIL	0.962	20	31.38
100cS 200 OIL	0.977	100	30.69
ISODODECANE	0.754	1	40.97

TABLE 4 - Test for Coalescence at Low Concentration

SPEED (rpm)	TIME	DROP DIAMETER (microns)
160	1 hr	110
130	2 hrs	154
160	3 hrs	117
130	4 hrs	151

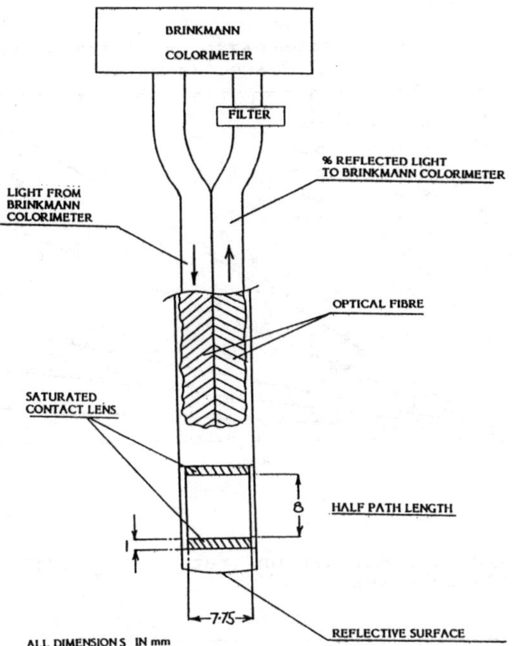

Figure 1: Light transmittance probe (Brinkmann colorimeter).

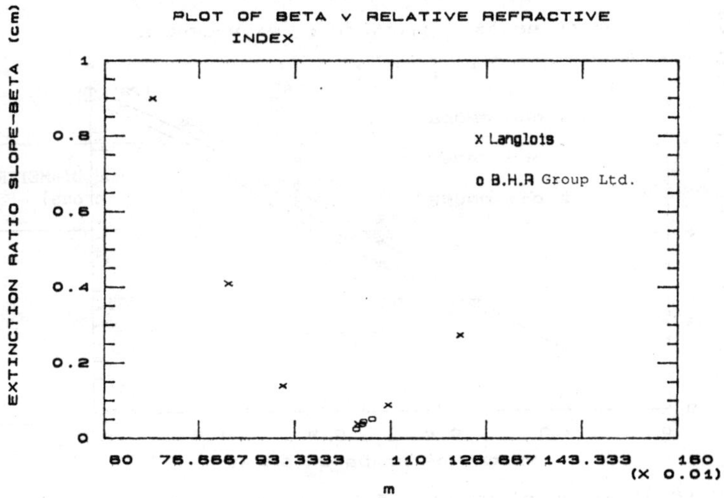

Figure 2: Graph of beta against relative refactive index.

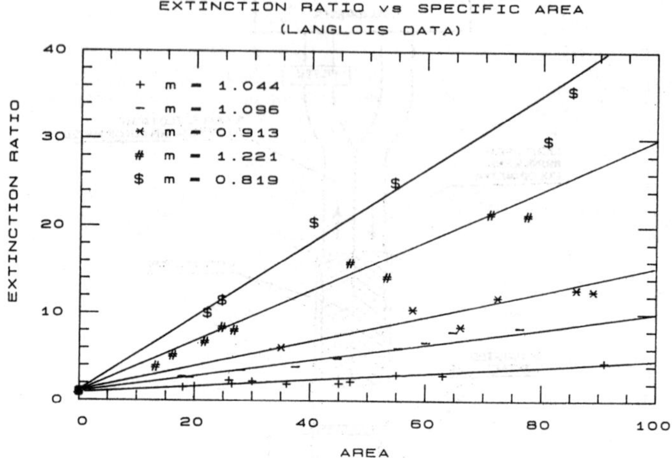

Figure 3: Graph of extinction ratio (I_o /I) v specific interfacial area.

	2.0 cm probe	1.6 cm probe	1.0 cm probe
Slope	2.68	2.61	2.57
R-squared	99.96%	99.33%	99.63%

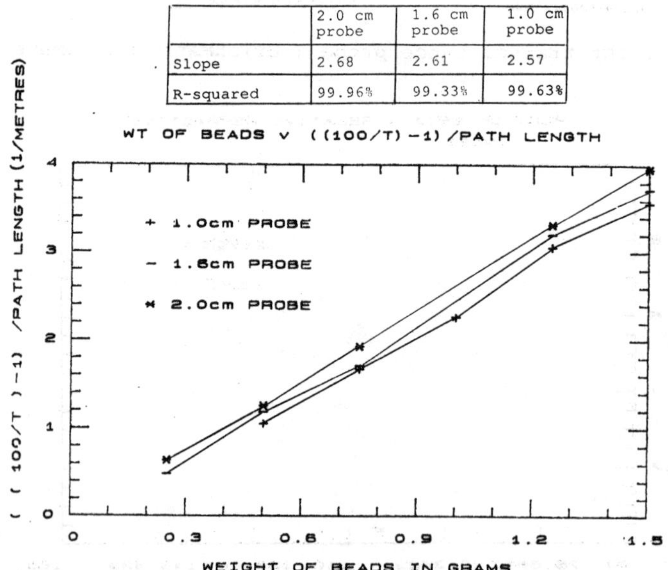

Figure 4: Graph of weight of Ballotini beads against $((I_o/I)-1)$ / path length of probe.

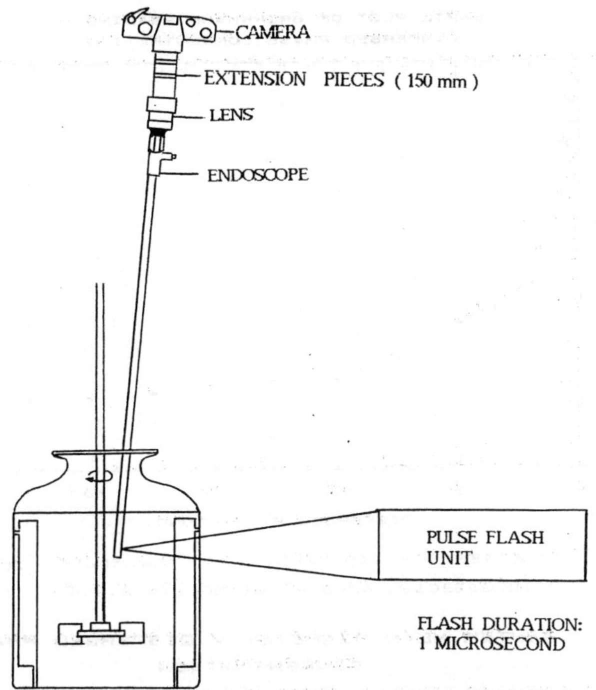

Figure 5: Experimental set-up for taking photographs.

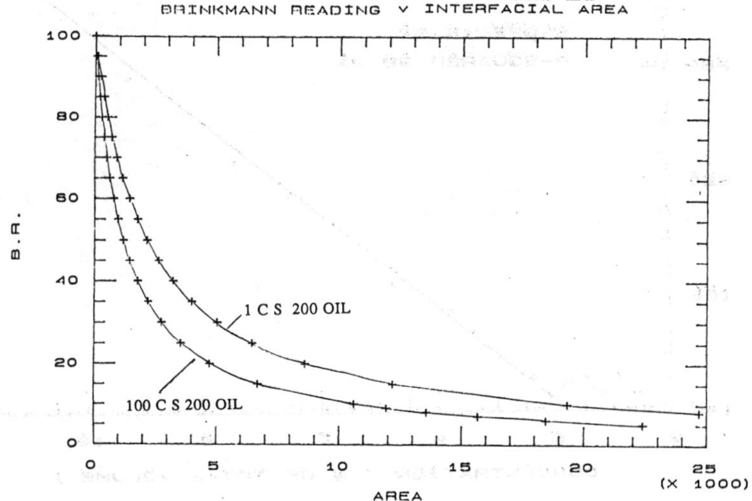

Figure 6: Theoretical graph of Brinkmann reading v interfacial area.

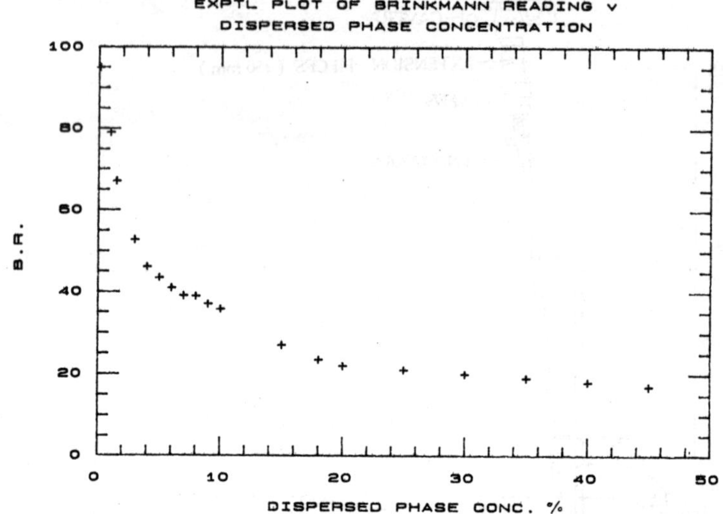

Figure 7: Experimental results for Brinkmann reading v interfacial area at speed 170 r.p.m.

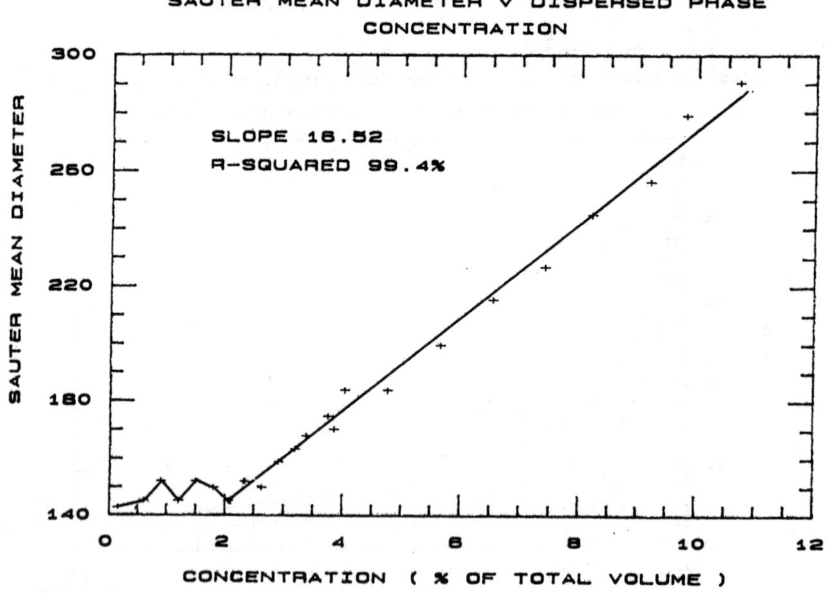

Figure 8: Variation of Sauter mean diameter with dispersed phase concentration at speed 170 r.p.m.

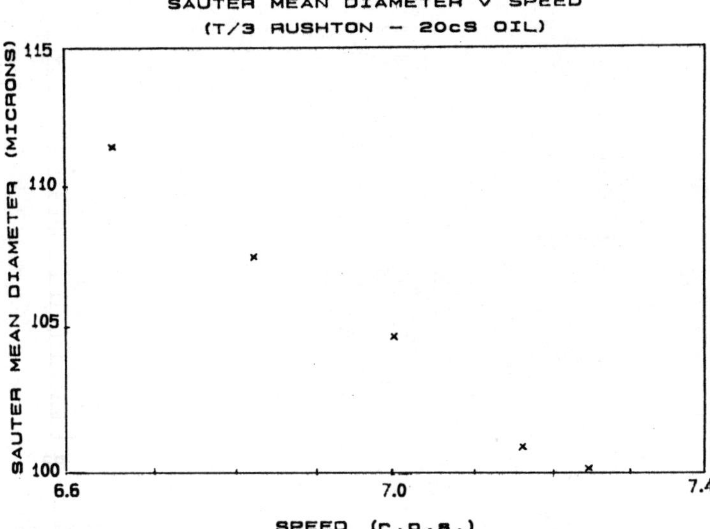

Figure 9: Dependence of Sauter mean diameter with impeller speed for Ø = 0.15%

LOCAL MIXING BEHAVIOUR OF AIRLIFT MULTIPHASE CHEMICAL REACTORS

A. Lübbert*, B. Larson*, L.W. Wan*, S. Bröring*

Detailed investigations of a pilot-scale airlift tower loop reactor led to new insights into the mixing behaviour of these reactors. Measurements in the air/water model system as well as during yeast cultivations have been performed. Gas residence time distribution measurements, enhanced in the high signal-to-noise ratio by pseudostochastical tracer addition, could be used to perceive detailed information on the gas circulation within similar reactors. The local heat-pulse-technique allows for measuring liquid velocities and the locally acting mixing mechanisms. It turned out that airlift loops cannot be described by means of the isotropic turbulence model, even not at high energy inputs, where the flow has a chaotic outer appearance. The main mixing mechanism in the riser seems to be the convective mechanism due to the transport of liquid in the bubbles' wakes.

INTRODUCTION

Airlift tower loop reactors have been investigated by various authors. Chisti (1989) investigated the gas residence time behaviour of airlift tower loop reactors in a conventional way. Verlaan (1987) investigated the liquid velocity through the downcomer of the reactor by means of an inductive flow meter, which at best integrates the flow velocity over the cross-section of the downcomer. Chisti (1989) used tracer techniques to investigate the mean liquid velocity, but could not resolve liquid velocity profiles. Weiland and Onken (1980) compared the velocity profiles of bubble columns and airlift loop reactors and stated that the profiles within the loops are flat, whereas the profiles in bubble columns are of a parabolic form.

The liquid circulation, which is the dominant hydrodynamical property of airlift tower loop reactors, has been investigated by several groups by means of tracer measurements. The common result of all investigations showed that the continuous liquid fluid elements, marked by the tracer molecules, do maintain their identity for up to more than 10 circulations of the fluid around the loop. This means that the airlift tower loop reactor is not a highly efficient mixer. A typical tracer response curve, as obtained in the pilot-scale airlift tower loop reactor of 4 m^3 total volume, chosen to be the accompanying example in this paper, is depicted in Figure 1. This reactor had a loop section of 14 m in length.

* Institut für Technische Chemie, Universität Hannover, D-3000 Hannover, FRG

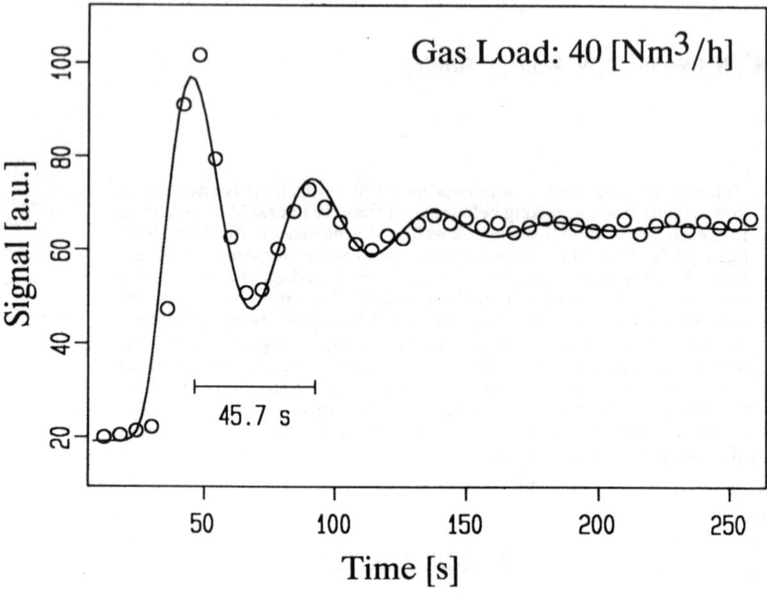

Fig. 1: Result from a circulation time measurement in an airlift loop reactor. The tracer response on a single-pulse addition of coumarin was measured by means of a fluorescence detector.

Since only 4 circulations could be resolved in this special case, the reactor used is a comparatively well-mixing airlift contactor. This fact should be kept in mind during the detailed study of its mixing behaviour.

GLOBAL GAS-PHASE MOTION IN AIRLIFT REACTORS

Since the gas holdup in the downcomers of airlift tower loop reactors is by no means negligible, it is reasonable to assume that there is a gas recirculation from the riser through the downcomer and back into the riser. Verlaan (1986) assumes that the gas fraction recirculated constitutes an insignificant fraction of less than 10% of the fresh gas input. However, in some systems, especially in biotechnological applications, the gas holdup in the downcomer reaches such high values that the amount of recirculated gas cannot possibly be that low.

A simple straightforward way to probe the gas circulation in an airlift tower loop reactor should be a gas residence time distribution measurement. If it would be possible to enhance the signal-to-noise ratio of this general measurement technique significantly, the structure of the residence time distribution curves, due to the internal gas circulation, should be observable.

Measuring Technique

Gas residence time measurements are classical measurement techniques in chemical engineering. Usually a gaseous tracer, e.g., helium, is added to the main gas input stream of an aerated reactor in form of a Dirac delta function. The concentration of the tracer molecules is measured as a function of time, where the gas leaves the dispersion. The detection can then be performed by means of a mass filter detector. In order to obtain sufficient tracer concentration at the detector, one is led to enhance the amount of gas shot into the reactor. This, however, is not possible, since a fast input of the tracer, as required to simulate a Dirac pulse, cannot be made without an additional mechanical energy input into the column and not without producing bubbles, which deviate significantly in size from those usually produced at the gas distributor. Then, the probed system would be different from its normal operation state.

A way around this difficulty is to use the concept of pseudostochastical test functions instead of single Dirac pulses. By this technique (Wilhelmi and Gompf 1970) it is possible to enhance the signal-to-noise ratio of the resultant residence time distribution density curves by more than two orders of magnitude. This proved to be sufficient to resolve a structure within the residence time distributions density functions at sufficiently small tracer additions.

Gas RTD Results

A typical result of such a gas residence time distribution measurement is depicted in Figure 2. It was obtained in a pilot-scale airlift tower loop reactor of 4 m^3 total volume during a yeast cultivation. It can be seen that the scatter of the data is low enough to allow for a fit of a mathematical model to the measured data.

The model fitted to the data was composed of a sum of two axial dispersion terms, one for the gas, which starts at the gas distributor and leaves the dispersion by a direct path through the riser section of the reactor, the second is due to the gas which circulates once around the loop and then leaves after a second run through the riser.

In this way, a model-supported measurement of some characteristic parameters of the gas-phase motion within the reactor can be obtained. The residence time τ_1 of the first term then gives information on the gas which passes the reactor on the direct way, the time τ_2 gives the corresponding time for tracer particles which moved around the loop. Their difference τ_1 - τ_1 gives the gas circulation time.

From the point of view of mixing the amount of gas recirculated is much more important. This amount is hidden in the integrals of the two components. The ratio of the integrals of the two model peaks in Figure 2 is directly related to the amount of gas which

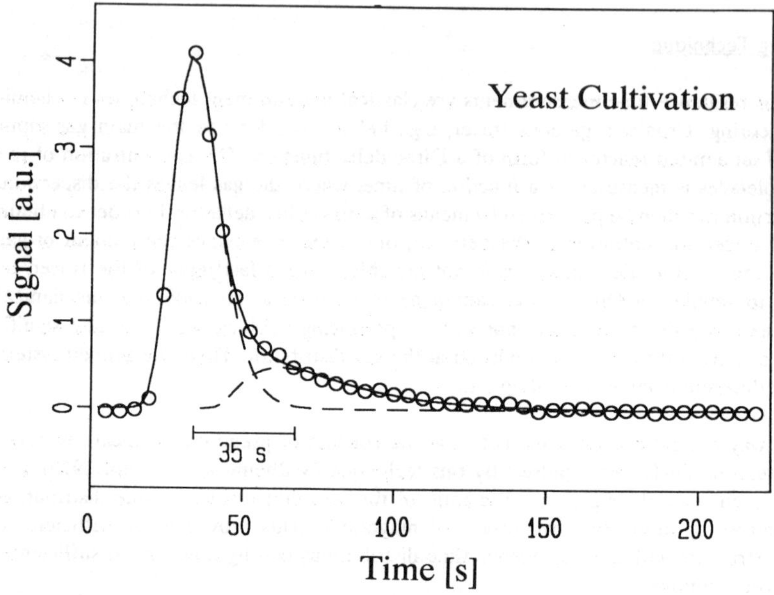

Fig.2: Typical gas residence time distribution obtained during a yeast cultivation in a pilot-scale airlift fermenter, if one adds helium pseudostochastically to the main gas input stream.

is recirculated. This fraction turned out to be more than 20% in the special case probed by the shown data. This result clearly demonstrates that there is a considerable backmixing of the gas phase in the reactor investigated. Similar amounts of gas recirculation were also found in other pilot-scale reactors of the same and of smaller sizes.

There are some hints on the mixing behaviour of these reactors, which additionally can be obtained from gas tracer techniques. If one introduces the tracer at different places into the reactor, then the highly resolved residence distribution data exhibit more details. The curves take on a slightly different form, as is shown in Figure 3.

The data of the two curves displayed were obtained in an air-in-water model system. It can be observed that the relative (!) amount of gas recirculated is significantly higher, provided the tracer is injected into the downcomer section at the place indicated in Figure 3. This result, which is surprizing at first glance, finds a simple explanation in the fact that the dispersion in the downcomer is different from that in the riser section. Only the smaller bubbles are carried down into the downcomer. Thus, small bubbles are selectively marked if the tracer is fed into the downcomer.

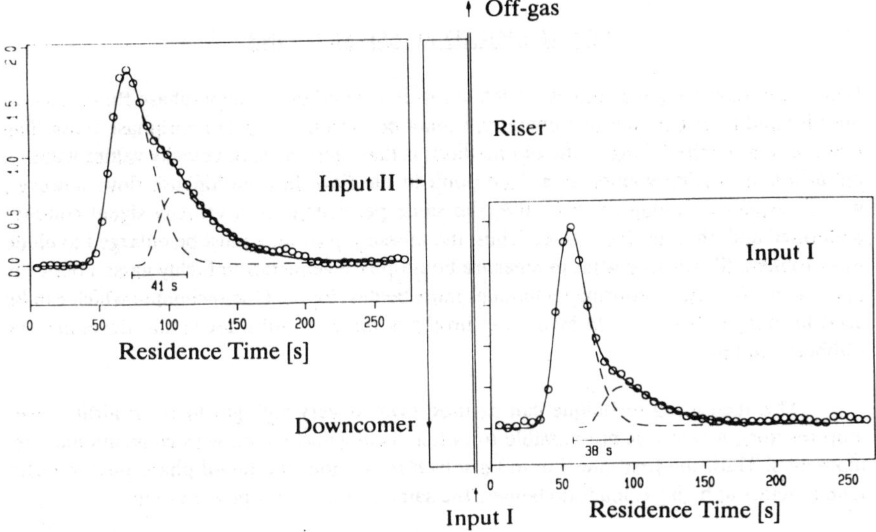

Fig.3: Comparison of gas residence time distributions, obtained from measurements with different tracer input locations.

Since the amount of small bubbles that are circulated around the loop is considerably higher than the amount of larger ones, this result is not surprizing. As compared to the residence time distribution curve depicted on the right hand side of Figure 3, one can notice a third very small peak in the left curve. This means that the fraction of recirculated smaller gas bubbles is so large that one even observes a very small number of bubbles circulated a second time. This third peak is not resolved in the right curve.

The measurements with different input places do not only show the different behaviour of different bubble composites, but they also give some hints on the gas-phase mixing by means of bubble coalescence. If we would have a strong coalescence within the riser of the loop reactor, a different behaviour would not be observable in such a clear way.

These examples show that highly resolved gas residence time distributions can result in a lot of usable local information on the gas-phase behaviour within the different sections of tower loop reactors. For spatially more resolved information, local measuring methods must be used.

LOCAL LIQUID-PHASE MOTIONS

Most local measuring techniques, which are well established in single-phase flows, such as hot film and laser Doppler anemometry, cannot be used in complex multiphase flows. The main reason for the failing of the old methods is that they measure velocity values within a point-like measuring volume at a fixed point in the flow. In a multiphase flow, however, with a dispersed holdup of, say, 20%, the same percentage of a velocity signal contains undefined and, thus, useless values. Thus, the measuring volume must be enlarged to elude this problem. Should one want to measure liquid-phase velocities in highly gassed optically opaque media, new measuring techniques must be developed. One technique which can be used in many real viscous fermentation broths is the heat pulse technique developed by Lübbert and Larson (1987).

This measuring technique can be used even at very high gas loads in airlift tower loop reactors, where it is not possible to obtain local velocity data with conventional anemometers. Thus, the first question to be solved is, whether the liquid phase flow in airlift loop reactors at high gas loads do behave the same way as at low power inputs.

Heat pulse technique

In the heat pulse technique, the time-of-flow technique is measured that takes a fluid element, started at a well-defined point in the flow to reach an adjacent detector. With the known probe distance, the average flow velocity component along the straight line, defined by the two probe locations, can be obtained. Moreover, from the shape of the time-of-flow curve, one can get information on the mixing mechanisms acting in the flow.

It is not possible to inject even small amounts of material tracers within a very short time into the dispersion without transferring a momentum into the fluid. In the vicinity of the probes, this will change the flow velocity to be measured. Therefore, heat was used to mark the fluid elements. If the temperature enhancement of the marked fluid elements is kept at or below about 1 K, buoyancy effects can be neglected. Concerning the signal-to-noise ratio, which is low if one uses a conventional Dirac pulse markation of the fluid elements, the same problems arise as with the residence time distribution measurement. Thus, to keep it at a high level, pseudostochastical signals are used as tracing signals. The time-of-flow curves are then obtained by cross-correlating the tracer signal with the measured tracer concentration curve.

To extract the characteristic parameters of the fluid flow, which is assumed to flow at constant mean velocity v from the heater to the detector, the following mathematical model was fitted to the measured data:

$$p(t;x) = (2\pi\sigma^2(t))^{-1/2} \exp(-0.5\,(t-\tau)^2 / \sigma^2(t)), \tag{1}$$

with $\tau = x/v$, x being the fixed distance between heating element and temperature sensor. The standard deviation $\sigma(t)$ is assumed to follow the simple relationship usually taken in literature (Richardson 1926, Levenspiel and Fitzgerald 1983, Großmann and Procaccia 1984):

$$\sigma(t) = \sigma_0\, t^{\beta} \tag{2}$$

σ_0 characterizes the strength of the mixing and β the mixing mechanism (Lübbert and Larson 1990). This model assumes that the developing tracer particle cloud can be described by a normal distribution which, by the influence of the different mixing mechanisms, becomes broader with time. A typical time of flow curve, as obtained in the riser of the probed airlift loop reactor, is depicted in Figure 4.

Results of pilot-scale airlift reactors

If one measures a mean liquid velocity profile within the riser of the airlift tower loop reactor, which serves as accompanying example in this paper, one obtains the profile depicted in Figure 5. The curve was obtained at a gas load of 40 Nm^3/h. The relatively flat profile was expected. If one doubles the gas load, the profile (Figure 6) becomes more pronounced, which was not expected. A further enhancement of the gas throughput up to 120 Nm^3/h led to an even more parabolic profile (Figure 7) which, after a fit of a simple mathematical model curve, gave rise to negative lobes near the walls of the riser. Thus, one is led to the conclusion that the downcomer of this reactor is not able to transport the large amount of liquid carried along the riser by the mammut pump effect, so that the flow within the riser section of the reactor evades into a bubble column-like flow with a countercurrent flow along the reactor walls. This example shows that airlift reactors may change their simple flow profiles, if the gas load gets higher.

Since a bubble column is known to be a better mixer than an airlift loop (Weiland and Onken, 1980), this leads to enhancement of the mixing strength within the riser part of the reactor, which can be extracted from parameters σ_0 defined in eq. 2. Quantitative measurements have been performed in laboratory scale concentric tube airlift reactors, where it is possible to remove the draft tube. The result of such measurements was that σ_0 was more than 50% larger in the bubble column (Larson and Lübbert 1990).

The exponent β in eq. 2, however, which characterizes the mixing mechanism, turnes out to stay below 1. This means (Lübbert and Larson 1990) that even at higher gas

loads within pilot-scale airlift tower reactors, local isotropic turbulence cannot be assumed to be dominating. The exceedingly dominating local mixing mechanism seems to be the transport of liquid in the bubbles' wakes, which theoretically should lead to an exponent $\beta = 1$.

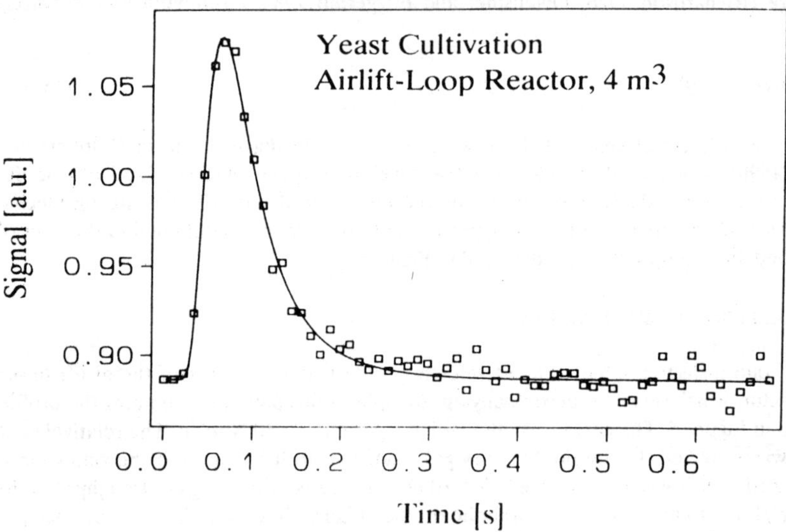

Fig.4: Typical time-of-flow curve as measured with the heat pulse technique in the riser of an airlift tower loop reactor during a yeast cultivation. The symbols are measured data, the full line a fit of equations 1 and 2 to the data.

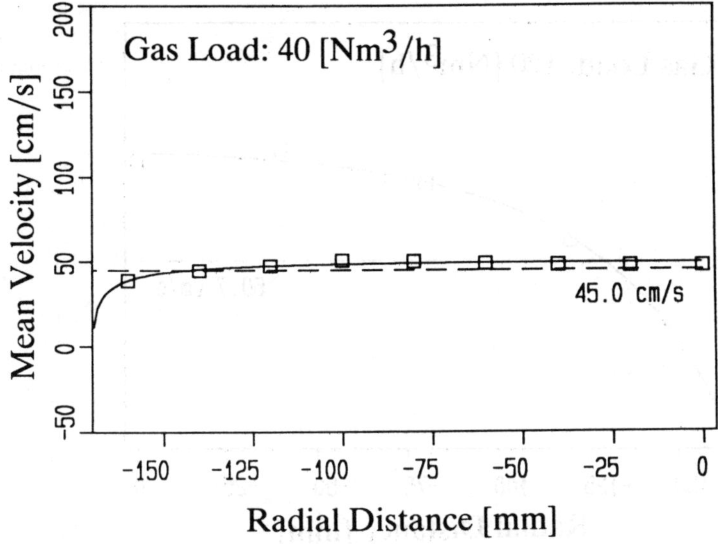

Fig.5: Liquid velocity profile, measured with the heat pulse technique in the riser of an airlift loop reactor at 40 Nm^3/h. The flat profile was expected from literature data.

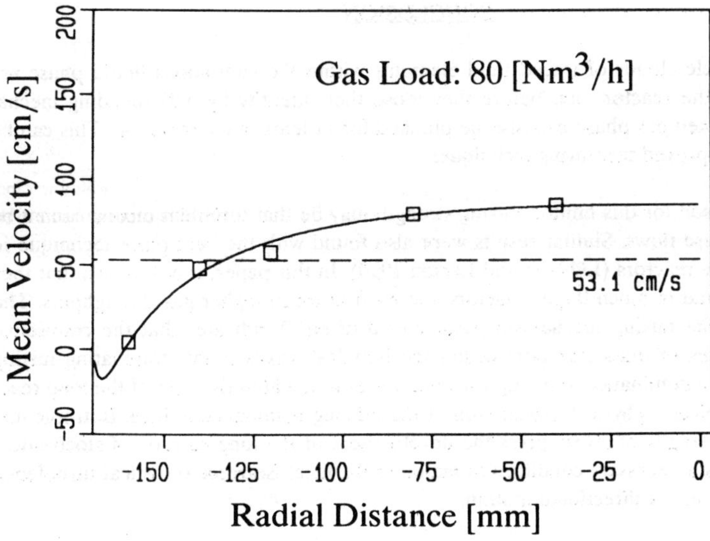

Fig.6: Liquid velocity profile, measured at the same place at a gas load of 80 Nm^3/h.

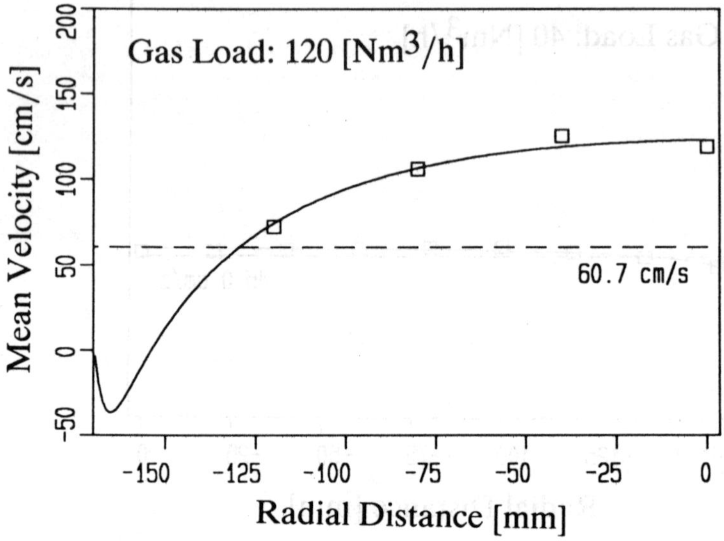

Fig.7: Liquid velocity profile, measured at the same place at a gas load of 120 Nm^3/h.

CONCLUSION

Not only particle clouds of traced fluid elements within the continuous liquid phase are swept around the reactor loop before they loose their identity by bulk-blending mechanisms, the marked gas phase can also be pursued for at least one circulation. This can be observed by improved measuring techniques.

The reason for this limited mixing strength may be that turbulent mixing cannot be observed in these flows. Similar results were also found with the heat pulse technique in laboratory-scale reactors (Lübbert and Larson 1990). In this paper, it was shown that they can also be found in much larger reactors and even at much higher gas throughputs. The estimation of the mixing mechanism parameter β of eq. 2 indicates that the convective mixing by means of mass transport within the bubbles' wakes is the dominating mixing mechanism. The dominance of the upward bubble motion within the riser of the loop reactors, likewise, gives a physical explanation of the missing dominance of local isotropic turbulence: The rising bubbles suppress the development of the long cascade of stochastical eddies, which is a necessary condition to maintain the local isotropic statistical turbulence theory, by their strong directional motion.

LITERATURE

1. Chisti, M.Y., 1989, "Airlift bioreactors", Elsevier Applied Science, London, England.

2. Lübbert, A., Fröhlich, S., Schügerl, K., 1987, "Characterization of bioreactors by mass spectrometry analysis", pp. 125-142 in: Mass Spectrometry in Biotechnological Process Analysis and Control, E. Heinzle, M. Reuss, eds., Plenum Publ. Corp., New York

3. Lübbert, A., Larson, B., 1987, Chem.Eng.Technol. 10, 27-32.

4. Lübbert, A., Larson, B., 1990, Chem.Eng.Sci., in press.

5. Verlaan, P. (1987), Modelling and characterization of an airlift-loop reactor, Doctoral Thesis, Waageningen.

6. Verlaan, P., Tramper, J., Riet, K. van't, Luyben, K.C.A.M. (1986), "Hydrodynamics and axial dispersion in an airlift loop bioreactor with two- and three-phase flow", pp. 15-17 in: Proc.Int.Conf. Bioreactor Fluid Dynamics, Cambridge, Elsevier, London.

7. Weiland, P., Onken, U., 1980, Chem.-Ing.-Techn. 52, 986-987, Synopse 858.

8. Wilhelmi, G., Gompf, F., 1970, Nucl.Instr.Meth. 81, 36-44.

MIXING PHENOMENA IN AIRLIFT LOOP REACTORS

J.A. Trilleros* and A. Recio*

Experimental research has been carried out in two reactors with an overall volume of one and four cubic metres, operating with the air-in water model system. The gas was introduced and dispersed by a gas jet and the water was lifted. One and more draft tubes have been used to guide the flow of the dispersion.

A qualitative analysis is set forth of the flow in the reactors and the mixing times for each reactor design. To such end, a heat zone has been used to mark the fluid elements instead of a material tracer, forming an upper area of hot liquid.

INTRODUCTION

Currently, within heterogeneous gas-liquid transfer phenomena, airlift-loop reactors are most frequently used in Biotechnology and in Chemical and Metallurgical processes; this is, among others, due to the fact that they have no mobile mechanical devices, they generate a high interfacial area and homogenization of the phases is favoured by both the turbulence and the liquid recycle phase.

The real flow problem in chemical reactors is related to the scaling up and frequently the factor that cannot be controlled is flow unsuitability. Ignorance and/or incorrect estimate of such factor may lead to significant errors when designing and/or operating the reactors.

Gas-liquid heterogeneous reactors have preferably been designed and used, by free gas dispersal with draught tubes or otherwise. This experimental work studies the flow behaviour in such reactors when a bundle of draught tubes is used working with flow in parallel, grounding our work on the experiments we made on their fluid dynamics made by ourselves at the pilot plant of the Department of Materials Science and Metallurgical Engineering of the Complutense University in Madrid, regarding reactors typical of processes beneficial for Metallurgy.

To obtain the flow type, the analogous air-water system was worked in, using the thermal tracer technique of hot zone

(*) Department of Materials Science and Metallurgical Engineering. University Complutense. Madrid. Spain.

formation, logging the system's responses, temperatures, at twenty-five different points of the reactor spaced out radially and axially. To such end a FLUKE data logging system was provided (HELIOS host computer interface data logger) connected to a personal computer, allowing the temperature fields to be obtained, both to make the flow analysis and to secure mixing times.

The experiments summed up hereinafter shall first of all deal with the problems and current state of airlift-loop reactors, followed by a description of the testing installations used, and the method used in tests. An analysis of the reactors' flow patterns and of the mixing times is finally set out.

AIRLIFT-LOOP REACTORS

All gas-liquid contact stages require a system for dispersing the more or less dense phase within the liquid phase, resulting in bubbles developing and moving up through the liquid. If this situation is analyzed heeding movement quantity transport criteria, a vertical field of estimates shall be established in a two-phase system, causing a liquid flow towards the free surface through gas entrainment and subsequent return of such liquid towards the starting zone.

The liquid's upward flow is due to the generation of a central biphasic column providing a significant turbulence, together with a downward liquid flow with no gas. Hydrostatic pressure difference is established between the system's central column and the outer zones causing liquid to be entrained by the gas.

The bibliography, (1) to (3), sets out the above situation together with the velocities profiles and the evolution of the gas old-ups in reactors.

If the gas phase distributor fails to do so uniformly throughout the column section, turbulent eddies are generated and grow when the column diameter does so, and sometimes such eddies change their position in the column, which results in chaotic movement which is significant in columns with a large diameter (4).

As opposed to conventional bubble columns, columns have been worked with draft tubes improving flow stabilities in columns with a large diameter reducing the turbulence provided by the eddies. Columns have also been designed with gas redistributors, as well as filling beds and static mixers, in order to reduce formation of large bubbles and improve the flow conditions (5).

The liquid flow established in bubble reactors brought about the need to include a central tube, to limit the space allocated to airlift, and where there is essentially a biphasic flow, together with another zone concentric with the former, through which the liquid returns. Bibliography (6) includes different designs of such reactors, with both internal and external loop.

The most significant differences between the bubble reactors and reactors with draught tube, lie in the different values of the liquid superficial velocities, reactors with draught tube being highest. Significant differences are also found in the gas old-ups, interfacial areas, bubbles distribution and axial and radial dispersal thereof, as well as the system's mixing times; (6) and (7).

The bibliography has excellent papers on fluid dynamics for this kind of reactors, in order to estimate the liquid lift flows, only the most significant in the last decade being included. (8) through (12).

Bubble reactors have been widely used in metallurgical processes. An important study was made throughout the nineteen seventies for Pyrometallurgy refining processes (13) through (16). Due to such processes' operating conditions, many of the research and development works were made using simulation techniques, both mathematical, and using models with analogous systems, and testing at a pilot plant. Proper combination of the information provided by such techniques has been useful to include different design and/or operation alternatives in the processes when the scaling up problem has been safely dealt with; (17) through (19).

Over the last years extensive information has been published on airlifts as bioreactors, with nuances that make them differ from airlifts as metallurgical reactors. In order not to render the bibliography too lengthy, two of these, (20) and (21), have been selected which, we believe, appropriately show aspects pertaining to bioreactors, moreover containing plentiful bibliography.

Airlift loop reactors work in a steady state, because of the load difference established in such system when inserting air through the far end, that is reversed in the liquid flow, in irreversible friction dissipation of the phases moving in the tube or otherwise, in directional changes and in the slipping phenomena arising in the two phase flow (22) through (24).

In these reactors what is most interesting is to know the turbulent flow in the gas-liquid system or otherwise, liquid flow in the stage, to know the real flow in the reactors and to determine mixing times. Experimental effort made in the last years has been significant, both working with real and analogous systems, (25) through (29).

Real flow deviations in the transfer operations are interesting, and this is mainly due to the formation of dead zones, by-pass and cross flow, which results in the transfer efficiency being smaller.

The real flow problem is closely linked to the scaling up and it is hence necessary to work at a pilot plant scale. The uncontrollable factor in the scaling up is frequently flow unsuitability, which results in significant errors regarding design and operation of the reactors.

Knowledge of the real flow is obtained from the residence

Time distribution (RTD), using to such end the tracer techniques. From the RTD, sufficient experimental information will be available to know the mixing times and on the overall homogenization attained.

The following tracer techniques, among others, have been used: calorimetric, following the evolution of the colourants concentration profiles in time; thermal, obtaining the medium's temperature fields in time; pH or electric conductivity measurements, logging the changes of either variable in time. There are obviously other techniques that are analyzed and summarized in the work of Ulbrect et al. (32).

EXPERIMENTAL APPARATA AND TECHNIQUES

The experimental study has been carried out on airlift loop reactor designs, taking bubble reactors as reference, to subsequently analyze the behaviour of reactors with simple draught tubes and other lift designs with several flow draught tubes, in parallel, forming a bundle.

Thermal tracer techniques have been used to analyze the flow in reactors and to determine the mixing times, using simulation techniques, the apparata being operated with the air-water system, in pilot plant experimental devices.

Apparata

Two apparata have been used, with different capacity reactors (1.6 and 4.1 m^3) and cylindrical geometry (1.30 m height by 1.26 m diameter), the former, and (1.60 m height by 2.00 m diameter), the latter.

Figure 1 sets out a flow diagram of the apparatus, showing the following parts:

a) Hot fluid (water) preparation circuit, from steam, to form the hot zone in the reactor, flowing through a distributor ring.

b) The air circuit, coming from the plant network, is verified and the flow values and air overpressures controlled, before sending to the reactor.

c) The reactor of cylindrical geometry and that is provided with a manifold position stainless steel plate, up to twelve jets, formed by capillary ducts with a diameter of 1.0 mm and a length of 100 mm.

The designs considered for the reactor were:

C-1) Airlift loop reactor with air bubbling through a single capillary and without draught tube.

C-2) Airlift loop reactor with air bubbling through a single capillary and with draught tube. Heights of 60 and 110 cm were tested for the draught tube.

C-3) Airlift loop reactor with air bubbling through 6 and 12 capillaries and with draught tubes. Tubular draught bundles with heights of 60 and 110 cm were tested.

In sections C-2 and C-3, testing used draught tubes with an inside diameter of 20 and 25 mm.

The air flow volumes used ranged between 10 and 400 Ncm^3/s.

Test method

The thermal tracer technique was used, forming an upper water hot zone in the reactor, temperature between 40 and 60 °C and height of 12 cm.

For temperature fields evolution, 25 K type (Cromel-Alumel) thermocouples were used, five in radial position, spaced out every 10 cm., and with the same number of longitudinal positions, at a height of 20 cm. each, yielding information on both the hot and the cold area, at four different height levels.

To log data, FLUKE brand equipment was used (HELIOS, Host Computer Interface Data-logger), figure 2, connected to an IBM P.C.

ANALYSIS OF THE FLOW PHENOMENA

The flow aspects studied are summarised in the following sections, with separate remarks for matters relating both to flow in the reactors and to mixing times, for each type of reactor design and for the scales used.

Thus, for apparatus no. 1, where less liquid is withheld, the behaviours of the different draught systems have been studied for internal ducting diameters of between 20 and 25 mm. In apparatus no. 2, the larger one, only the behaviour of the different draught systems for internal diameters of 25 mm, after verifying the small apparatus data.

Mixing times

When using tracer techniques for flow analysis, mixing time is defined as the time in which the variation in the tracer concentration fluctuates by a very small percentage around the average value.

Okamoto et al. (33) set out a theoretic analysis, pointing out that geometric variables, such as momentum transport phenomena, and the energy dispersed in the mixing process, vary potentially and affect the length of the mixing time, depending on the kind of motion in the system.

Dissipation energy in mixing mainly refers to the using up of energy in gas expansion in order to entrain the liquid and make the mixture both due to loop of the latter and to turbulent dissipation as the gas is checked by the liquid.

Frequently, the specific energy - - is taken as

reference, taking as reference the system's volume unit (34) and (35). Murthy et al. (36), moreover introduced the kinetic energy interchange, which term must be considered in systems with small liquid old-up.

In airlift loop reactors, the mixing time is preferably correlated with the superficial gas velocity or with the specific energy density. Reference is also made to geometric design variables such as the diameter-height ratio, the elevation between the discharge point and the liquid level and the ratio of the reactor's and the draught tube's diameters.

The simplest form of relating both variables is:

$$\tau \propto \dot{E}_b^{\,n} \quad (1)$$

Test values of the exponent -n-, that have been furnished in experimental works, are set out hereunder:

-n- value	Reference
0.23	(37) Lehrer
0.25	(27) Szekely
0.31	(38) Haida
0.33	(39) (40) (41) Brodkey, Kipke, Mori
0.40	(31) Nakanishi
0.45	(42) and (43) Nakanishi, Kato

The mixing times have been obtained from the time temperature data, logged in each test, with the twenty-five thermocouples. Figure 3 shows that the lift profiles for the cold temperatures are more reliable than those for the heat area when determining the mixing times values.

Tables I and II group the mixing times values for each type of reactor design and for different values of the air flows used.

It can be observed that the mixing times decrease inversely with the air flows in each type of reactor. Furthermore, it is verified that when the reactor works with free gas dispersal, least overall homogenization times are required within the tested flow intervals.

The mixing times are very significantly lowered in reactors with tubular draught bundles, this trend being more noticeable the larger the number of tubes in the bundle. It was also experimentally verified that decreasing the diameter of the draught tube does not imply that the mixing times decrease for the same air flows despite the fact that kinetic energy in the biphasic flow discharged increases, as does the turbulence in such area.

It has also been verified that the reactor design with the half draughts, always require less homogenization time as opposed to full length draughts, the number of ducts in the bundle being very significant.

The effects of the scaling up in the mixing times is summarized hereinafter for the interval used.

For the smaller apparatus, only in the reactors with tubular bundle, having twelve half draft tubes, do the homogenization times grow close to those attained when the reactor works with free dispersal. This trend is more noticeable for the larger reactor, where the tank with a twelve tubes tubular bundle needs mixing times of the same size as when free dispersal is worked with, homogenization times being smaller when the reactor is provided with twelve half draft tubes.

The mixing times have been correlated with specific gas expansion energy - -, using the expression given by Themelis (33). The result of linear regression is given in table III. It can be observed that the energy slope ranges between 0.24 and 0.65, being at a maximum for the reactor with free expansion and close to the tanks with twelve half draft tubes, decreasing in the other type of airlift loop reactors.

Flow in the reactors

The flow analysis in the reactors shall be made from the temperature time profiles in the cold area. In order to set out the said profiles in the simplest possible form, figures 4 through 24 only have the evolution of the longitudinal temperatures of the thermocouples closest to the reactor's external wall, i.e. those with a greater delay in the evolution of the flow conditions.

It can be observed that the tanks operating with free gas dispersal have a significant contribution of mixing flow, mixing flow contribution being far less the smaller the height of the liquid on the reactor bottom. In fact, it can be verified that there is a delay in homogenization in the last area of thermocouples, that corresponds to 20% of the reactor volume, for there are situations where up to one half of the mixing time is required to attain total homogenization of the liquid; figures 4 through 7.

In the tanks with only one draught tube, it may be observed that there is a significant plug flow contribution, homogenization being mainly attained thanks to the loop. When a half length draught tube is worked with, it may be verified that the mixing contribution in the surface area of the reactor increases significantly. All of this can be summed up indicating that in the first situation, no delays take place in the homogenization of the far ends of the reactor, while in the second there are delays, since such area is far from the upper reactor turbulence; figures 8 through 12.

When the reactors operate with tubular bundles, it may be verified that the mixing flow gradually increases, as does the number of bundle tubes and the gas flow, figures 13 through 18. The temperature profiles in the tanks with half length bundles are similar to when the reactor works with free gas dispersal. The appearance of dead areas or reduction thereof due to the greater loop contribution, follows the lines already set out in the above two paragraphs. Figures 19 through 24.

FINAL CONSIDERATIONS

From the experimental studies made and summed up in the previous paragraphs, the following considerations can be inferred regarding both the thermal tracer technique used and the different reactors tested and the scaling up.

Thermal tracer technique

The thermal tracer technique used to form a hot zone, is precise and sensitive, producing no turbulence whatsoever in the system's momentum transport. The hot zone's temperature evolution profiles are essentially useful to study the mixing flow in the upper reactor zone. However, the information supplied by the cold zone temperature profiles are useful to inform of the contributions of the mixing and plug flows at different levels and of the tank, furnishing data on the existence of dead zones.

Design of loop reactors

If the values obtained in the mixing times and the specific energy consumptions in the reactors are considered, such tanks as are provided with a twelve-tubes bundle are competitive as opposed to reactors working with free dispersal.

If the specific flow type in reactors is also considered, the tubular bundle tank for the lifting has more favourable flow conditions for when mass transfer phenomena take place with chemical reaction or otherwise, for it has a surface area with a strong turbulence, following by significant liquid loop.

Scaling up

It has been verified that the flow type may have small variations within the scale interval that has been tested; nevertheless in the mixing time adjusted functions, specific expansion energy, no different behaviours are noticed, the values of the exponents of such energy are close to those furnished by other investigators, for different systems and scales.

NOMENCLATURE

Q = Normal flow rate of gas, (m^3/s)
T = Temperature, (°C)
n = Parameter of equation no. 1
t = Time, (s)
$\dot{\varepsilon}_b$ Mixing power input per unit bath volume ($Kgm/m^3.s$)
τ Mixing time, (s)

AIRLIFT LOOP REACTOR

Gas dispersion	Key
Jet	DL
Jet and draft tube	ET
Jet and half draft tube (O/ = 20 mm)	MT
Jet and six draft tubes (O/ = 25 mm)	E6m
Jet and six draft tubes (O/ = 25 mm)	E6r
Jet and six half draft tubes (O/ = 20 mm)	ME6m
Jet and six half draft tubes (O/ = 25 mm)	ME6r
Jet and twelve draft tubes (O/ = 20 mm)	E12m
Jet and twelve draft tubes (O/ = 25 mm)	E12r
Jet and twelve half draft tubes (O/ = 20 mm)	ME12m
Jet and twelve half draft tubes (O/ = 25 mm)	ME12r

REFERENCES

1.- Joshi J.B., Shaima M.M., 1979, Transt. Inst. Chem. Eng., 57, 244.

2.- Hills J.H., 1974, Transt. Inst. Chem. Eng., 52, 1.

3.- Ueyama K., Mianchy T., 1979, AIChE Journal, 14, 258.

4.- De Nevers N., 1968, AIChE Journal, 14, 122.

5.- Schügerl K., 1980, Chem. Ing. Tech., 52, 951.

6.- Weiland P., Oukeng U., 1981, German Chem. Eng., 4, 174.

7.- Weiland P., 1984, German Chem. Eng., 7, 373.

8.- Hsu U.C., Dudukovix M.P., 1980, Chem. Eng. Sci., 35, 135.

9.- Merchuk J.C., Stein Y., 1981, AIChE Journal, 27, 377.

10.- Bello R.A., Robinson C.W., Moo-Toung M., 1984, Can., J. Chem. Eng., 62, 573.

11.- Jones A.G., 1985, Chem. Eng. Sci., 40, 449.

12.- Clark N.N., 1984, Mineral and Metallurgical Processing, Nov., 226.

13.- Scaninject I and II, 1977 and 1980, "Proceeding of the International Conference on Injection Metallurgy", MEFOS, Lulea, Sweden.

14.- Wilson W.G., McLean A., 1980, "Desulfuration of Iron and Steel and Sulfite Sherpe Control" Iron and Steel Society. AIME, Pittsburgh.

15.- Balkar R., Normaton A.S., Pemally G.D., Atteinson R.A., 1980, Ironmaking and Steelmaking, 7, 227.

16.- Saigusa M., Nagai J., Suro F., Bara H., Tamara S., 1980, Ironmaking and Steelmaking, 7, 242.

17.- Robertson A.D., Sheridan A.T., 1970, J. of the Iron and Steel Inst., July, 625.

18.- McNallam M.J., King T.B., 1982, Metallurgical Transaction, 13B, 165.

19.- Johansen S.T., Engh T.A., 1985, Scand. J, of Metallurgy, 14, 214.

20.- Blenbe H., 1979, Adv. Biochemical Eng., 13, 121.

21.- Chisti M.Y., 1989, "Airlift Bioreactors", Elserier Applied Science, London, England.

22.- Ramírez F. Trilleros J.A., Otero J.L., 1987, Anales RSEQ, 83A, 420.

23.- Ramírez F. Trilleros J.A., Otero J.L., 1988, Anales RSEQ, 84A, 221.

24.- Lombardero L., López F., Trilleros J.A., Otero J.L., 1989, Extraction Metallurgy Symposium, London, 361.

25.- Szekely J., Wang J.Y., Kiser K.M., 1976, Met. Trans., 7B, 287.

26.- Deb-Roy T. Majundar A.K., Spalding D.B., 1978, Appl. Math. Mod., 2, 146.

27.- Szekely J., Schener T., Chag C.W., 1979, Ironmaking and Steelmaking, 6, 285.

28.- Tse-Chang H., Lehner T., Kjellberg V., 1980, Scand. J. Met., 9, 105.

29.- Grevert J.H., Szekely J., El-Kaddah N.H., 1982, Int. J. Heat Mass Transfer, 25, 185; 1984, 27, 116.

30.- Deb-Roy T., Majundar A.K., 1981, J. Met., Nov., 421.

31.- Nahanisshi K., Fujii T. Szekely J., 1975, Ironmaking and Steelmaking, 3, 193.

32.- Ford D.E., Mashelkar R.A., Ulbrect J., 1982, Proc. Tech. Inst., 17, 803.

33.- Asai S., Okamoto T., He J., Muchi I., 1985, Trans. ISIJ, 23, 43.

34.- Bhabaraju S.M. Russele T.W.F., Blanch H.W., 1978, AIChE Journal, 24, 454.

35.- Themelis N.J., Goyal P., 1983, Can. Met. Quaterly, 22, 320.

36.- Murthy G.G., Mehrotra S.P., Gosh A., 1988, Metallurgical Transaction, 19B, 839.

37.- Leher L.H., 1988, IEC. Proc. Des. Dev., 7, 226.

38.- Haida O., Emi T., Yamada S., Sudo F., 1980, Scaninject II, 2nd Int. Conf. on Injection Metallurgy, Mefos, June.

39.- Brodkey R., 1979, Mefos Scan. Lancers Club. 1st Meeting. Stockholm.

40.- Kipke K., 1970, Verfahrenstechenische Fortschritte beina Mischem, VDJ., 21.

41.- Mori K., Sano M., 1980, The 19th Committee (Steelmaking) Japan So. for the Promotion of Science, May.

42.- Nakanishi K., Fujii T., 1973, Testsu-to-Hagane, 59, 11.

43.- Kato T. Okamoto T., 1979, Denkseiko (Electric Furnace Steel), 50, 128.

TABLE nº 1

Liquid old up= 1.6 m^3

KEY	$Q \times 10^{-6}$ (m^3/s)	τ (s)
DL	9.570	1940
↓	40.26	575
	102.3	320
	345.1	260
ET	9.763	4730
↓	42.47	2140
	103.5	1700
	352.6	1160
MT	127.1	860
↓	355.7	675
E6m	186.7	700
E6r	49.79	900
↓	96.7	665
	193.8	540
E12m	183.3	475
E12r	97.81	575
↓	189.3	350
ME6m	96.26	590
↓	186.7	355
ME6r	95.67	475
↓	183.8	300
ME12m	94.72	425
↓	179.6	290
ME12r	94.96	370
↓	179.1	235

TABLE nº 2

Liquid old up =4.1 m^3

KEY	$Q \times 10^{-6}$ (m^3/s)	τ (s)
DL	24.77	2400
↓	41.03	1600
	59.20	1300
	81.42	955
	129.5	590
	243.1	500
	393.8	340
MT	102.6	700
↓	156.6	590
	220.6	445
E6r	98.76	1485
↓	141.1	1030
	197.7	1000
E12r	96.86	835
↓	137.5	630
	181.6	625
ME6r	97.69	935
↓	139.1	655
	189.3	600
ME12r	95.91	720
↓	137.0	530
	181.1	385

TABLE nº 3

KEY	n
DL	- 0.646
ET	- 0.387
MT (liq. old up 1.6 m^3)	- 0.235
MT (liq. old up 4.1 m^3)	- 0.585
E6m + E6r	- 0.470
ME6m + ME6m	- 0.583
E12m + E12r	- 0.425
ME12m + ME12r	- 0.614

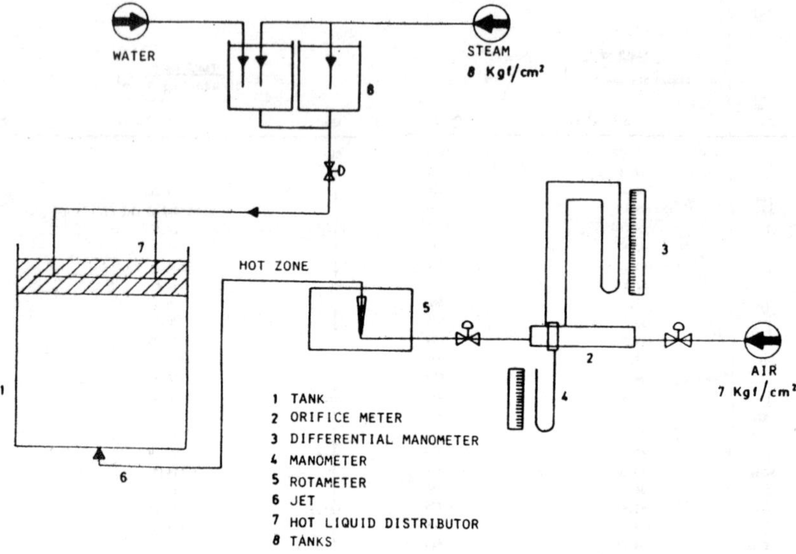

Fig. 1.- Flow diagram of installation.

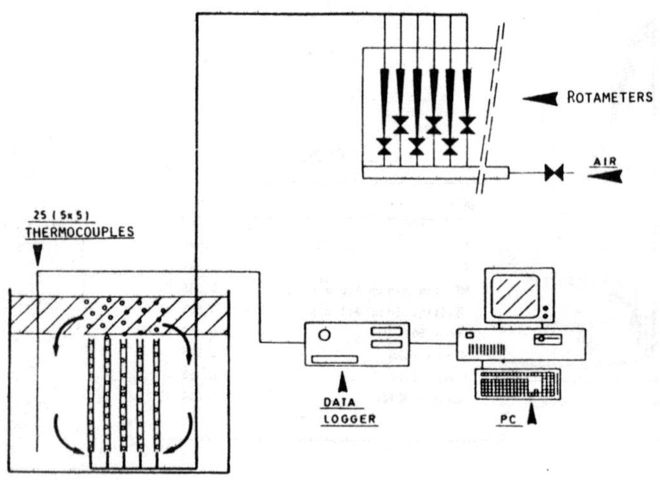

Fig. 2.- Thermal tracer technique.

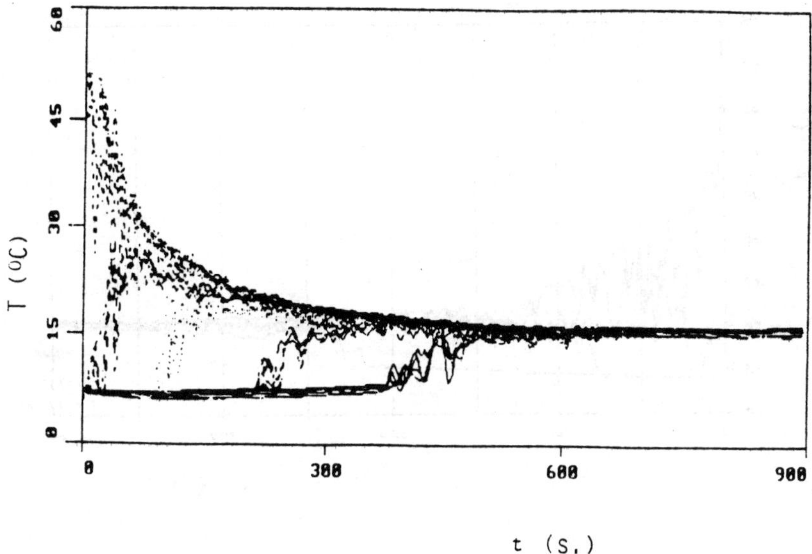

Fig. 3.- Temperatura-time data for 25 thermocouples.

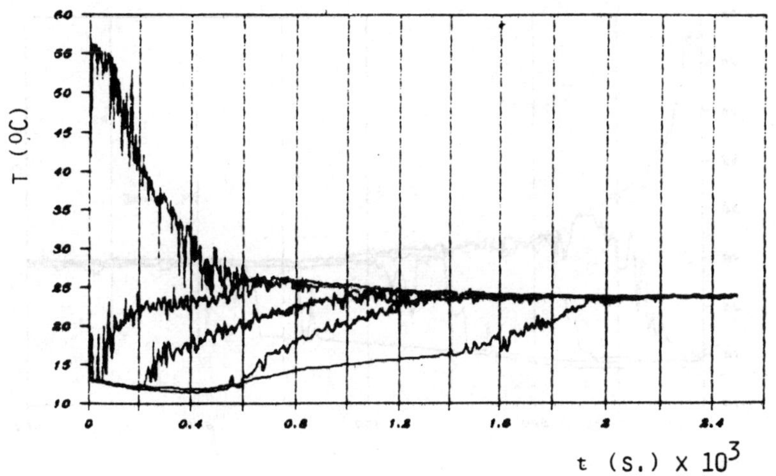

Fig. 4.- Key:DL, Tank = 1.6 m^3, Q = $9.57.10^{-6}$ m^3/s.

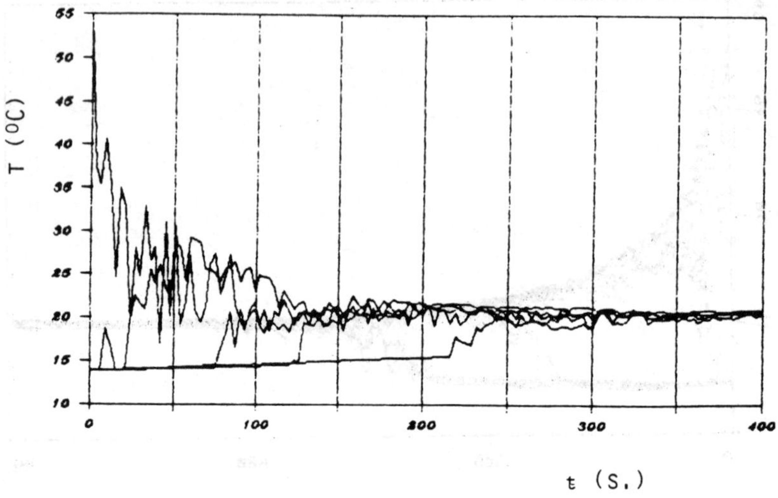

Fig. 5.- Key:DL, Tank = 1.6 m^3, Q = $102.3.10^{-6}$ m^3/s.

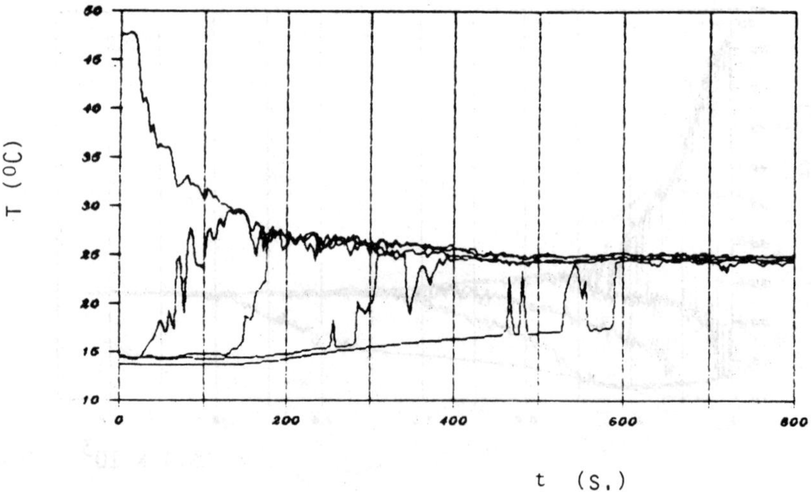

Fig. 6.- Key: DL, Tank = 4.1 m^3, Q = $129.5.10^{-6}$ m^3/s.

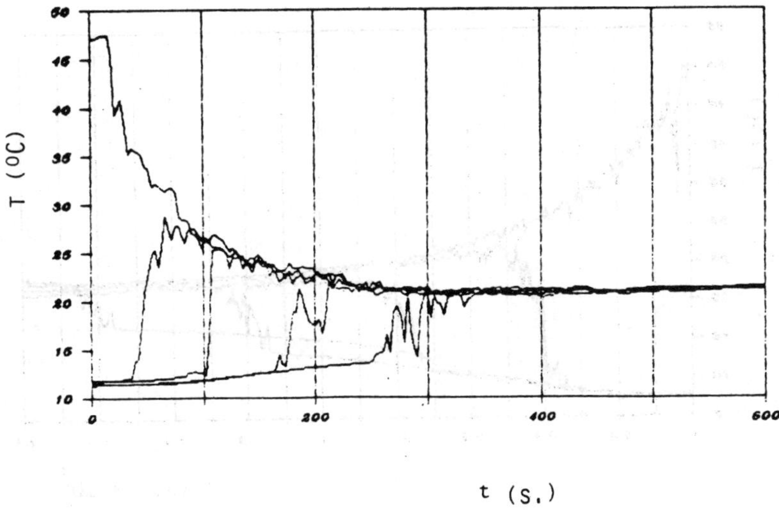

Fig. 7.- Key: DL, Tank = 4.1. m^3, Q = $393.8.10^{-6}$ m^3/s.

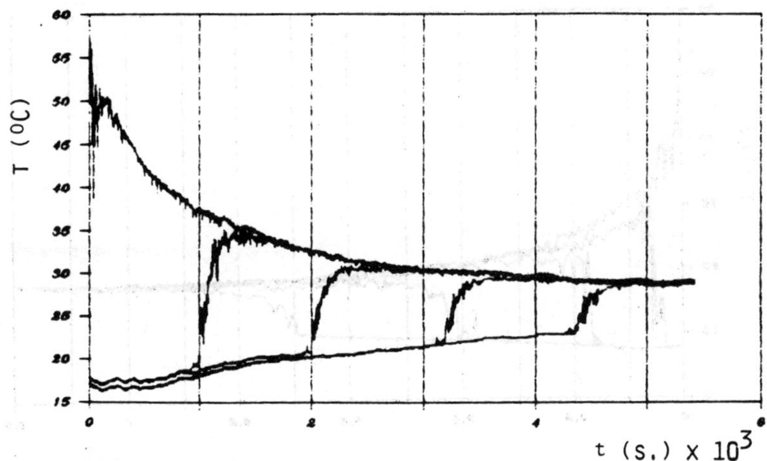

Fig. 8.- Key: ET, Tank = 1.6 m^3, Q = 976.10^{-6} m^3/s.

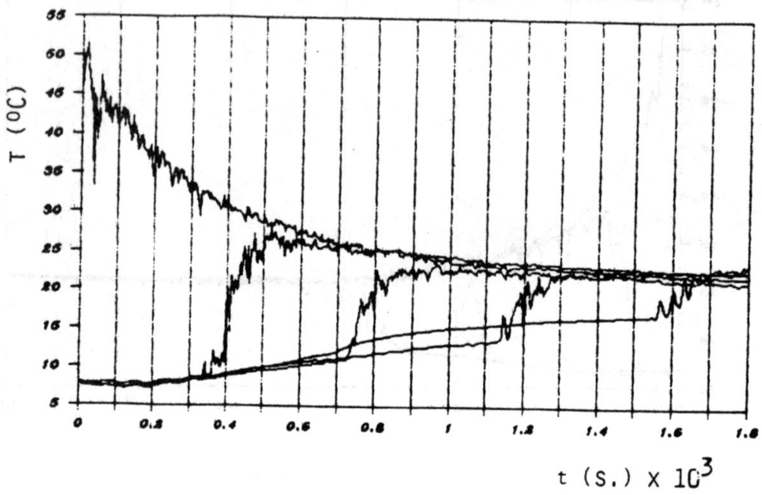

Fig. 9.- Key: ET, Tank = 1.6 m^3, Q = $103.5.10^{-6}$ m^3/s.

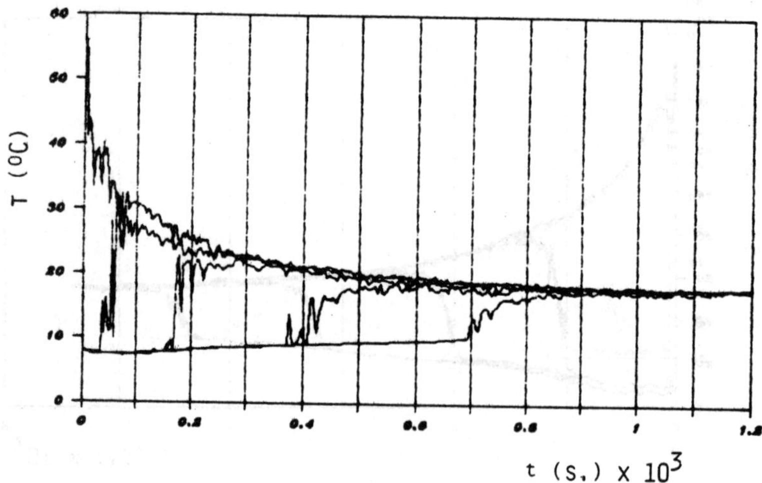

Fig. 10.- Key: MT, Tank = 1.6 m^3, Q = $127.1.10^{-6}$ m^3/s.

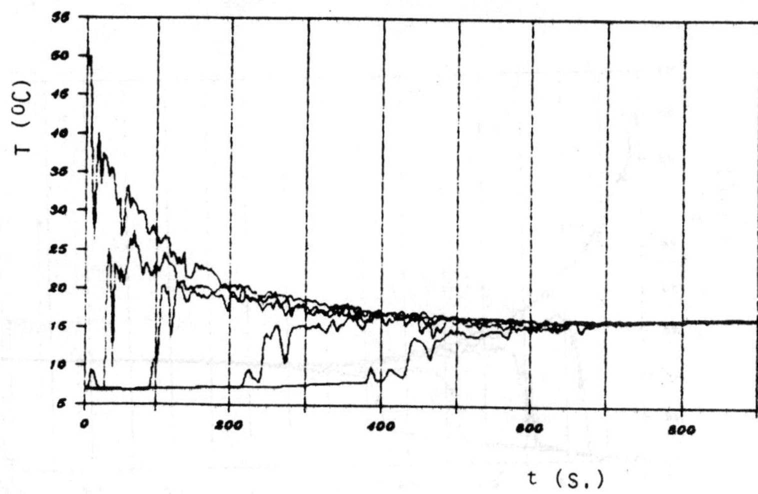

Fig. 11.- Key:MT, Tank = 1.6 m^3, Q = $355.7.10^{-6}$ m^3/s.

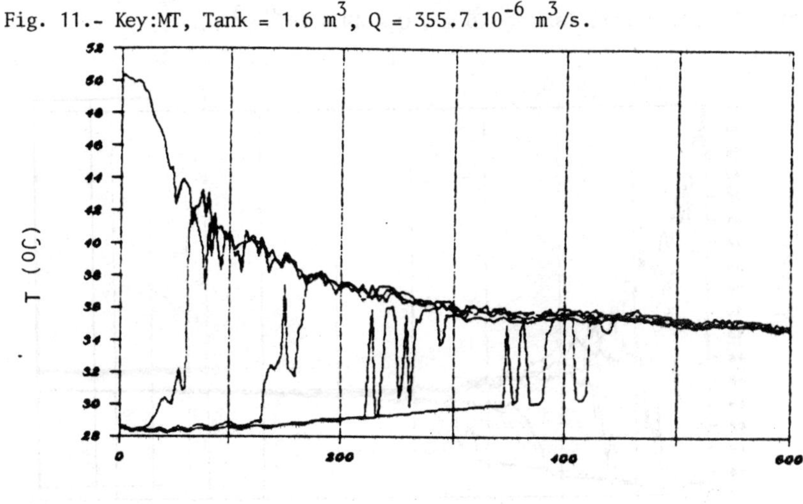

Fig. 12.- Key: MT, Tank = 4.1 m^3, Q = $220.6.10^{-6}$ m^3/s.

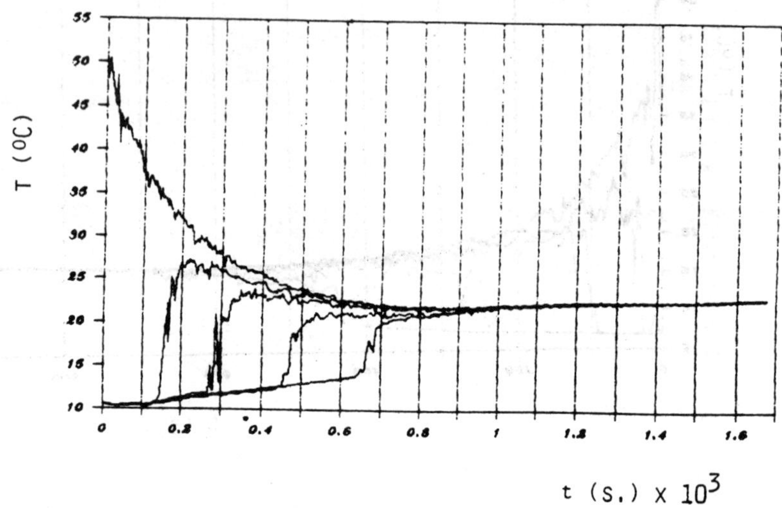

Fig. 13.- Key: E6r, Tank = 1.6 m^3, Q = $49.8.10^{-6}$ m^3/s.

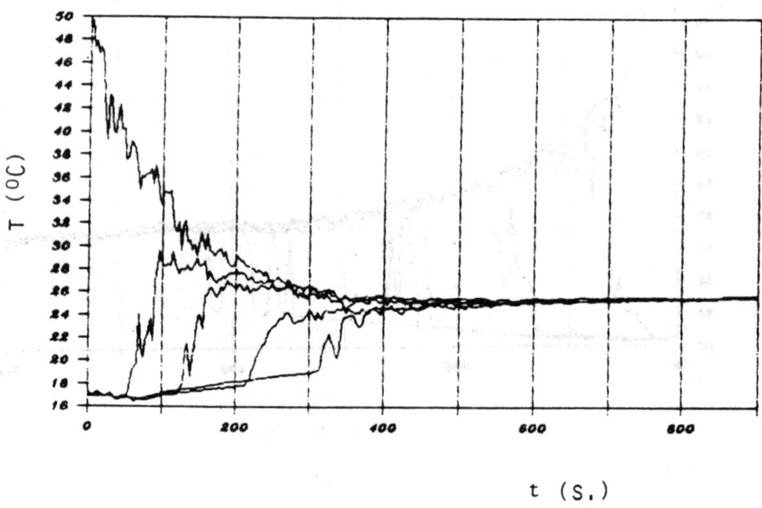

Fig. 14.- Key: E6r, Tank = 1.6 m^3, Q = $193.8.10^{-6}$ m^3/s.

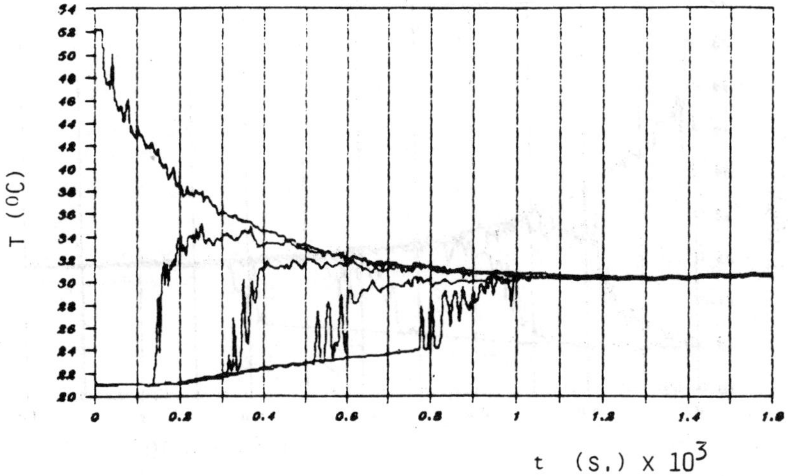

Fig. 15.- Key: E6r, Tank = 4.1 m^3, Q = $98.6.10^{-6}$ m^3/s.

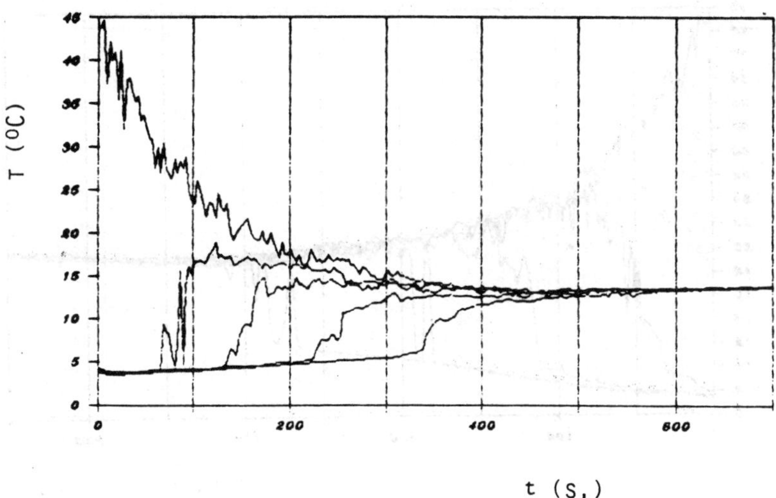

Fig. 16.- Key: E12r, Tank = 1.6 m^3, Q = $97.8.10^{-6}$ m^3/s.

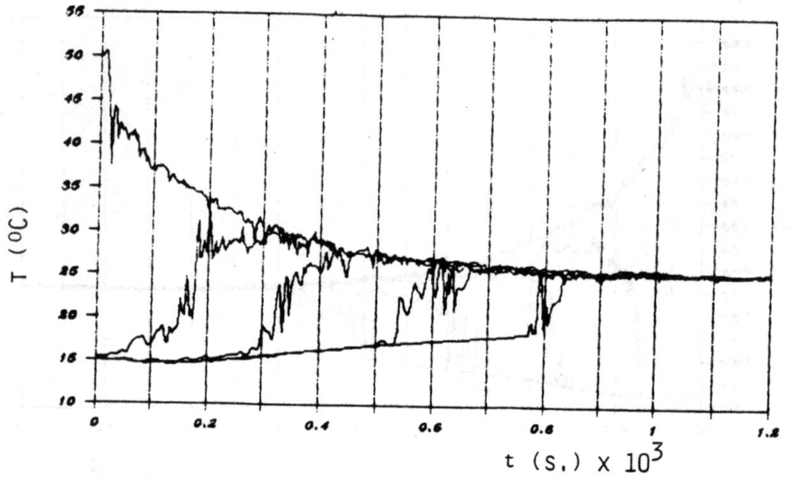

Fig. 17.- Key: E12r, Tank = 4.1 m^3, Q = $96.9.10^{-6}$ m^3/s.

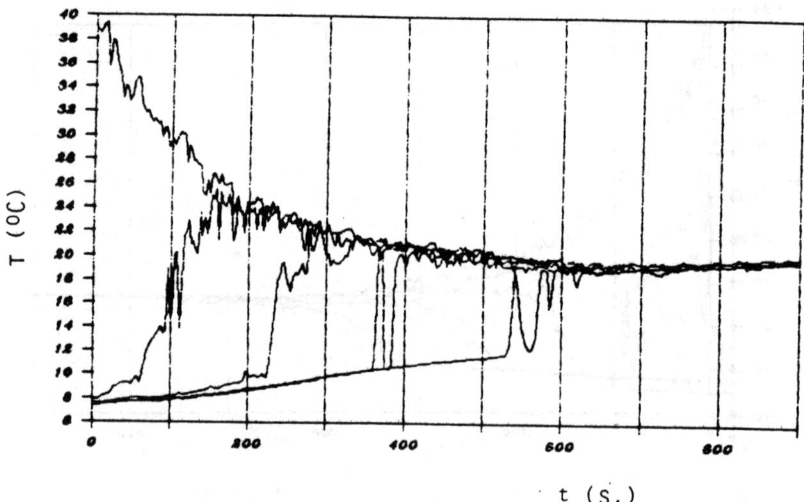

Fig. 18.- Key: E12r, Tank = 4.1 m^3, Q = $181.6.10^{-6}$ m^3/s.

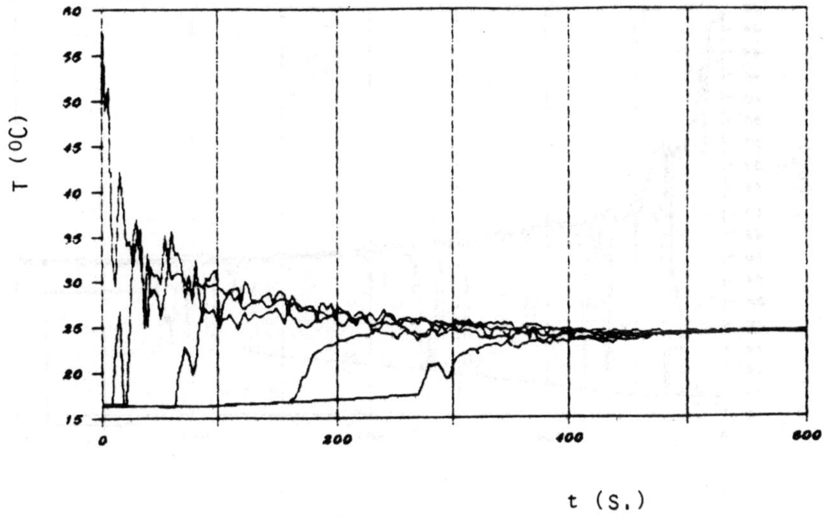

Fig. 19.- Key: ME6r, Tank = 1.6 m^3, Q = $95.7.10^{-6}$ m^3/s.

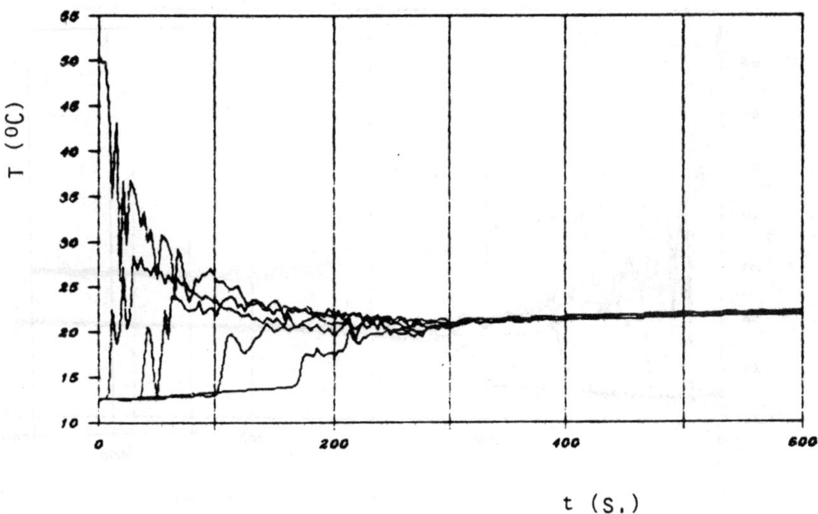

Fig. 20.- Key: ME6r, Tank = 1.6 m^3, Q = $183.8.10^{-6}$ m^3/s.

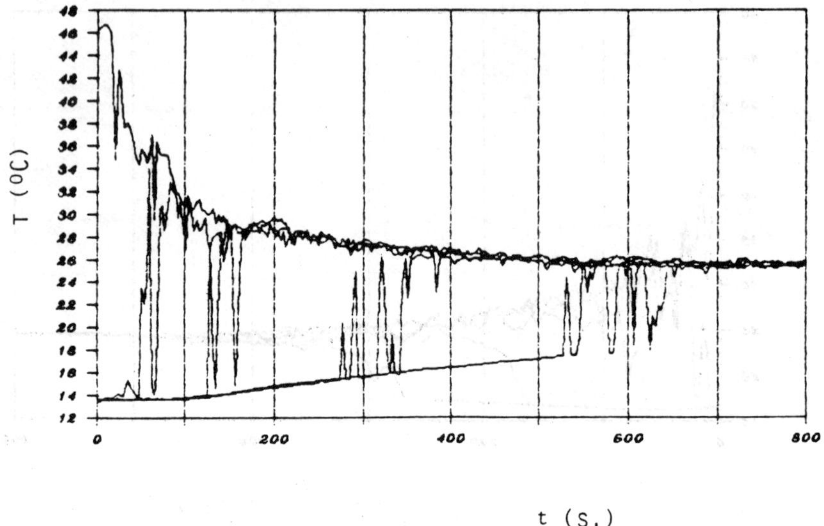

Fig. 21.- Key: ME6r, Tank = 4.1 m^3, Q = $139.1.10^{-6}$ m^3/s.

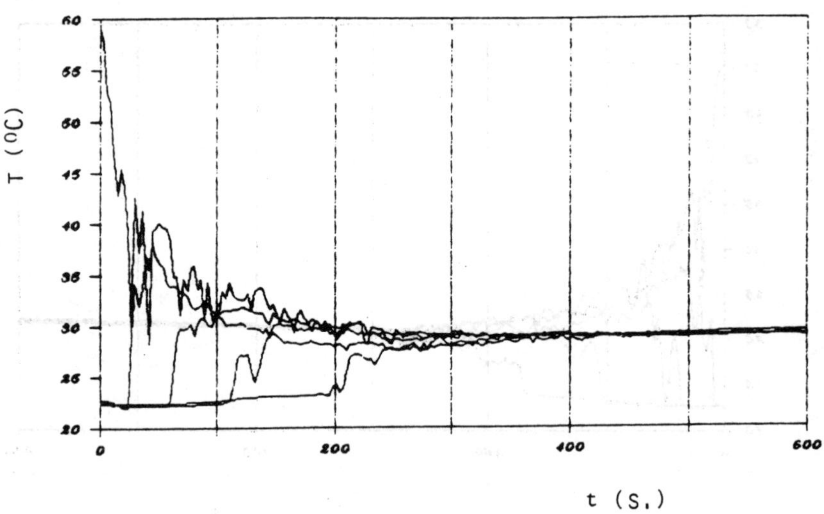

Fig. 22.- Key: ME12r, Tank = 1.6 m^3, Q = $94.9.10^{-6}$ m^3/s.

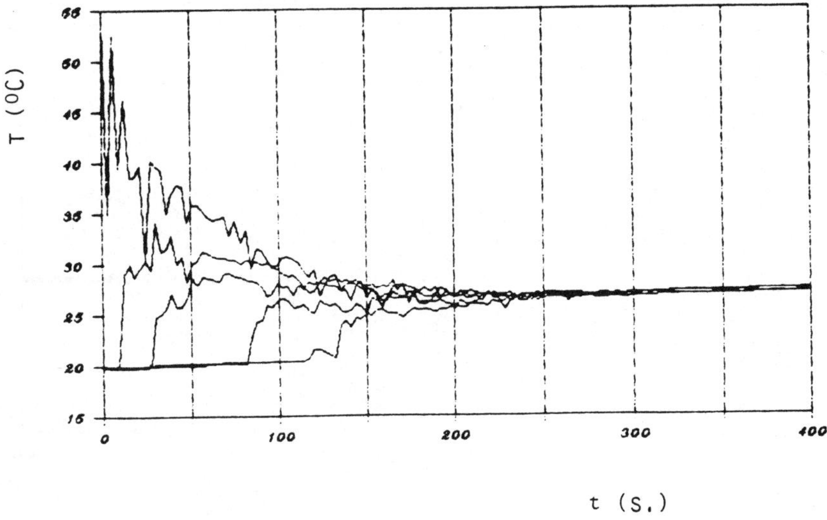

Fig. 23.- Key: ME12r, Tank = 1.6 m^3, Q = 179.1.10^{-6} m^3/s.

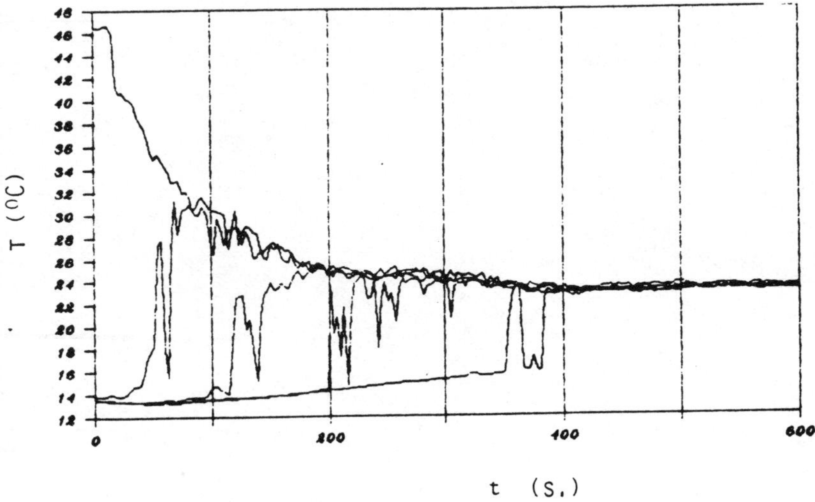

Fig. 24.- Key: ME12r, Tank = 4.1 m^3, Q = 181.1.10^{-6} m^3/s.

"MIXING IN A STIRRED SEMI–BATCH REACTOR: PARTIAL SEGREGATION FOR A PAIR OF COMPETING REACTIONS ANALYSED VIA NETWORKS–OF–ZONES"

Y.D.Wang* and R.Mann*

Partial segregation inside a stirred vessel gives rise to spatially non–uniform concentration fields which evolve in time under semi–batch operation. The extent and character of this non–uniformity may have profound effects on the reaction pathways and selectivity to desired products. An application of the network–of–zones model to the following reaction scheme

$$A \xrightarrow{k1} S$$

$$A+B \xrightarrow{k2} R$$

is described. The predicted internal concentration fields under axi–symmetry are examined for a 20 litre vessel stirred at speeds ranging from 12 to 302 rpm. Results are presented as coloured sectional image reconstructions. The predicted yield behaviour closely matches the experimental observations.

INTRODUCTION

Mixing inside batch and semi–batch stirred reactors continues to be poorly understood. Most efforts seem to have concentrated on mixing in continuous flow reactors, despite the fact that stirred batch reactors are clearly more prevalent than their continuous flow counter parts. Part of the reason for this state of affairs is that the internal mixing processes in a batch vessel are intrinsically unsteady as fed reagents tend to accummulate and charged reactants disappear by being consumed in the reactions. In contrast, for a flow reactor modelled deterministically, the internal processes are steady in time and as a consequence concentrations are steady at any internal position.

The experimenal equivalance to the well developed stimulus response technique for a flow reactor is to inject a pulse of an inert tracer into a batch stirred vessel as indicated in Fig. 1. Typically such an experiment gives the timewise concentration response at some fixed position, usually the impeller. The concentration response curve contains almost all the information necessary to quantify the internal mixing, but there are so far no generally satisfactory methods for quantitative interpretation.

SIMPLE ZONES–IN–LOOPS MODELS

Until now most interpretations for batch vessels have used very simple mixing models based on single loops of backmixed zones as shown in Fig. 1. Khang and Levenspiel [1] were the first to relate the peaks and troughs of the mixing curve at the impeller to the oscillation responses predicted by theory. Subsequently, Sasakura et al [2] applied a similar model and were the first to recognize the significance of measuring the mixing curve at different internal positions. However it is clear that very simple loops of zones are unable

* Dept. Chemical Engineering, UMIST, Manchester, M60 1QD.

to represent the internal fluid mechanics, except in a most superficial way. The total internal flow generated by an impeller can be accommodated, but it is impossible to relate the small number of mixing zones to internal position, local velocities, role of baffles etc.

THE PROBLEM OF PARTIAL SEGREGATION

Improved models are especially important for characterising partially segregated stirred reactors. Whenever reactions are reasonably fast (which is desirable for productivity reasons) there will tend to be at any instant of time internal concentration fields giving rise to areas which are rich in various reagents. This is depicted schematically in Fig 2 for a reactor being fed with reagents A and B. If complex reactions are taking place, in which intermediate products compete to react with A and B, product selectivity is likely to be sensitive to the quality of mixing produced by the stirrer. The mixing process itself is typically a complex blend of fluid mechanical convection and turbulence which at first sight seems hopelessly formidable.

When reactions accompany mixing, it is always necessary to allow for the impact of micro–mixing. This process governs the rates of reaction at a molecular level. In the past, the combination of the partial differential eqns for local rates of micro–mixing with the macroscopic flow patterns has proved extremely difficult.

Quite recently Baldyga and Bourne [3] have shown how the partial differential eqns of micromixing can be converted to ordinary differential equations. This involves a simplification of the engulfment–deformation–diffusion model under conditions where the deformation–diffusion can be neglected. The simplification amounts to the fluid engulfement being the rate determining step. Baldyga and Bourne [4] then combine the simplified micro–mixing description with a representation for the stirred reactor based upon a plug flow element and a perfectly backmixed element. This configuration can be used to model semi–batch operation as shown in Fig 3.

A related approach based upon the so called "interact and exchange with the mean" model (acronymically IEM) has been thoroughly developed by Villermaux and co–workers [5,6]. The IEM model also avoids the complexity of partial differential equations. As shown in Fig 4 it can provide a reasonably thorough representation for the complex internal mixing and reaction processes in semi–batch operation.

However, both these approaches by Bourne and Villermaux's groups simplify the circulatory fluid mechanics into a small number of zones, within which it is difficult to encompass the processes of flow and turbulence as they are distributed throughout the whole fluid inside a stirred vessel.

THE NETWORK OF ZONES MODEL

In an attempt to evaluate the interaction of flow, turbulence and chemical reactions within a recognisable fluid mechanical description for a stirred vessel, a model based upon networks of backmixed zones has been developed by the present authors and previous co–workers [7,8,9,10,11].

Because the model is based upon an assembly of backmixed zones, the unsteady timewise behaviour in a batch vessel is governed by odes for which convenient library integration software is available. As depicted in Fig 5 and previously described in detail [7,8,9], the network is constructed of nested flow loops around focii of circulation above and below the impeller. Each flow loop introduces an element of circulatory mixing. In addition, the model incorporates lateral equal and opposite exchange flows between zones adjacent to one another. This mechanism mimics tubulent exchange and the model thereby provides a simultaneous dispersive radial mixing.

The computational algorithms for a general network of size 2 x (NxN) are particularly simple if the flow q is equal in each loop. Any velocity profile may then be created by adjustment of the zone volumes $V_{i,j}$ as previously described [8], although it should be noted that the lateral dispersive mixing attenuates the effect of differing local velocities during circulatory flow. Thus equal volume zones throughout the vessel gave a close approximation to the mixing with typical velocity profiles [8].

The specific configuration of the zone 4N shown circled in Fig 5 which has i=4 and j=N is shown in Fig 6. The magnitude of the lateral exchange flows (which are equal and opposite) are ratiod to the main flow through the backmixed zone using the exchange parameter β given by

$$\beta = \frac{\text{exchange flow with adjacent zones}}{\text{main flow through zone}} \qquad \text{....(1)}$$

For a 2x(20x20) network with equal sized zones, the set of 800 zones appears as shown in Fig 7. The foci of circulation appear equidistant between the centre line of the impeller and the vessel base and fluid surface for a centrally located impeller. These foci positions are almost exactly coincident with the recent laser measurement of velocity profiles reported by Costes and Couderc [12].

The total internal circulatory flow induced by the rotating impeller is usually expressed as

$$Q = KN^1D^3 \qquad \text{....(2)}$$

This total flow should not be confused with the pumping flow within the confines of the impeller periphery. For a Rushton turbine, Revill [13] gave a K for impeller flow of 0.75, whereas for the total internal circulation (or turnover) a value of 2.75 has been reported [14]

TRACER TEST INTERPRETATIONS

The validity of the networks–of–zones approach has been tested against experimental results using pulse tracer tests with and without reaction.

For the case of an inert tracer, for a perfectly mixed i,j zone, the unsteady material balance for the tracer is given by

$$V_{i,j} \frac{dC_{i,j}}{dt} = q\,[C_{i-1,j} - (1+2\beta)\,C_{i,j} + \beta(C_{i,j-1} + C_{i,j+1})] \qquad \text{....(3)}$$

when the flow directions are as shown in Fig 5. The initial conditions for the integration of the set of odes are then

$$C_{i^*,j^*} = C^\circ \quad \text{for } i=i^* \text{ and } j=j^*$$

$$C_{i,j} = 0 \quad \text{for } 1<j<N \text{ and } 1<j<2N \qquad \text{....(4)}$$

where i*, j* refer to the zone into which the initial injection pulse is made.

If C° is chosen such that

$$C^\circ = 2N^2$$

then the calculated values of $C_{i,j}$ are normalised with respect to a final dimensionless concentration of unity.

Fig 8 then shows a simulation of the dispersal of a nigrosine dye tracer pulse for a 2x(20x20) configuration of zones. The injection has been made into zone (1,1). The shading of concentration uses 1 line for $C_{i,j}=1$ (the final concentration) with one line of shading for each increment of concentration up to $C_{i,j}=10$. Dimensionless concentrations above 10 use the maximum of 10 lines. By comparison with experimental visualisations captured on video, the appropriate extent of dispersion accompanying the convective flow is achieved by a value of $\beta=0.2$ [10]. Fig 8 refers to a 30 litre laboratory vessel fitted with a standard Rushton turbine rotating at 100 rpm.

The same vessel has also been used to study experimentally the dispersion and reaction of an axisymmetric pulse of acid into alkali. The experiments used hydrochloric acid and caustic soda solutions. With methyl red indicator, acid shows as a vivid crimson colour with alkali as a bright yellow. Experimental results captured using a motorised camera at a frame rate of 0.32s compare almost exactly with the sectional computer graphic reconstructions shown in Fig 9. The initial growth of the red acid pulse by convection and dispersion and its division after the impeller and continuing recirculation are closely simulated. The subsequent shrinking of the acid cloud, because of the original deficit of acid injected compared to the alkali in the vessel, is also closely reproduced. The final complete disappearance of red colouration exactly matches the visualisations. Complete details have been reported previously [11]. Using the subscript A to refer to acid and B to alkali, the initial conditions for integration are

$$C_{A\,i^*,j^*} = C_A^\circ \quad \text{for } i=i^* \quad \text{and } j=j^*$$

$$\text{with } C_{A\,i,j} = 0 \quad \text{for } 1<i<N \quad \text{and } 1<j<2N \qquad ....(4)$$

$$\text{and } C_{B\,i,j} = C_B^\circ \quad \text{for } 1<i<N \quad \text{and } i<j<2N$$

The corresponding zone material balances are then

$$V_{i,j}\frac{dC_{A\,i,j}}{dt} = q\left[C_{A\,i-1,j} - (1+2\beta)\,C_{A\,i,j} + \beta\,(C_{A\,i,j-1} + C_{A\,i,j+1})\right] - k\,C_{A\,i,j}\,C_{B\,i,j} \quad ....(5)$$

$$V_{i,j}\frac{dC_{B\,i,j}}{dt} = q\left[C_{B\,i-1\,j} - (1+2\beta)\,C_{B\,i,j} + \beta\,(C_{B\,i,j-1} + C_{B\,i,j+1})\right] - k\,C_{A\,i,j}\,C_{B\,i,j} \quad ....(6)$$

Equations (5) and (6) employ a second order reaction between the acid (A) and alkali (B) and implicity assume that each zone is perfectly micro–mixed.

SEMI–BATCH OPERATION WITH TWO REACTIONS

ICI in a CASE project with Birmingham University (Drain [15]) have developed an experimental test system based upon a pair of competing parallel reactions of the form

$$A \xrightarrow{k_1} S \quad \text{by product}$$

$$A + B \xrightarrow{k_2} R \quad \text{desired product}$$

The component A is a diazonium salt which undergoes a first order decomposition. A will also couple with B, in this case pyrazolone. The coupling reaction produces a desired product R whereas the parallel decomposition forms an unwanted by–product S. The coupling reaction has so–called elementary reaction kinetics, being first order in A and B.

In typical semi–batch operation the vessel is charged with a solution of A and a solution of B is continuously fed from a suitable feed vessel. The initial situation at 0.0s is as depicted in Fig 10.

It is necessary to calculate the concentrations of each of the reagents A and B and the products R and S as the semi–batch stirred addition proceeds. The initial conditions are

$$\begin{aligned} &C_{A}{}_{i,j} = C_{A}^{\circ} \\ &C_{B}{}_{i,j},\ C_{R}{}_{i,j},\ C_{S}{}_{i,j} = 0 \end{aligned} \qquad \text{for } 1<i<N \text{ and } 1<j<2N \qquad \text{....(7)}$$

The individual zone material balances are then, for the general i,j zone

$$V_{i,j}\frac{dC_{A}{}_{i,j}}{dt}=q\left[C_{A}{}_{i-1,j}-(1+2\beta)C_{A}{}_{i,j}+\beta(C_{A}{}_{i,j-1}+C_{A}{}_{i,j+1})\right]-k_1C_{A}{}_{i,j}-k_2C_{A}{}_{i,j}C_{B}{}_{i,j} \qquad \text{....(8)}$$

$$V_{i,j}\frac{dC_{B}{}_{i,j}}{dt}=q\left[C_{B}{}_{i-1,j}-(1+2\beta)C_{B}{}_{i,j}+\beta(C_{B}{}_{i,j-1}+C_{B}{}_{i,j+1})\right]-k_2\,C_{A}{}_{i,j}\,C_{B}{}_{i,j} \qquad \text{....(9)}$$

$$V_{i,j}\frac{dC_{R}{}_{i,j}}{dt}=q\left[C_{R}{}_{i-1,j}-(1+2\beta)C_{R}{}_{i,j}+\beta(C_{R}{}_{i,j-1}+C_{R}{}_{i,j+1})\right]+k_2\,C_{A}{}_{i,j}\,C_{B}{}_{i,j} \qquad \text{....(10)}$$

$$V_{i,j}\frac{dC_{S}{}_{i,j}}{dt}=q\left[C_{S}{}_{i-1,j}-(1+2\beta)C_{S}{}_{i,j}+\beta(C_{S}{}_{i,j-1}+C_{S}{}_{i,j+1})\right]+k_1\,C_{A}{}_{i,j} \qquad \text{....(11)}$$

For the zone into which reagent B is to be fed, i=i* and j=j*, the unsteady state material balance for Eqn.9 has to be modified to account for the feeding of B over the designated addition period. If the time of addition extends for t_a and if the volumetric addition rate of B is linear, the modified differential equation for zone i*, j* is

$$V_{i^*,j^*}\frac{dC_{B}{}_{i^*,j^*}}{dt} = q\left[C_{B}{}_{i^*-1,j^*}-(1+2\beta)\,C_{B}{}_{i^*,j^*}+\beta(C_{B}{}_{i^*,j^*-1}+C_{B}{}_{i^*,j^*+1})\right] - k_2\,C_{A}{}_{i^*,j^*}\,C_{B}{}_{i^*,j^*} + q_fC_{B}^{f} \qquad \text{....(12)}$$

The product term $q_fC_B^f$ is the molar addition rate of B into the feed zone, where q_f is the volumetric flow rate and C_B^f is the concentration of B being fed semi–batchwise.

Eqn. 12 then applies over the addition period $0<t<t_a$. Thus when $t>t_a$, the addition term in Eqn. 12 is deleted and Eqn 12 reverts to Eqn. 9, so that the addition or injection zone i*,j* behaves exactly as all the other zones once addition is complete.

PREDICTIONS OF VESSEL CONCENTRATION FIELDS

The relevant kinetic data and experimental details are presented in Table 1. These data correspond to the ICI/University of Birmingham experiments on a 20 litre vessel. The time of addition of B was 15 seconds.

Results at a low stirring speed of 12 rpm

Fig 10 shows a sectional image reconstruction using horizontal line coloured shading in each cell to represent the concentration of B. The stirring rate in Fig 10 is 12 rpm which implies a relatively low flow and corresponding rate of mixing. The shading convention uses linear shading with up to 5 horizontal lines per cell. An absence of a line indicates a concentration between 0 and 0.2 C_A°. The number of lines for shading are then 5 for a concentration equal to the maximum of C_A°. All other concentrations have been normalised to C_A° and use the same shading, except for the use of a colour to indicate a specific component.

Fig 11 then shows an assembly of sectional image reconstructions which use green shading for A, red shading for B and blue shading for the desired product R. At t=0, the vessel contains just the reagent A at its initial concentration. Neither B nor R are present at first.

The time sequences in Fig. 11 are for each 15s up to 60s and for each 30s from 60s to 180s. The first shaded section at 15s shows the situation at the moment the addition of reagent B has been completed. The speed of the A–B coupling is such that A and B do not coexist in significant amounts so that there is a clear segregation into a B enriched region growing from the point of addition (i*,j* = 1,1), with the remainder of the vessel volume containing the reagent A. It is clear that the rate at which A and B react is controlled by the rates of convection and lateral mixing engendered in the loops of zones which form the overall network. It needs to be noted that although reaction takes place in a thin swathe of zones that adjusts outwards as addition continues, the product R, having been continuously produced within the emerging swathe of reaction zones, actually occupies a somewhat larger sub–set of zones. Thus the cloud of product R extends somewhat beyond the cloud of reagent B having been subjected to more convective and dispersive mixing.

This trend for the R rich zones to extend beyond the B rich region can be seen to continue after the completion of addition of B. In fact the trend is more exaggerated because the R produced being inert, continues to undergo mixing, whereas the reagent B after completion of addition occupies smaller shrinking volumes as it is consumed by reaction with A. This can be readily observed after times of 30, 45 and 60 seconds.

In respect of reagent B after completion of addition, the 'cloud' of B begins to circulate around the vessel. At 45s it has divided after passing through the impeller and is located mostly along the walls, being slightly more advanced in its circulation in the upper part of the vessel. After 60s the B can be seen to be returning towards the impeller via the uppermost and lowermost parts of the vessel. At 90s, circulation has just taken place through the impeller and a small residual 'patch' of high concentration is just approaching the walls. Circulation and reaction continue, but after 180s all of the reagent A has either reacted with B or decomposed. At this point, where A vanishes by reaction, some unreacted B is still left in the vessel. This unreacted B reflects the yield loss which arises from the by–product decomposition of

A to S. In the event that there was no by–product loss of A, all the B and A would be mutually consumed.

This yield shortfall of R is also manifest in the shading for R after 180s. A small light patch which indicates a lower R value in part of the vessel on completion of reaction can be seen. Ultimately for longer times the R (and of course also the B) would become homogenised throughout the vessel.

Results at a high stirring speed of 302 rpm

The results for the relatively high speed of 302 rpm are presented in Fig 12. As in the slow stirring case, the addition of B takes place over 15 seconds. Fig 12 presents coloured sectional image reconstructions over 5 second intervals up to 25s.

As the addition of B proceeds, the higher relative rate of circulation and internal mixing succeeds in dispersing B much more effectively throughout the vessel. At 5s B is dispersed into about 50% of the volume and after 10s B is present at significant concentrations throughout some 90% of the vessel volume. On completion of addition after 15s, there is a measurable B concentration everywhere. The increased mixing rates result in the local co–existence of A and B throughout most of the vessel by the time all the B has been added. On completion of addition, the vessel contents are almost uniformly mixed so that after 20 seconds the A–B reaction proceeds in close accordance with an assumption of perfect mixing. However, it is neverthless quite clear that even at this high rate of stirring the conditions do not approach perfect mixing during the semi–batch addition phase.

OVERALL YIELD PERFORMANCE

Whilst the individual concentration fields for reagents and products pictorialised in Figs 10, 11 and 12 provide a picture of how the mixing process and chemical reaction rates interact with one another, those involved in designing improved semi–batch operation are more interested in the yield performance of a batch.

It is, however, the case that the local relative yields rely upon the local concentration values. The sum total of all the individual local yields over both space and time then produce the overall yield behaviour at the termination of the semi–batch process.

Since R is the desired product, yields need to be defined with respect to this component. However, yields may then be defined either with respect to A or B. Since A participates in both reactions, yields of R with respect to A will be used here, although it should be noted that the yield of R with respect to B can be deduced via the stoichiometry of the pair of reactions. The local (zone) instantaneous yield of R with respect to A, following the terminology adapted by Levenspiel [16] based on the pioneering work of Denbigh [17], is defined by

$$\varphi\left[\frac{R}{A}\right] = \frac{\text{rate at which R is produced}}{\text{total rate at which A is consumed}}$$

so that for the i,j perfectly backmixed zone

$$\varphi_{i,j}\left[\frac{R}{A}\right] = \frac{k_2\, C_{A\,i,j}\; C_{B\,i,j}}{k_1\, C_{A\,i,j} + k_2\, C_{A\,i,j}\; C_{B\,i,j}} \qquad \ldots(13)$$

Because the zone by zone calculations implicity material balance the reagents and products, the overall yield is manifest as the amount of R present in the vessel relative to the amount of A consumed. The overall yield evolves in time, and at any instant is given by

$$\Phi_t\left[\frac{R}{A}\right] = \frac{\text{mols of R produced}}{\text{total mols of A consumed}}$$

$$\Phi_t\left[\frac{R}{A}\right] = \sum_{i=N,j=1}^{i=N,j=2N} C_R i,j/(C_A^\circ - \sum_{i=1,j=1}^{i=N,j=2N} C_A i,j) \qquad \text{....(14)}$$

The ultimate yield applies when all the A has been consumed (so that all reactions have ceased) and is given by

$$\Phi\left[\frac{R}{A}\right] = \sum_{i=1,j=1}^{i=N,J=2N} C_R i,j/C_A^\circ \qquad \text{....(15)}$$

Without resorting to the integrals (or summations) associated with the overall yields, it can be discerned from Eqn.13 that the local instantaneous yield is optimal when the local product of $C_A i,j$ $C_B i,j$ is maximised and the local value of $C_A i,j$ is minimised.

Thus any portion of the vessel that is rich in A will be giving rise to a low yield. In other words, A and B should be mixed together to the greatest extent, thereby reducing the tendency for A to decompose to S. It follows that the overall yield of R is maximised by having the maximum possible stirrer speed.

The predicted ultimate overall yields for a 2x(20x20) network are presented in Fig 13, together with the experimental results for a 20 litre vessel obtained by Drain [15]. The agreement is good. At lower stirrer speeds, the ultimate overall yield of R starts to fall off rapidly below 50 r.p.m. Above 50 r.p.m. the yield improvement is almost linear with stirrer speed and improves from 88% at 100 r.p.m. to 92% at 300 r.p.m. This trend of a slowly increasing yield is exactly reflected by the experimental results.

CONCLUSIONS

1. The network–of–zones model for flow and mixing inside a stirred vessel has been successfully applied to model a semi–batch process with a pair of competing reactions.

2. The use of a 2(20x20) set of 800 zones provides a good level of spatial discrimination inside a stirred vessel. Partial segregation between reagents can be graphically presented through the batch using shaded sectional image reconstructions.. Reagents and products can be distinguished by using different colours.

3. The changes in partial segregation occasioned by different rates of stirring are readily predicted by the model. Using a lateral mixing coefficient value of β=0.2 and a total internal flow coefficient of K=2.75 provided good agreement with the experimentally observed yields for the two parallel reactions $A+B \longrightarrow R$ and $A \longrightarrow S$

NOMENCLATURE

$C_{i,j}$	concentration of inert tracer in zone i,j
C°	initial tracer concentration in zone i*,j*
$C_{I\,i,j}$	concentration of component I in zone i,j (I=A, B, R or S)
C_A°	initial concentration of A in vessel
C_B°	initial concentration of B in pulse addition
C_B^f	concentration of B in semi–batch feed stream
i	column position of zone (radial)
j	axial position of zone (axial)
K	total internal volumetric flow coefficient
N	size of network of zones
N^1	stirrer rotation speed
Q	total internal circulatory flow in vessel
q	flow in circulation loops
q^f	flow rate of addition of reagent B .
t	time
t_a	semi–batch addition time of reagent B
$V_{i,j}$	volume of i,j zone ($V_{i,j} = V/2N^2$)
β	lateral exchange flow coefficient
$\varphi_{i,j}\left[\frac{R}{A}\right]$	instantaneous yield of R relative to A in zone i,j
$\Phi_t\left[\frac{R}{A}\right]$	overall yield of R relative to A after time t
$\Phi\left[\frac{R}{A}\right]$	overall yield of R relative to A on completion of reaction

Initial concentration of A	$C_{A_0} = 6.0 \times 10^{-5}$ kmol m^{-3}
Initial concentration of B	$C_B^f = 7.4 \times 10^{-2}$ kmol m^{-3}
Semi–batch addition rate	$q_f = 1.1 \times 10^{-3}$ m^3 s^{-1}
Semi–batch addition time	$t_a = 15$ s
Volume of coupler B	$V_b = 17$ ml
Volume of vessel	$V = 0.02$ m^3
Standard Rushton turbine	$D = 0.102$ m
First order rate constant	$k_1 = 0.001$ s^{-1}
Second order rate constant	$k_2 = 7000$ l mol^{-1} s^{-1}

Table 1: Experimental and Simulation Parameters

REFERENCES

[1] Khang S.J. and Levenspiel, O., "New Scale–up and Design Method for Stirrer Agitated Batch Mixing Vessels", 1976 Chem. Eng. Sci., 31, 569, .

[2] Sasakura, T., Kato, Y., Yamamuro, S. and Ohi, N., "Mixing Processes in a Stirred Vessel," 1980 Intl. Chem. Eng., 20, 251,

[3] Baldyga, J. and Bourne, J. R., "Simplification of Micromixing Calculations. I. Derivation and Application of New Model", 1989 Chem. Eng. Jnl., 42, 83,

[4] Baldyga J. and Bourne J. R., "Simplification of Micromixing Calculations II. New Applications", 1989 Chem. Eng. Jnl., 42, 93,

[5] Klein, J.P., David, R. and Villermaux, J., "Interpretation of Experimental Liquid Phase Phenomena in a Continous Stirred Tank Reactor with Short Residence Times", 1980 Ind. Eng. Chem. Fund., 19, 373,

[6] David, R. and Villermaux, J., "Interpretation of Micromixing Effects on Fast Consecutive–Competing Reactions in Semi–Batch Stirred Tanks by a Simple Interaction Model," 1987 Chem. Eng. Commun., 54, 333,

[7] Mann, R. and Mavros, P., "Analysis of Unsteady Tracer Dispersion and Mixing in a Stirred Vessel Using Interconnected Networks of Ideal Flow Zones", 1982 Proc. 4th Europ. Conf. on Mixing (BHRA), 35,

[8] Mann, R and Knysh, P., "Utility of Networks of Backmixed Zones to Interpret Mixing in a Closed Stirred Vessel." 1984 I.Chem.E. Symp. Series 89, 127,

[9] Mann, R, Knysh, P., Rasekoala, E. A. and Didari, M., "Mixing in a Closed Stirred Vessel: Use of Networks of Zones to Interpret Mixing Curves Acquired by Fibre Optic Photometry," 1987 I. Chem. E. Symp. Series, 108, 49,

[10] Mann, R., Knysh, P., Rasekoala, E. A and Motlagh, M. D. K., "Mixing of a tracer in a Batch Stirred Vessel: Visualisation Interpreted by Networks of Zones" Submitted to Trans I. Chem. Engrs.

[11] Mann, R., Pillai, S.K. and Wang, Y. D., "Mixing of Inert and Reactive Tracers in a Batch Stirred Vessel: An Image Reconstruction Approach", Submitted to A. I. Ch.E. Jnl.

[12] Costes, J. and Couderc, J.P., "Study by Laser Doppler Anemometry of the Flow Induced by Rushton Turbine in a Stirred Tank: Influence of the size of the units, 1988 Chem. Eng. Sci., 43, 2751,

[13] Revill, B.K., "Pumping Capacity of Disc Turbine Agitators – A Review" 1982 Proc. 4th Europ. Conf. on Mixing (BHRA)., 11,

[14] Oldshue, J.Y., 1983 "Fluid Mixing Technology", p 174, McGraw Hill,

[15] Drain, S., 1987 Ph.D. Thesis "The Development of a Competing Reaction Scheme and Its Application to the Study of Mixing in Stirred Tanks", Birmingham University.

[16] Levenspiel,O., 1972 "Chemical Reaction Engineering", 2nd Ed, Wiley,

[17] Denbigh, K.G., 1966 "Chemical Reactor Theory : An Introduction" p111, Cambridge University Press, 1966.

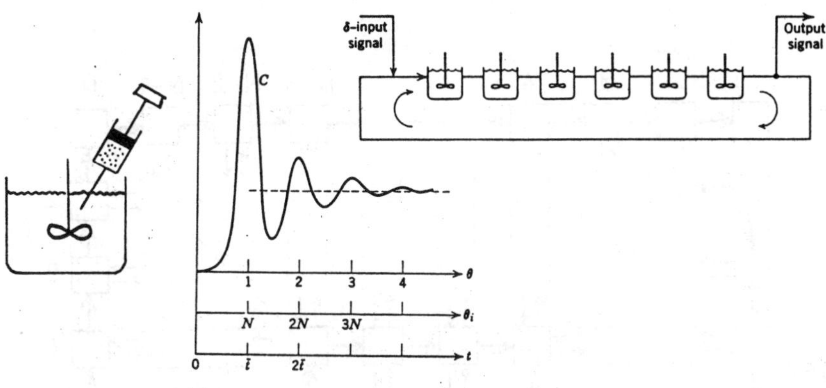

Fig. 1 Mixing Curve for a Batch Stirred Vessel

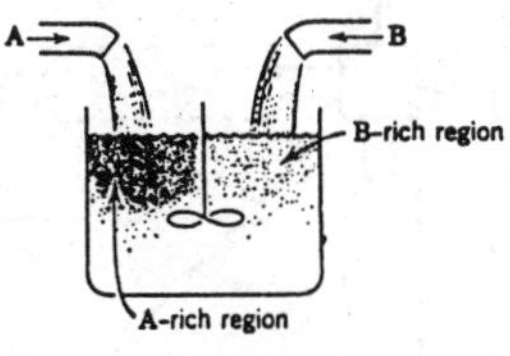

Fig. 2 Partial Segregation in Semi–Batch Operation

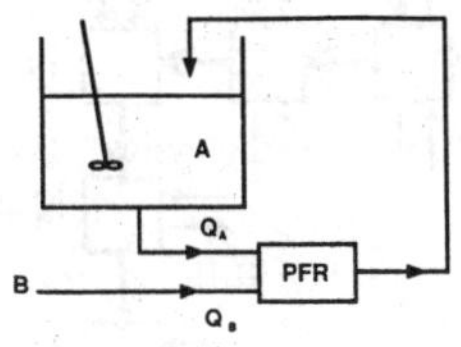

Fig. 3

Semi–Batch Operation with a Plug Flow Reaction Zone [4]

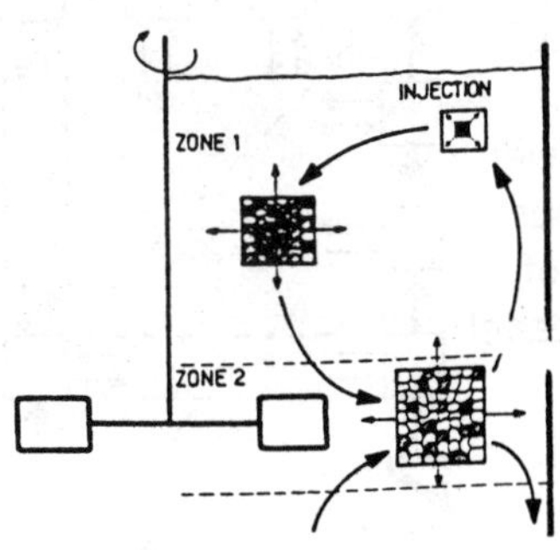

Fig. 4

"Interaction and Exchange with Mean" Model [5,6]

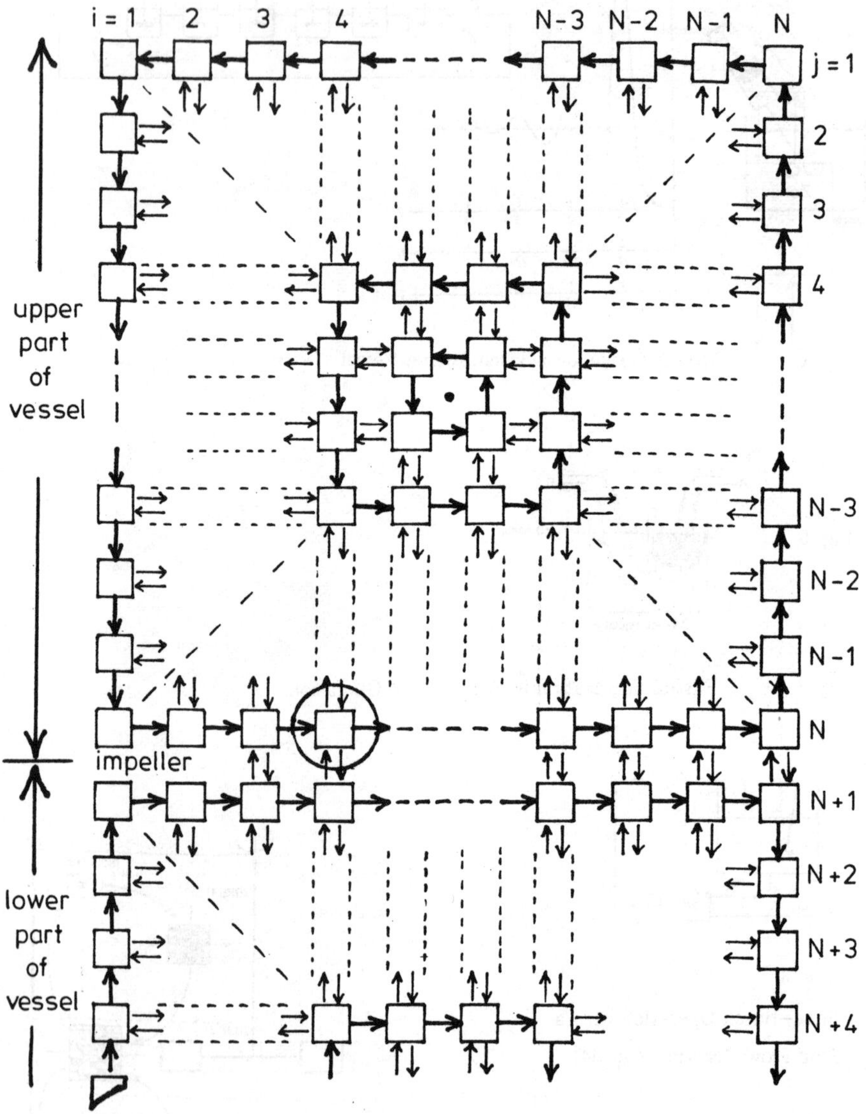

Fig. 5 Network of zones of general size 2x(NxN)

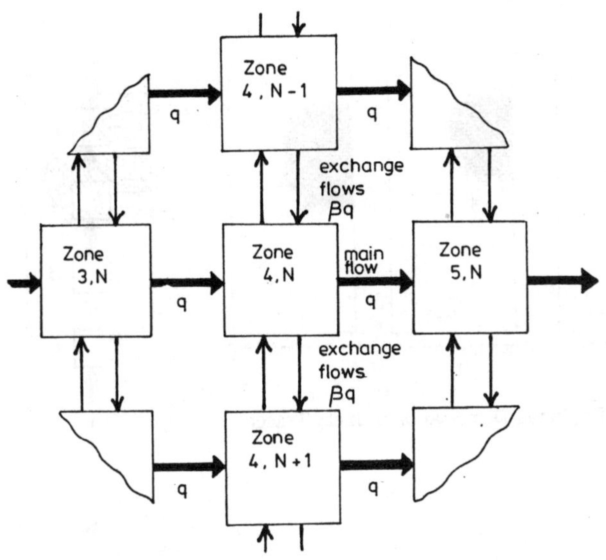

Fig. 6 Main Flow and Exchange Flows for Zone (4,N)

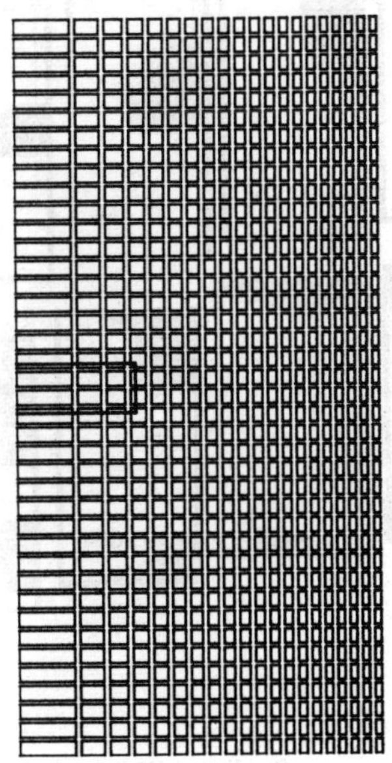

Fig. 7

Set up of 800 Equal Volume Zones for 2x(20x20) Configuration

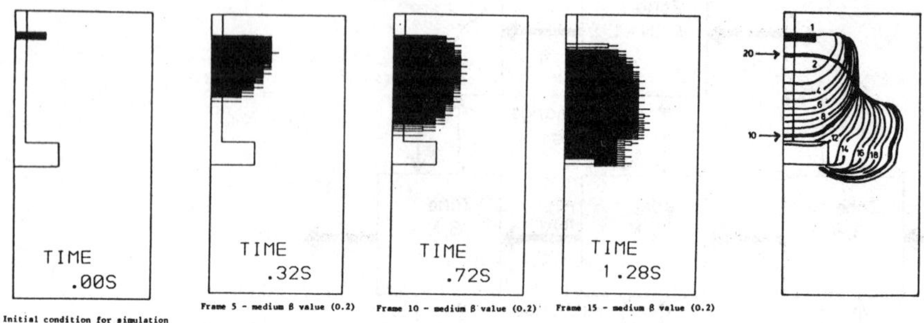

Fig. 8 Predicted Dispersal of an Inert Tracer

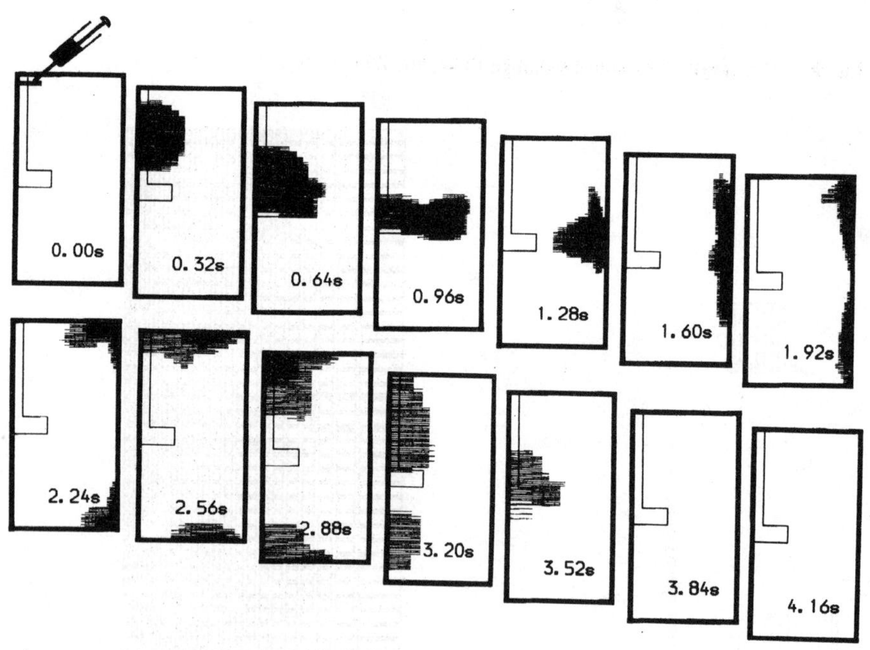

Fig. 9 Predicted Dispersal of an Acid Pulse in Alkaline Vessel

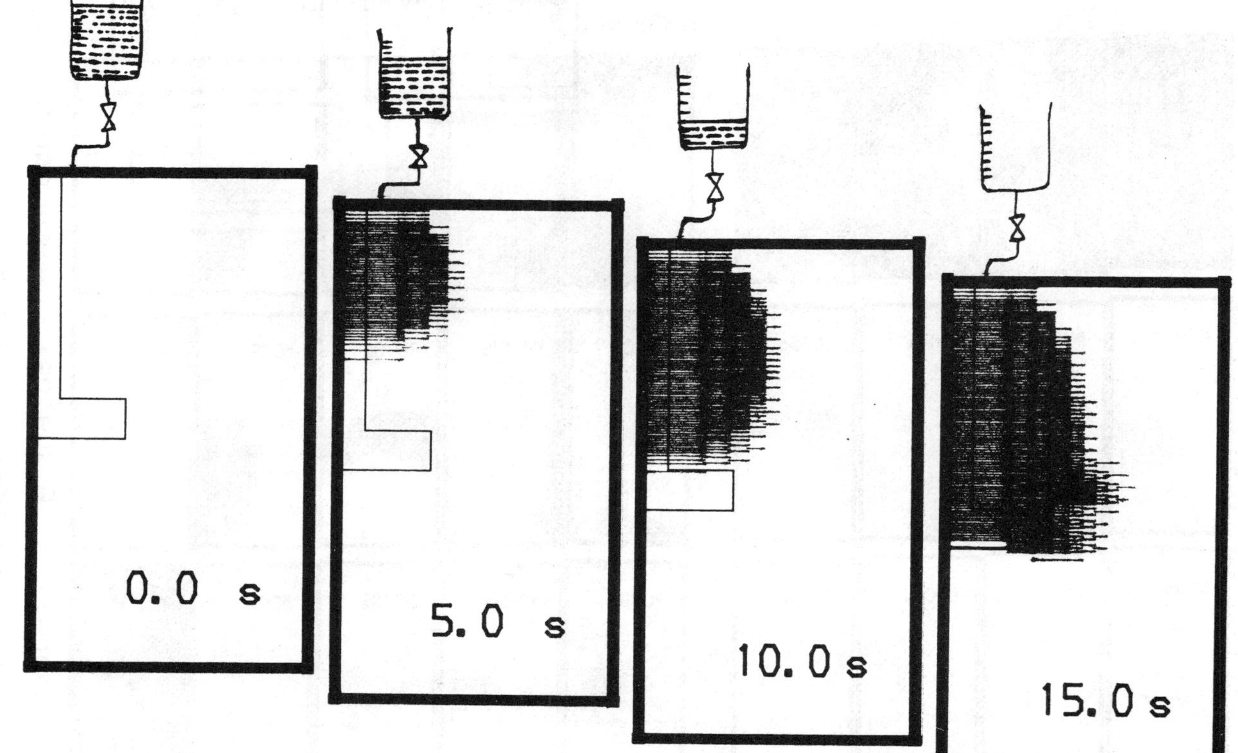

Fig. 10 Semi–Batch Addition of Reagent B to Stirred Vessel

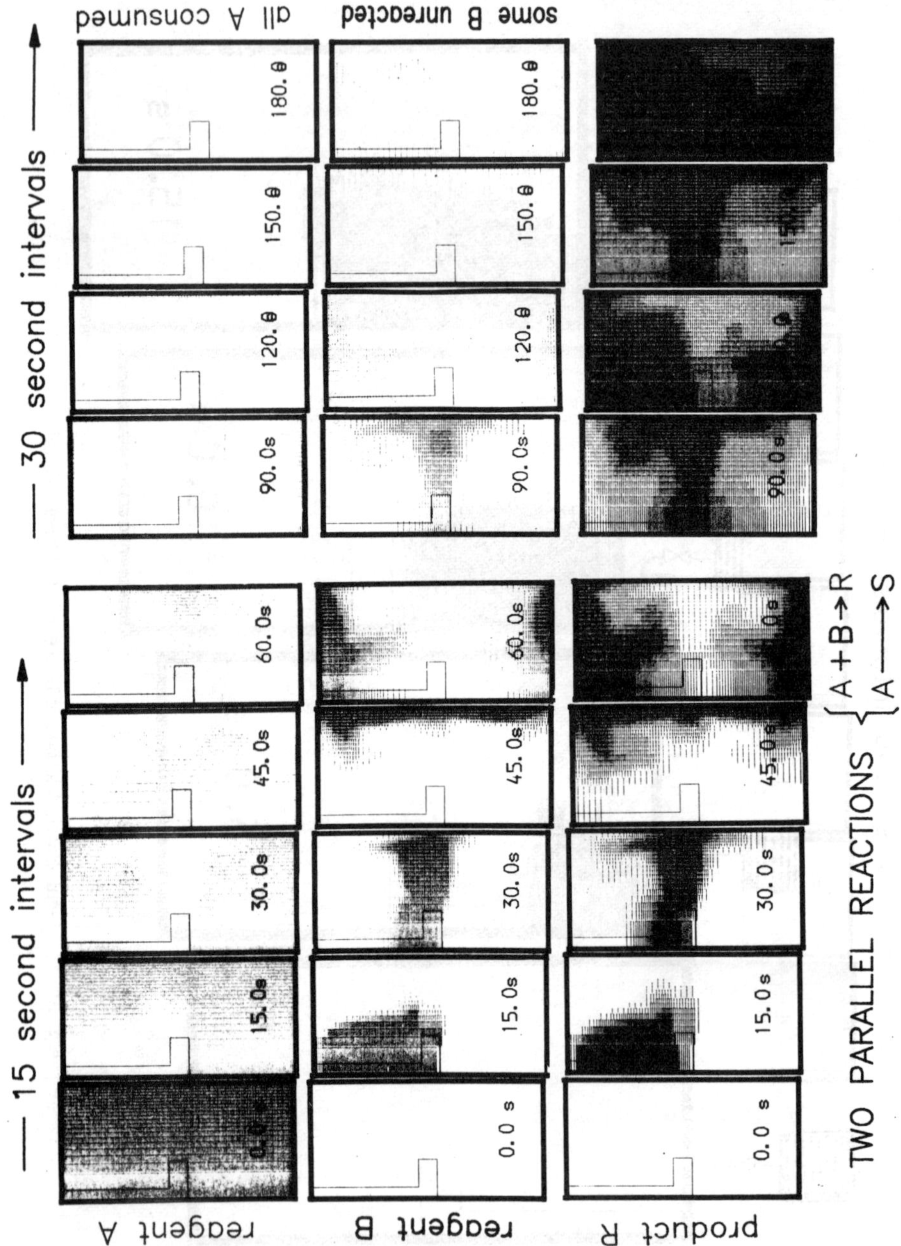

Fig. 11 Sectional Image Reconstructions of Semi–Batch Operation (N^1=12 rpm)

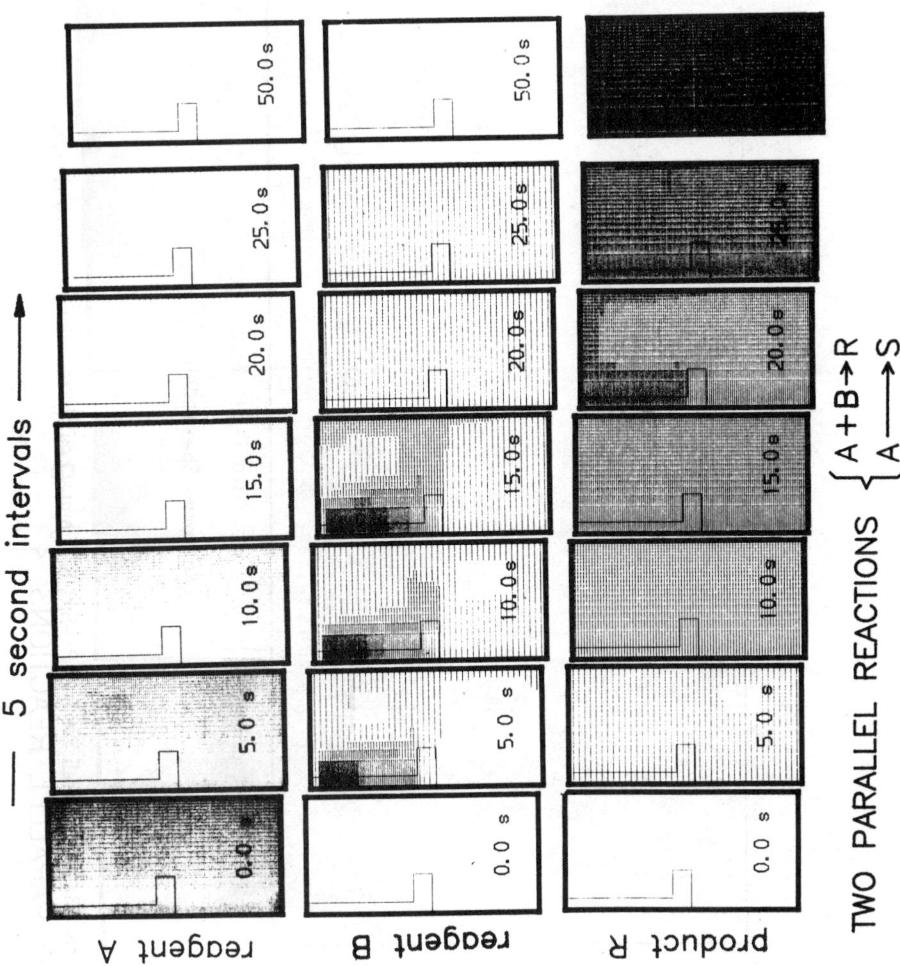

Fig. 12 Sectional Image Reconstructions of Semi–Batch Operation (N^1=302 rpm)

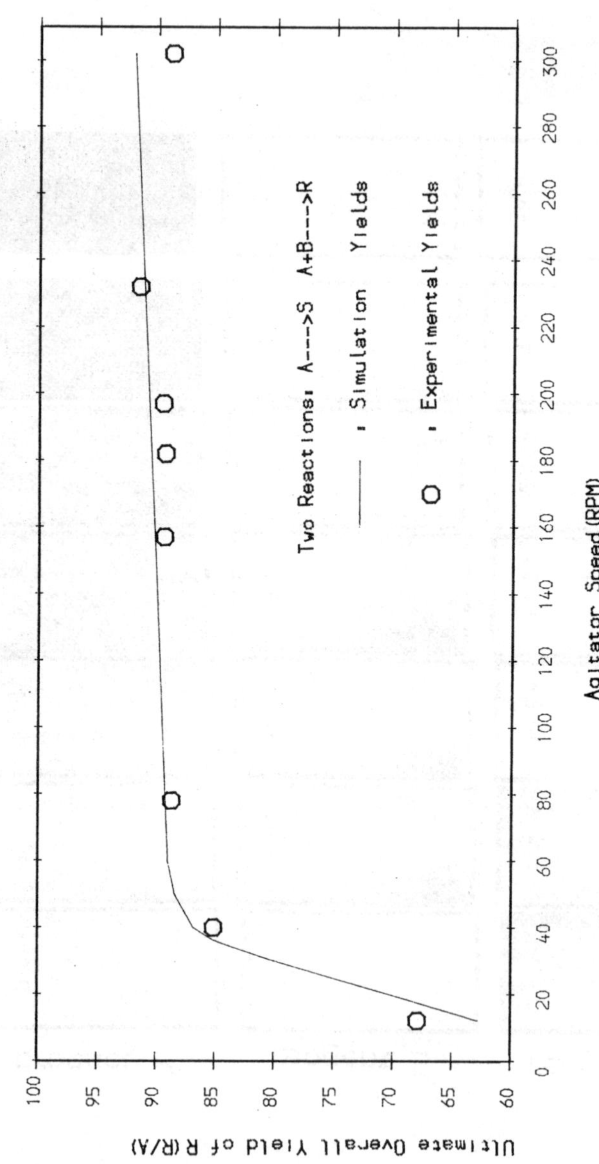

Fig. 13 Overall Yields of Product R : Comparison of Predictions with Experiments

HYDRODYNAMICS, MIXING AND OXYGEN TRANSFER IN LIQUID-IMPELLED LOOP REACTORS*

H.M. van Sonsbeek, P. Verlaan and J. Tramper

The liquid-impelled loop reactor (LLR) is a new type of bioreactor in which the density difference of two immiscible liquid phases provides the mixing. Two of the many configurations are discussed here and a general model is presented that gives the dispersed-phase hold-up and the continuous-phase circulation velocity, both as function of dispersed-phase input flow rate. Furthermore, the mixing per reactor section is characterized and some results of $k_l a$ measurements are presented.

INTRODUCTION

In the process industry biotechnological processes are becoming more popular in comparison to chemical processes due to the milder process conditions and the high specificity of the reactions. Biocatalytic processes involve enzymes or living organisms such as bacteria, yeasts, moulds or plant and animal cells. These biocatalysts are sometimes immobilized for protection or facilitated continuous operation and biocatalyst separation. The medium in which the reaction takes place used to be an aqueous solution, but in the last decade many processes have been developed that use media significantly less polar then aqueous solutions (1). The pertinent biocatalytic research aims to use water-immiscible organic solvents as a solution to the restrictions imposed by aqueous solutions. An example of this is the possible better solubility of substrate and/or product. Furthermore substrate and/or product inhibition or hydrolysis may be reduced. However, some disadvantages may be encountered when using organic solvents, for instance biocatalyst denaturation or inhibition and increased complexity of the process.

* Patent pending

Food and Bioprocess Engineering Group, Agricultural University, P.O. box 8129, 6700 EV Wageningen.

These developments in biocatalysis induce new developments in reactor engineering. The most basic demands on a bioreactor are good mixing and mass transfer properties. The stirred vessel, usually with air sparging, is the simplest form of a bioreactor, that can satisfy these conditions. For shear-sensitive biocatalysts the air-lift loop reactor (ALR) is an attractive alternative as it forms a good compromise between good mixing and low shear forces and gives a uniform distribution of (immobilized) cells. A logical extension of this development is a loop reactor in which two immiscible liquid phases are used in the same way as gas and water in an ALR (2). This new type of bioreactor, the so called liquid-impelled loop reactor (LLR), consists of two vertical tubes: a main tube and a circulation loop. These tubes have an open connection at the bottom and the top and are filled with the continuous phase. The dispersed phase is injected at one end of the main tube: in the case shown in figure 1 the dispersed phase enters at the base of the column. After reaching the top of the column, phase separation and dispersed-phase carry-off takes place at the top of the column. Circulation of the continuous phase is induced by the density difference between the contents of the main tube and the contents of the circulation loop, which does not contain the dispersed phase. Many different configurations can be designed using the same principle. The dispersed phase can be in up flow or down flow , the circulation loop may be internal or external, a combination of up flow and down flow may be constructed, even a combination with aeration is possible. Furthermore the aqueous phase can be either the dispersed phase or the continuous phase. Two possible configurations are given in Figures 1 and 2. The suitability of the LLR to carry out biotechnological processes was shown for an up-flow and a down-flow configuration, both with an external loop (3,4).

As mentioned before, mixing and mass transfer are key aspects of a reactor. In an LLR, both characteristics depend directly on the flow, i.e. the circulation velocity and the dispersed-phase hold-up in the main tube. A model description of the flow in an LLR is presented here. It gives estimations of both quantities starting with the reactor geometry and the imposed dispersed-phase flow rate. In the chapter on theory some details about this model are given. Furthermore some preliminary experiments on mixing and oxygen mass transfer are presented and discussed.

THEORY

Hydrodynamic model

The continuous-phase circulation velocity and the dispersed-phase hold-up are interdependent and are controlled by the dispersed-phase flow rate. For reactor and process design it is important to be able to give a good estimation of both quantities as a function of the dispersed-phase flow rate. The velocity of dispersed-phase drops is estimated with the two-phase drift-flux theory of Zuber and Findlay (5) :

$$v_d = C(v_{sd} + v_{sc}) + v_{d,\infty} \tag{1}$$

where v_d is the velocity of the dispersed-phase drops and v_{sd} and v_{sc} are the superficial velocities of the dispersed and continuous phase in the main tube, respectively. C is the distribution parameter and $v_{d,\infty}$ is the terminal velocity of a single drop in an infinite volume of the continuous phase. If the concentration of dispersed phase is uniform over the cross section of the tube, the value of C is 1. If the concentration of dispersed phase at the wall of the tube is zero and the velocity profile pronounced parabolic, C has a value of about 1.5. With this estimation the hold-up α can be calculated as follows :

$$\alpha = \frac{v_{sd}}{v_d} \tag{2}$$

Combining equation 1 and 2 gives the relationship between hold-up (α) and the circulation velocity (v_{sc}). A second relationship is necessary to calculate the actual values of α and v_{sc} for a given dispersed-phase flow rate. For that, the circulation flow in a loop reactor is considered as one-phase flow in a pipe with appendages. The pressure difference between both columns can then be regarded as the pressure drop in the whole pipe system and then the following equation applies (6) :

$$\alpha \Delta\rho g H = \frac{1}{2} K_f \rho_c v_{scc}^2 \tag{3}$$

where $\Delta\rho$ is the density difference of dispersed and continuous phase, g the gravitational constant, H the height of the reactor, ρ_c the density of the continuous phase, K_f an overall friction coefficient and v_{scc} the superficial continuous-phase velocity in the circulation loop. Substitution of equations 1 and 2 gives :

$$\frac{v_{sd} \Delta\rho g H}{C(v_{sd} + v_{sc}) + v_{d,\infty}} = \frac{1}{2} K_f \rho_c v_{scc}^2 \tag{4}$$

For the continuous-phase flow in the main tube and in the circulation loop, the superficial velocities v_{sc} and v_{scc} are interrelated. The value of C is derived from the experimental results and the value of $v_{d,\infty}$ is derived from physical properties (see Materials and Methods). This makes equation 4 suitable for calculation of the circulation velocity (v_{sc}), and from that the hold-up, both as a function of the dispersed-phase flow rate.

Mixing

Mixing in the main tube, the circulation loop, and top and bottom section can be described by the axial dispersion model:

$$\frac{\delta \hat{c}}{\delta \theta} = \frac{1}{Bo} \frac{\delta super2 \hat{c}}{\delta x^2} - \frac{\delta \hat{c}}{\delta x} \tag{5}$$

where $\hat{c}$ is the dimensionless concentration, θ the dimensionless time, x the dimensionless axial coordinate and Bo the Bodenstein number, defined as:

$$Bo = \frac{vL}{D_{ax}} = \frac{convective \quad transport}{axial \quad dispersion} \tag{6}$$

with v the average velocity in the tube, L the length of the tube and D_{ax} the axial dispersion coefficient.

Oxygen transfer

The solubility of oxygen in FC40, the organic solvent that is used for oxygen transfer measurements, is about 12 times higher than in water and therefore the mass transfer is mainly limited by the mass transfer coefficient in the water phase. Oxygen transfer from FC40 to the continuous water phase can thus be described by the following equation :

$$\frac{dC(t)}{dt} = k_l a(C^* - C(t)) \tag{7}$$

where k_l is the water-phase mass-transfer coefficient, a is the specific exchange area, C^* is the oxygen saturation concentration in the water phase and $C(t)$ is the actual oxygen concentration in the water phase.

MATERIALS AND METHODS

For the experiments two up-flow configurations were used: a lab-scale and a pilot plant LLR with volumes of 0.004 m^3 and 0.165 m^3, respectively. One down-flow configuration was also used: a lab-scale LLR with a volume of 0.004 m^3 and height of 0.55 m. In all cases water was the continuous phase. For the up-flow configurations hexane ($\rho = 660$ kg/m^3) was used as dispersed phase, while for the down-flow configuration this was the perfluorcarbon FC40 ($\rho = 1870$ kg/m^3). The reactors are equipped with liquid spargers, designed to give a uniform drop size. In the lab-scale reactors 12 pipes of 0.2 m length and in the pilot-plant reactor 49 pipes of 0.3 m length are evenly spaced over the cross section of the main tube. The pipes are of stainless steel and have an inner and outer diameter of 1.4 and 2 mm, respectively. A chamber at the inlet of all pipes gives a uniform pressure distribution over the pipes. The hold-up was derived from pressure-difference measurements between top and bottom of the main tube of the reactor. The circulation velocity of the continuous phase was measured by means of an inductive flow meter, located in the circulation loop. The drop size of dispersed phase drops was estimated with the theory of Scheele and Meister and photographically verified (6,7). This estimated drop size was used to derive the terminal rise or fall velocity in an infinite medium from a graphical correlation given by Hu and Kintner (8). The value of parameter C, which characterizes the flow, was derived from experimental results. Mixing experiments were carried out by injecting an acid tracer in the continuous water phase and the detection was by means of a pH probe. Both up-flow LLR's were used for these experiments. Oxygen transfer from the dispersed phase to the continuous phase was measured in the down-flow LLR using the dynamic method. For this nitrogen-saturated FC40 was used to deoxygenate the reactor. After switching to air-saturated FC40 the oxygenation rate was measured as a function of the time.

RESULTS AND DISCUSSION

Hydrodynamic model

Using equation 3 and the experimental results of α and v_{scc}, the overall friction coefficients are derived and are found to be constant in the measured flow range (see Figure 3). Using these overall friction coefficients, the value of 1.6 for parameter C gave the best model description of the experimental results (see for example figure 4 and 5). This value indicates a pronounced parabolic concentration and velocity distribution, and is rather high compared to the value of 1.07 for an ALR found by Verlaan et al (9), which may be due to the considerable lower flow range of the experiments.

Mixing

In figure 6 experimental Bodenstein values for the main tubes of the lab-scale and the pilot-plant up-flow LLR are shown and appear to be about constant (12 and 18, respectively) at the flow rates used in this work. The average value of the Bodenstein numbers for the circulation loop and the top section for the reactors on both scales are found to be 30 and 5, respectively. For further modelling this means that the circulation tube can be characterized by plug flow, while the top section is rather well mixed. The main tube is plug flow with a small influence of axial dispersion.

Oxygen transfer

In figure 7 the preliminary results of the dynamic $k_l a$ measurements are shown. An increase is seen due to an increased hold-up at higher dispersed-phase flow rates. The k_la values are one order of magnitude lower than k_la values for an ALR found by Verlaan (10).

The average k_l values are estimated to be 1.3 * 10^{-4} s^{-1}, which is in good agreement with literature values. Gas bubbles with a diameter above 1.5 mm have a k_l value of 4 * 10^{-4} s^{-1}. For a smaller bubble size the k_l value decreases to 1.3 * 10^{-4} s^{-1} due to a more rigid liquid-gas interface (11). The experimentally detemined k_l values for FC40 drops, having a mean diameter of 2.7 mm, indicate a more rigid liquid-liquid interface compared to the gas-liquid of gas bubbles.

CONCLUSIONS

The experimental results can be described by the two-phase drift-flux theory, using a distribution parameter of 1.6. The flow in both tubes can be characterized by plug flow, with some axial dispersion in the main tube. The top section is well mixed. Compared to ALR results, preliminary $k_l a$ measurements in the LLR gave rather low values, but the k_l is comparable to literature values for gas bubbles.

NOMENCLATURE

a = specific exchange area (m^{-1})
Bo = Bodenstein number (-)
C = distribution parameter (-)

$C(t)$ = oxygen concentration in the water phasemol (m^{-3})

C^* = oxygen saturation concentration in the water phasemol (m^{-3})

D_{ax} = axial dispersion coefficient ($m.s^{-2}$)

$\hat{c}$ = dimensionless concentration (-)

g = gravitational constant ($m.s^{-2}$)

H = distance from nozzle opening to separation level (m)

K_f = overall friction coefficient (-)

k_l = mass-transfer coefficient ($m.s^{-1}$)

L = tube length (m)

t = time (s)

v_{sd} = superficial dispersed phase-velocity in the main tube ($m.s^{-1}$)

v_d = velocity dispersed-phase drops in the main tube ($m.s^{-1}$)

v_{sc} = superficial velocity continuous phase in the main tube ($m.s^{-1}$)

v_{scc} = superficial velocity continuous phase in the circulation loop ($m.s^{-1}$)

$v_{d,\infty}$ = terminal velocity of a single drop in an infinite medium ($m.s^{-1}$)

v = average velocity ($m.s^{-1}$)

x = dimensionless axial coordinate (-)

α = hold-up (-)

ΔP = pressure difference (Pa)

$\Delta\rho$ = density difference dispersed and continuous phase ($kg.m^{-3}$)

ρ = density ($kg.m^{-3}$)

ρ_c = continuous-phase density ($kg.m^{-3}$)

θ = dimensionless time (-)

REFERENCES

[1] Brink, L.E.S., Tramper, J., Luyben, K. Ch. A. M., and van 't Riet, K., Enzyme Microb. Technol. 10, 736.

[2] Laane, C., Tramper, J. and Lilly, M.D., 1987 "Biocatalysis in organic media", Elsevier, Amsterdam.

[3] van den Tweel, W.J.J., Marsman, E.H., Vorage, M.J.A.W., Tramper, J., and de Bont, J.A.M., 1987, "The application of organic solvents for the bioconversion of benzene to cis-benzeneglycol", International conference on bioreactors and biotransformations, Gleneagles, Scotland, UK.

[4] Buitelaar, R.M., Vermuë, M.H., Schlatmann, J.E., and Tramper, J., "The influence of various organic solvents on the metabolism of free and immobilized cells of Tagetes minuta", submitted for publication.

[5] Zuber, N., and Findlay J.A., 1965, J. Heat Transfer 87, 453.

[6] van Sonsbeek, H.M., Verdurmen, R.E.M., Verlaan, P., and Tramper, J., "Hydrodynamic model for liquid-impelled loop reactors", submitted for publication

[7] Meister, B.J, and Scheele, G.F., 1969, AIChE J. 15, 700.

[8] Hu, S., and Kintner, R.C., AIChE J. 1, 42.

[9] Verlaan, P., Tramper, J., van 't Riet, K., and Luyben, K.Ch.A.M., 1986, Chem. Eng. J. 33, 43.

[10] Verlaan, P., "Modelling and characterization of an airlift-loop bioreactor", Ph.D. Thesis, Agricultural University Wageningen, The Netherlands.

[11] van 't Riet, K., 1983, Trends in biotechnology 1, 113.

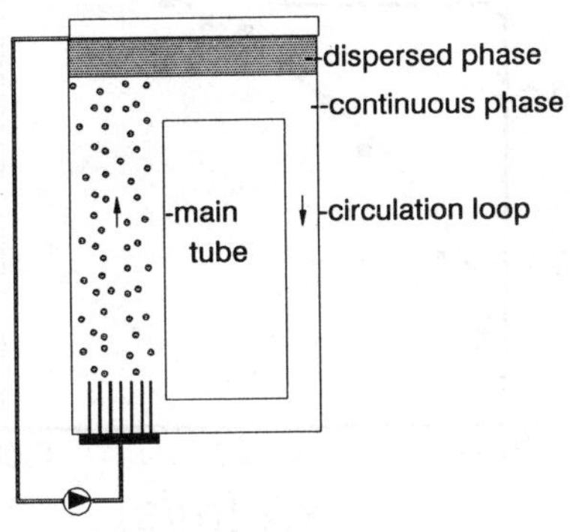

Figure 1 : outline of an up-flow LLR

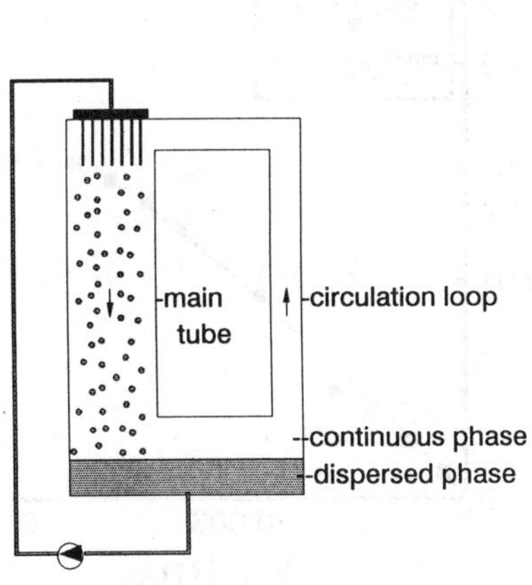

Figure 2 : outline of a down-flow LLR

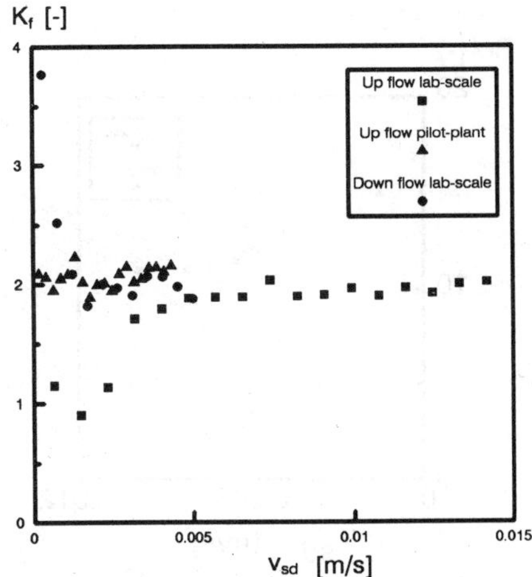

Figure 3 : overall friction coefficient as function of the superficial dispersed-phase velocity

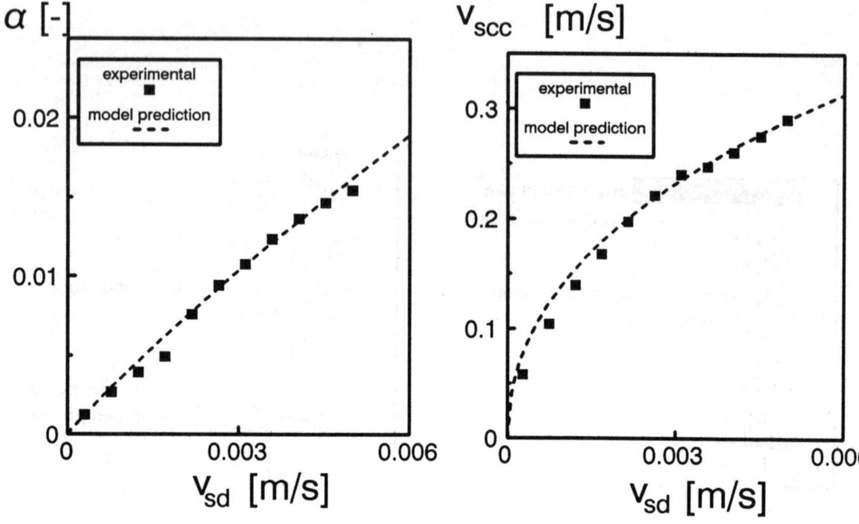

Figure 4 : the dispersed-phase hold-up in the main tube of the down-flow LLR

Figure 5 : the dispersed-phase hold-up in the main tube of the down-flow LLR

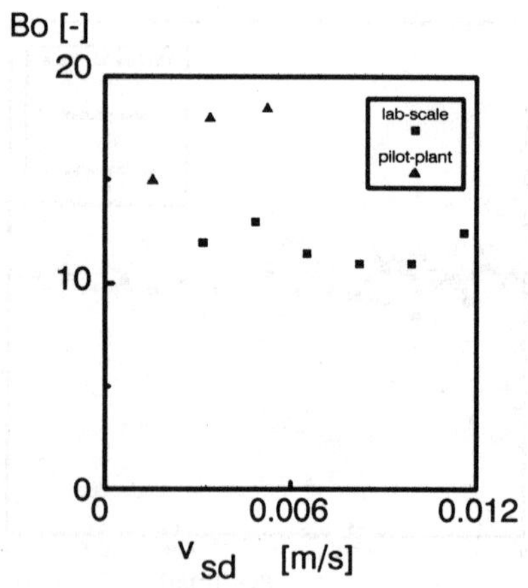

Figure 6 : Bo numbers for the main tube of the up-flow LLR.

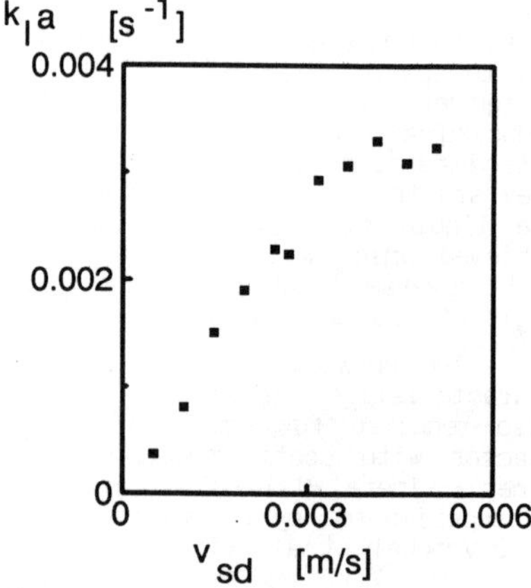

Figure 7 : K_l a-values in the main tube of the down-flow LLR

AXIAL MIXING STUDIES IN A CONTINUOUS REVERSED FLOW JET LOOP REACTOR

K. Remananda Rao* and G. Padmavathi

Mixing behaviour of individual sections of a reversed flow jet loop reactor was studied for water and air-water systems. The liquid circulation time and Bodenstein number (Bo) were calculated from the response curves and curve fitting of the experimental data with the axial dispersion model. The results of the effects of different equipment parameters in water system and flow rates of fluids in air-water system on mixing behaviour are discussed.

INTRODUCTION

Loop reactors are replacing other types of gas-liquid contactors mainly because of their simple construction, ease of operation and satisfactory mixing behaviour. In these reactors mixing is done either by a liquid jet or by different densities in the communicating spaces. The former type is designated as jet loop reactor(JLR) and the latter as airlift loop reactor (ALR). These reactors are particularly suited in application with demands for rapid and uniform distribution. Most of the work done in jet loop reactor were concerned with the upflow of liquid and gas using a two-fluid nozzle located centrally at the bottom of the reactor in which gas can escape unused due to upward rise velocity of bubbles. Wachsmann et al. (1) modified the design of the jet loop reactor by locating the two-fluid nozzle at the top, inducing a down flow in the draft tube and an upflow in the annulus. By this arrangement, the mean residence time of the gas bubbles increases as they are forced to move in a direction opposite to their buoyancy. By varying the liquid flow rate, complete circulation of the two phase flow can be achieved which will provide efficient primary dispersion and redispersion. Such reversed flow jet loop reactors are useful in processes where high ratio of liquid to gas is required.

The knowledge of mixing bahaviour in JLR is important because mixing characteristics influence the yield in many chemical reactions. It is also required for modeling of biotechnological processes. In a jet loop reactor with continuous operation, two mixing processes superimpose each other. There will be longitudinal mixing during each circulation and this is superimposed by backmixing due to recirculation. Several authors [Blenke (2), Voncken et al. (3), Murakami et al. (4), Verlaan et al. (5)] used axial

* Department of Chemical Engineering, Indian Institute of Technology, Madras - 600 036, India.

dispersion model to visualise mixing process in loop reactors. Mathur and Weinstein (6) developed expressions for the moments of the RTD as functions of recycle ratio and Peclet Number (Pe). Fields and Slater (7) and Verlann et al. (5) investigated the contributions of individual sections to mixing in a concentric tube and external loop type air lift reactors respectively. Weiland and Onken (8) have noted that the empirical mixing time for an air lift loop reactor changes little upon the addition of 0.25% CMC to water. However, Fields and Slater (9) carried out mixing studies in a concentric tube air lift loop reactor using an improved radio pill flow follower and reported the influence of pseudoplastic non-Newtonian fluid upon hydrodynamic performances and mixing parameters. Bello et al.(10) reported that the liquid mixing capability of an external circulation loop airlift contactor is considerably better than a bubble column and concluded that airlift columns are 2.5 times better than bubble column in terms of mixing efficiencies. Warnecke et al.(11) developed a model and reported a procedure to determine the main model parameters for the description of mixing behaviour of loop reactors. This model discriminated different mixing sections and considered the concentration profile caused by the recycle.

So far no work has been reported on the mixing behaviour of a continuous reversed flow jet loop reactor. This paper presents an experimental investigation aimed to study the influence of various equipment parameters on mixing behaviour and evaluate the contribution of individual sections of the equipment to mixing. This study also includes the effect of gas and liquid flow rates in air-water system on mixing characteristics of the reactor.

EXPERIMENTAL

A schematic diagram of the reactor used in this work is shown in Figure 1. The dimensions of the reactor and the range of operating variables studied are given in Table 1.

TABLE 1 - Reactor Dimensions and Range of Variables studied

1.	reactor diameter D(mm)	140
2.	reactor length H(mm)	1000
3.	bottom spacing H_B(mm)	45
4.	draft tube diameter D_d(mm) (D_d/D)	65.8 (0.47) 85.4 (0.61) 93.8 (0.67)
5.	draft tube length D_L(mm)	650
6.	immersion height of the two fluid nozzle into draft tube H_N(mm)	40 80 140
7.	liquid nozzle diameter d_N (mm)	7 10 12
8.	aeration tube diameter (d_{Gi}-d_{Go}) (mm) (inner-outer)	2.5-3.0
9.	liquid flow rate (l/h)	200-3000
10.	gas flow rate (l/h)	275-1200

The major parts of the reactor consists of a reactor tube and an interchangeable draft tube arranged concentrically inside the reactor tube on an impact plate. Liquid is withdrawn from the reactor through the outlet located under the impact plate. A two fluid nozzle installed at the top of the reactor consists of a liquid nozzle at the centre of which a stainless steel tubing is rigidly fixed by means of a guide system. Liquid from a storage tank is pumped and metered through calibrated rotameters and then passed through the liquid nozzle. Air from a compressor is metered through a calibrated rotameter before it is introduced to the aeration tube located at the centre of the nozzle. Tracer solution is introduced at the liquid inlet jet by means of a syringe. Three conductivity cells are installed each at the end of the riser, at the end of downcomer and at the outlet as shown in the Figure. The cells are connected to a conductivity meter and the output of the conductivity meter is connected to a strip chart recorder.

Impulse response method was used to study the mixing characteristics. 20 ml of 3 M KCl solution tracer was injected at a location just below the liquid inlet jet and the variation of conductivity of tracer with time was recorded at all the locations of the conductivity cells.

RESULTS AND DISCUSSION

Theoretical Consideration for Data Processing

Figure 2 shows a typical experimental record at the end of the downcomer, subsequent to the addition of the tracer. The tracer concentration C at time t and position x in the loop is given by

$$\frac{\partial c}{\partial t} = D_{ax} \frac{\partial^2 c}{\partial x^2} - U \frac{\partial c}{\partial x} \qquad (1)$$

Blenke (2) and Voncken (3) found a solution to the above equation by summing up the tracer response at every interval of the loop and applying the open vessel boundary conditions as

$$C_r = \frac{C}{C_\infty} = \sum_{j=1}^{\infty} \sqrt{\frac{Bo}{4\pi\Theta}} \exp\left[-\frac{(j-\Theta)^2 Bo}{4\Theta}\right] \qquad (2)$$

To take the dilution effect into consideration for a continuous process Blenke (2) modified the above equation as

$$C_r = \frac{C}{C_\infty} = \sum_{j=1}^{\infty} \sqrt{\frac{Bo}{4\pi\Theta}} \exp\left[-\frac{(j-\Theta)^2 Bo}{4\Theta}\right] \left[\frac{n_U - 1}{n_U}\right]^j \qquad (3)$$

$$\text{where } \frac{n_U - 1}{n_U} = \frac{\text{Circulation flow}}{\text{Total flow}} \qquad (3a)$$

$$Bo = \frac{U.L}{D_{ax}} \quad (3b)$$

$$\Theta = t/t_c \quad (3c)$$

The Bodenstein number (Bo) and $(n_U-1)/n_U$ were obtained by curve fitting of the experimental data with equation (3).

Circulation Time

Figure 2 shows a typical experimental record of conductivity as a function of time. As soon as the tracer was injected, for a short period the conductivity cell measured a very high conductivity peak and a second peak was produced because of the redistribution of the tracer due to recirculation. The circulation time t_c is defined as the time required for a tracer element to pass through the reactor once. This time was measured directly from the average distance between two consecutive peaks of the experimental record of tracer response. Each experiment was repeated four times and each response graph consisted of four peaks. Hence t_c calculated is the average of atleast ten observations.

Figure 3 gives circulation times t_c for the three draft tube diameters with varying liquid flow rate. Circulation time for single phase flow decreased with increasing draft tube diameter and increased liquid flow rate. Increase in draft tube diameter reduces annulus cross sectional area which in turn increases the superficial velocity of the liquid. Increase in the superficial velocity enhances liquid circulation which results in reduced values of t_c. The dependence of circulation time on nozzle diameter for various liquid flow rates is shown in Figure 4. This Figure shows that the increasing nozzle diameter increases t_c for the same liquid rate. The lower the diameter of the liquid nozzle, relatively higher liquid velocity results from the jet which in turn reduces t_c. But the immersion height of the two-fluid nozzle into the draft tube did not show any significant influence on t_c. These results clearly indicate that the enhanced liquid circulation induced by the liquid jet contributes to the reduced values of t_c. Results obtained on circulation time for single phase flow were correlated as follows:

$$t_c = 6.1\,(d_N/D)^{0.54}\,(D_d/D)^{-0.815}\,(H_N/U_L)^{0.89}$$

Mixing in Single Phase flow

Bodenstein number (Bo) which is the ratio of convective to diffusive transport rates, characterizes the degree of mixing in loop reactors. A decreased value of Bo indicates enhanced backmixing in the reactor. From the results of the present work Bodenstein number was obtained through curve fitting of the experimental data with equation (3) which gives the solution of the axial dispersion model. Figure 5 shows a comparison of the experimental response curve with the predicted one using equation (3) for different experimental conditions for the riser, downcomer and entire reactor section respectively using single phase liquid flow. The Figure clearly shows that the experimental data were in close agreement with the predicted values at higher values compared to lower Θ values. However, the model used matches the experimental data of the present work with a maximum root mean square deviation of 0.09. Figure 6 gives the variation of standard deviation with Bodenstein number for curve (a) shown in Figure 5. From Figure 6 it can be seen that according to the optimal value of the Bodenstein number, the estimation method is less sensitive at Bo values ranging from 10 to 15.

The dependence of Bo on reactor parameters, namely D_d/D, H_N and d_N is shown in Figures 7,8 and 9 respectively for single phase liquid flow. Figure 7 shows the dependence of Bo number on mean Reynolds number Re_m based on mean diameter of the reactor with draft tube diameter as parameter. Bo increased with increased Re_m and D_d/D ratio. Among the three immersion heights H_N studied, Bo values obtained for H_N = 4 cm were smaller compared to those obtained for H_N of 8 cm and 14 cm, as indicated in Figure 8. However no significant change in Bo values was observed in the case of H_N = 8 cm and H_N = 14 cm on the performance of the downcomer. For the riser, variation of Bo values with immersion height is found to be negligible.

Figure 9 shows the dependence of the Bo number on Re_m with nozzle diameter as the parameter for each of the three sections of the reactor. Figure 10 shows the dispersion coefficient calculated using equation (3b) corresponding to conditions shown in Figure 9. In Figure 9 it can be seen that Bodenstein number decreased with increasing nozzle diameter for all the three sections of the reactor. Increase in Re_m increased Bo values. Similar trends were observed in literature (12) also. However from Figure 10 it is seen that the dispersion coefficients increased with increase in Re_m and nozzle diameter. The circulation velocity of small diameter nozzle is always higher than larger nozzle diameter due to high velocity. Therefore, at constant Re_m, liquid input is less for smaller diameter nozzle due to which turbulence caused by flow is reduced resulting in lesser dispersion than with higher diameter nozzle. From both these Figures 9 and 10 it is seen that downcomer exhibits maximum dispersion because of high velocity and jet action obtained through two fluid nozzle located at the top. However smaller coefficients were obtained for the riser where the liquid velocity and intensity of turbulence is lower compared to the downcomer. These results clearly demonstrate that turbulence induced by liquid velocity contributes significantly to the dispersion in the reactor. Similar trend was observed for higher nozzle diameters studied.

In the present work, the Bo values obtained ranged from 5-12 for the downcomer, 5-24 for the riser and 5-14 for the reactor section. In literature (12), Bo values reported ranged from 5-20 and 35-60 for the downcomer and riser respectively. But the range of Bo numbers reported by Warnecke et al. (12) were much higher than the Bo values obtained in the present work. The improvement in the mixing performance observed in the present work can be attributed to the difference in the type of flow input in these two types of reactors. Warnecke et al. (12) used a JLR with liquid introduction at the bottom whereas in the present work a reversed flow jet loop reactor was employed which provided enhanced dispersion in the downcomer and redispersion in the riser. The value of n_U obtained in this work increased with decreasing Bo numbers which agreed well with the results of Blenke (2).

Mixing in two phase flow

In two phase flow of air-water system, experiments were conducted using a D_d/D of 0.61 and d_N of 12 mm. Figure 11 shows the dependence of Bo number on superficial liquid velocity U_l for all the three sections of the reactor namely downcomer, riser and total reactor section. The Figure shows that Bo number in the downcomer and total reactor section remains relatively constant throughout the range of U_l studied indicating that increasing U_l does not influence the mixing performance of downcomer and riser. However, a sharp increase in Bo values with an increase in U_l is observed for the riser. But increase in U_g as shown in Figure 12 indicates a gradual decreasing trend in Bo values for the downcomer and total reactor section where as a sharp decrease is observed in the case of riser signifying an improvement in riser performance with increasing superficial gas velocity. The values of Bo varied from 10-14, 20-60 and 15-20 for the downcomer, riser and total reactor section respectively indicating that the performance of the downcomer and total reactor section is not influenced significantly with flow conditions whereas the flow in the riser tends to behave like plug flow with superimposed dispersion (5).

The dependence of dispersion coefficient on U_g and U_l is shown in Figure 13. From the Figure it is seen that the downcomer exhibits high dispersion compared to riser and total reactor section due to relatively high liquid velocity and turbulent intensity in the downcomer. Presence of air bubbles in the downcomer has minor influence on dispersion coefficient D_{ax}. These results indicate that turbulence induced by liquid contributes to the liquid dispersion in downcomer and total reactor section whereas turbulence promoted by gas velocity improves the mixing performance of the riser.

CONCLUSIONS

From the results obtained the following conclusions can be drawn. In the case of water system, the mixing characteristics of the downcomer is found to be superior to the riser section whereas the performance of the total reactor section is somewhat similar to the downcomer.

For air-water system the flow behaviour in the riser can be considered as plug flow with superimposed dispersion. The range of Bo numbers obtained in the present work suggest that this type of reactor can be used for different processes by selecting the variables based on the requirements of mixing.

SYMBOLS USED

Bo = Bodenstein number ($U.L/D_{ax}$) (-)

C = tracer concentration (kg m^{-3})

C_∞ = mean tracer concentration for total tracer mass distribution in reactor (kg m^{-3})

D_{ax} = axial dispersion coefficient ($m^2 s^{-1}$)

D = reactor diameter (m)

D_d = draft tube diameter (m)

D_L = draft tube length (m)

d_G = aeration tube diameter (m)

d_m = logarithmic mean diameter (m)

d_N = liquid nozzle diameter (m)

H = reactor length (m)

H_B = bottom spacing (m)

H_N = immersion height of the two fluid nozzle into draft tube (m)

L = length of circulation path (m)

n_U = circulation number (-)

Re_m = mean Reynolds number $\left(\frac{U\, d_m \rho}{\mu}\right)$ (-)

t = time (S)

t_c = circulation time (S)

U_g = superficial gas velocity (m S^{-1})

U_l = superficial liquid velocity (m S^{-1})

U = circulation velocity (m S^{-1})

Greek Letters

ρ = density of water (kg m^{-3})

μ = viscosity of water (P)

θ = time (-)

REFERENCES

1. Wachsmann, U., Rabiger, N. and Vogelpohl, A., 1985, Ger. Chem. Eng., 8, 411.

2. Blenke, H., 1979 "Advances in Biochemical Engineering", 13, 121, Springer - Verlag, W. Germany.

3. Voncken, R.M., Holmes, D.B. and Den Hartog, H.W., 1964, Chem. Eng. Sci., 19(1), 209.

4. Murakami, Y., Hirose, T., Ono, S. and Nishijima, J., 1982, Jl. of Chem. Eng. of Japan, 15(2), 121.

5. Verlaan, P., Van Eijs, A.M.M., Tramper, J., Van't, R.K. and Luyben, K.Ch.A.M., 1989, Chem. Eng. Sci., 44, 5, 1139.

6. Mathur, V.K. and Weinstein, H., 1980, Chem. Eng. Sci., 35, 1452.

7. Fields, P.R. and Slater, N.K.H., 1983, Chem. Eng. Sci., 38 (4), 647.

8. Weiland, P. and Onken, U., 1982, Ger. Chem. Eng., 4, 42.

9. Fields, P.R., Mitchell, F.R.G. and Slater, N.K.H., 1984, Chem. Eng. Commun., 25, 93.

10. Bello, R.A., Robinson, C.W. and Moo-Young, M., 1981 "Advances in Biotechnology ", Vol.1, 547, Pergamon Press, Toronto.

11. Warnecke, H.J., Pruss, J. and Langemann, H., 1985, Chem. Eng. Sci., 40 (12), 2321.

12. Warnecke, H.J., Pruss, J., Leber, C. and Langemann, H., 1985, Chem. Eng. Sci., 40(12), 2327.

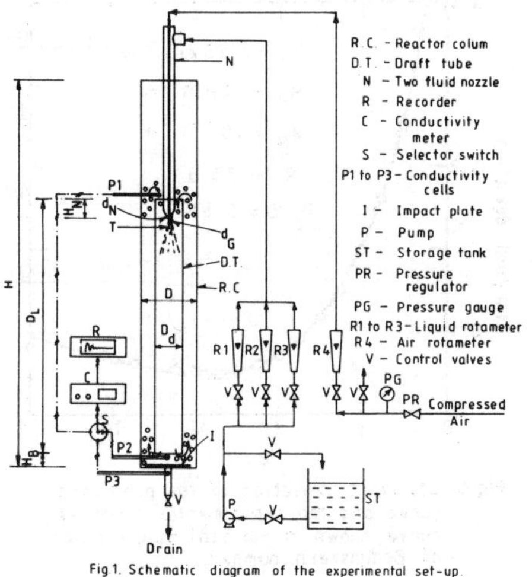

Fig 1. Schematic diagram of the experimental set-up.

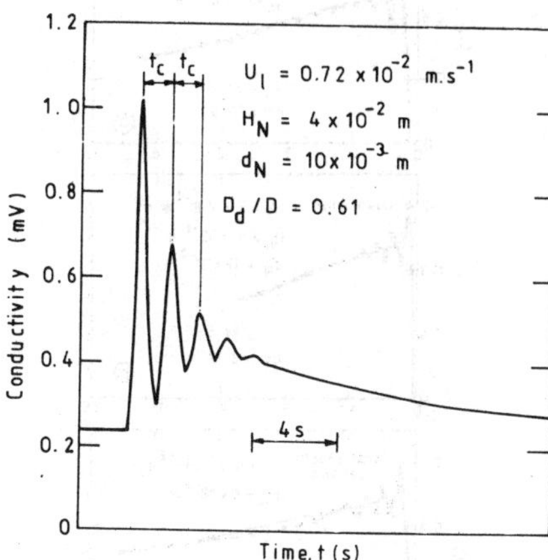

Fig. 2. Typical tracer response at the exit of the reactor downcomer subsequent to an impulse input.

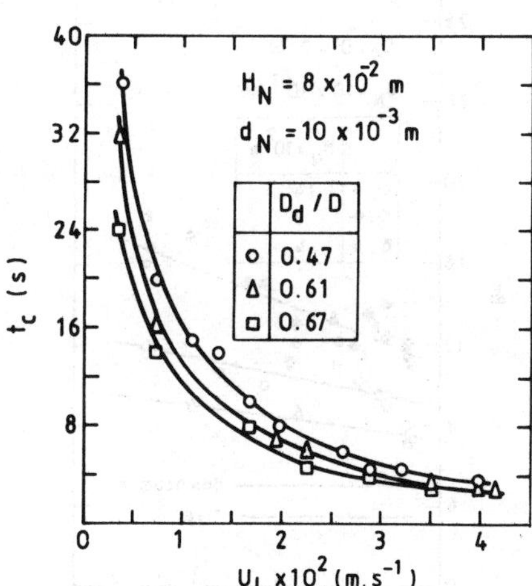

Fig.3. Dependence of circulation time on superficial liquid velocity for various draft tube diameters.

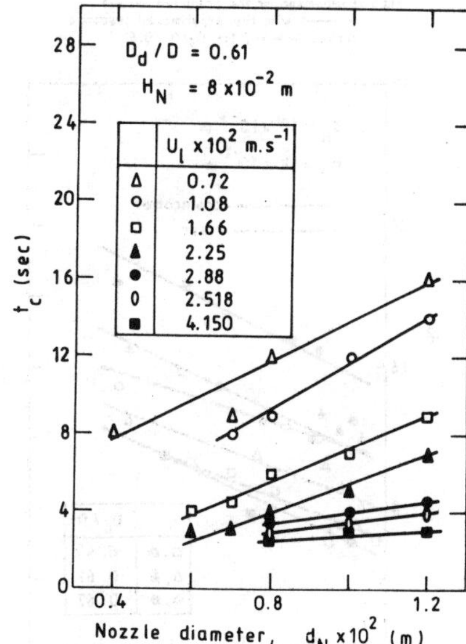

Fig. 4. Dependence of circulation time on nozzle diameter for various liquid flow rates.

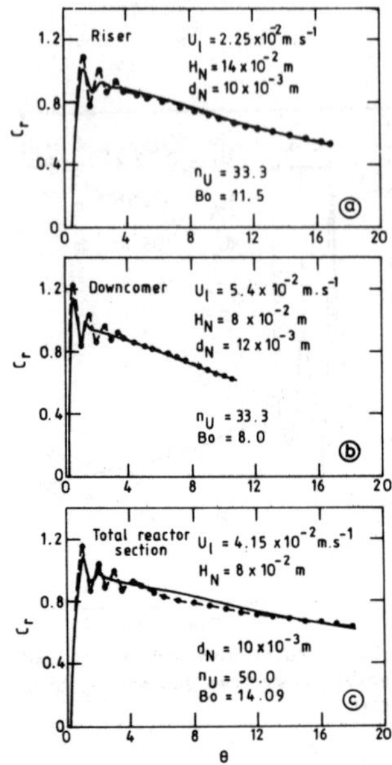

Fig.5. Comparison of the predicted curves (———) with the experimental response curves (-•-•-) for $D_d/D = 0.61$

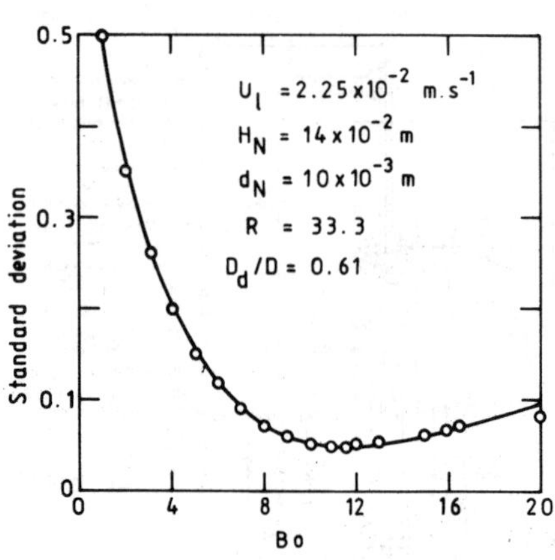

Fig.6. Standard deviation of the predicited curve and the experimental response curve shown in Fig.5(a) as a function of Bodenstein number.

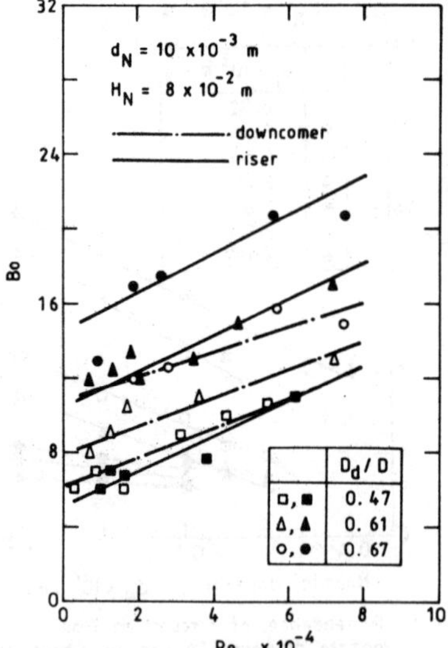

Fig.7. Dependence of the Bo number on the mean Reynolds number of the circulation Re_m for various draft tube diameters.

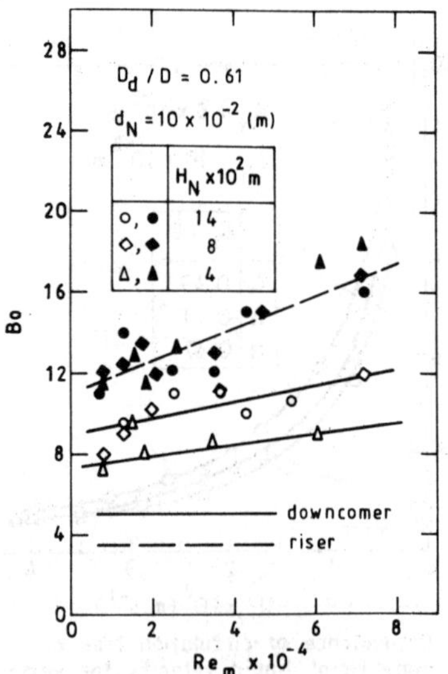

Fig.8. Dependence of the Bo number on the mean Reynolds number of the circulation Re_m for various immersion heights of nozzle into draft tube.

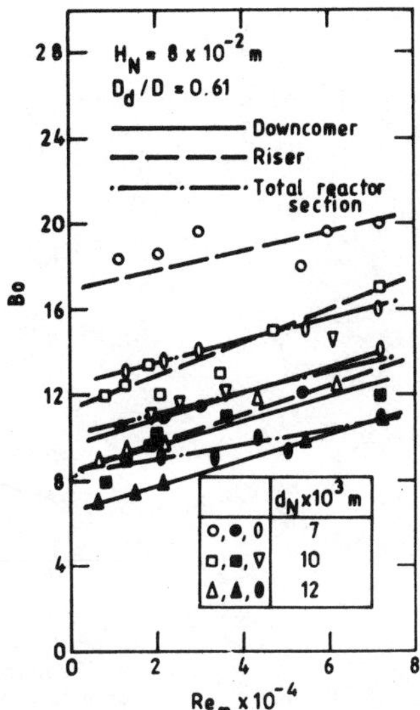

Fig.9. Dependence of the Bo number on the mean Reynolds number of the circulation Re_m for various nozzle diameter.

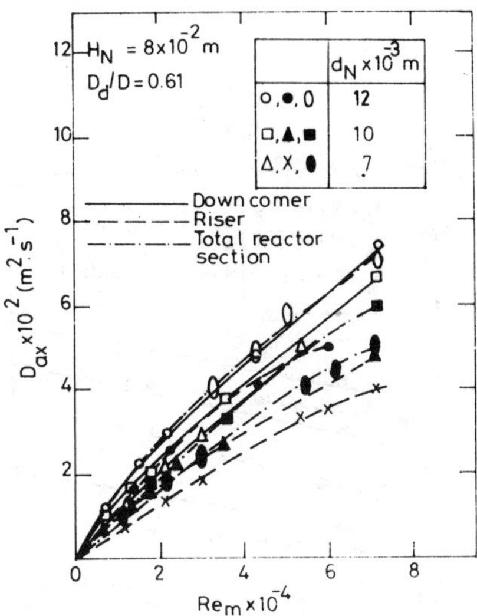

Fig.10. Dependence of axial dispersion coefficient on the mean Reynolds number of the circulation Re_m for various nozzle diameters

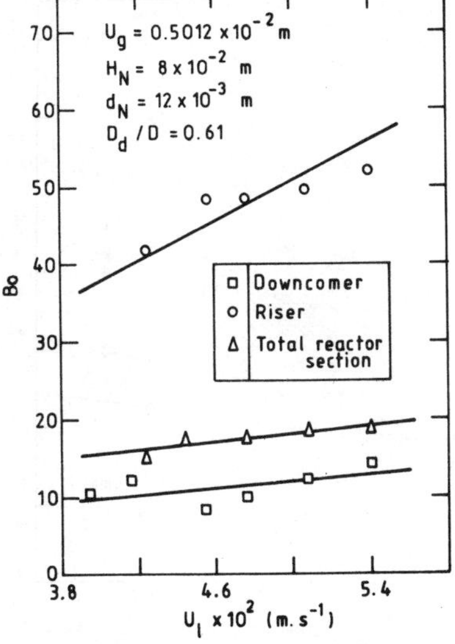

Fig.11. Dependence of Bo number on superficial liquid velocity (U_l) for a constant superficial gas velocity (U_g).

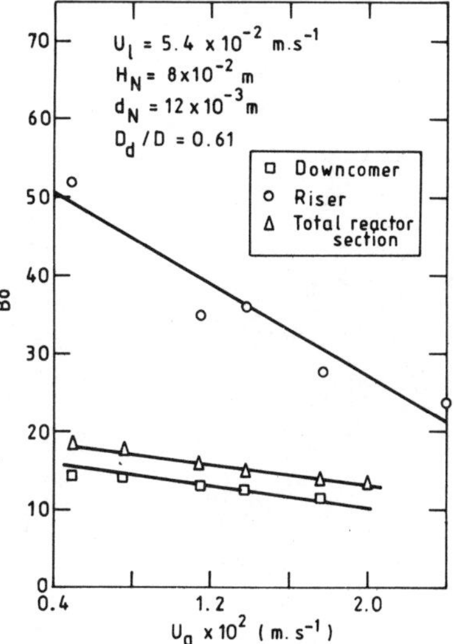

Fig.12. Dependence of Bo number on superficial gas velocity (U_g) for a constant superficial liquid velocity (U_l)

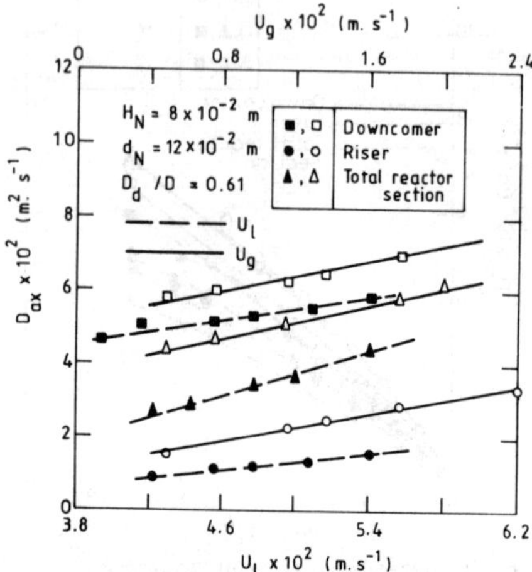

Fig.13. Dependence of axial dispersion coefficient on superficial gas and liquid velocities.

ON PARTICLE MOTIONS WITHIN ROTATING FLOW ANNULAR VESSELS

D. A. Janes* and N. H. Thomas**

This paper describes the ability to disperse slurrying particles in four configurations of rotating flow annular vessels, which can be applied widely in the process industries. In a Taylor-Couette system with a horizontal axis, a complete suspension of calcium alginate pellets in water could be achieved at lower speeds compared with when the axis was upright. At higher mass loadings there were reductions in the number of toroidal vortices, an effect which depended on the orientation of the vessel's axis. Regions of flow reversed against the rotor's direction of spin were manifested only in the horizontal case and partially filling the horizontal annulus reduced the effectiveness of the particle suspension. In an air-sparged vertical annulus the bubbles encouraged entrainment, but bubble capture by the moving rotor had a detrimental effect on the suspension.

INTRODUCTION

Probably because *prima facie* evidence (1) suggests that the Taylor-Couette class of flows in a cylindrical annulus (Fig. 1) offers an environment that is favourable for orderly dispersions, it is not surprising to find that over the past 70 years, double cylinder vessels, with the inner drum acting as a rotor, have been developed for a variety of process applications (Table 1), many involving multiphase fluids. During this period, extensive studies on hydrodynamic stability and flow regime transition (2), have furnished us with secure guidelines for single phase Taylor-Couette flows, along with information on the mass and heat transfer (3) and power consumption (4) characteristics of these flows. Of particular practical interest is the possibility of a mixing vessel with characteristics similar to plug flow when an axial flow is superposed on the Taylor-Couette flow (5). However, our knowledge of the two-phase dispersion characteristics of these vortex flows is quite limited. A few studies (1,6) have revealed: the importance of the effects of gravity on the system, and phenomena which cannot be accounted for by existing theories of homogeneous Taylor-Couette flow.

* Now at the Flour Milling and Baking Research Association, Chorleywood, Herts WD3 5SH.
** F.A.S.T. Team, School of Chemical Engineering, University of Birmingham, P.O. Box 363, Birmingham B5 2TT, with whom one should correspond.

Our curiosity in these annular reactor configurations derives from our interest in bioprocessing equipment and also how flow phenomena contribute to the accumulation of biomass and desirable products, particularly in plant cell cultures. A number of rotating annular bioreactors have been proposed for cell suspensions (7,8,9,10, 11,12). A drawback of most of these designs is that aeration of the culture by sparging adversely affects the toroidal vortices. An exception is our "bubble-free annular-vortex membrane" bioreactor (12), in which oxygen permeates through a membrane on the rotor and there are no bubbles in the working volume. Another bubble-free concentric cylinder bioreactor (13), provides for external oxygenation via an axial recirculating flow along the inner cylinder, but this is a fixed-film bioreactor.

In the modified Taylor-Couette flow within an incompletely filled vessel with horizontal axis, gases and vapours can be exchanged between the liquid film and the head space above. This mechanism is exploited in a "thin-layer rotating drum" bioreactor (10). An experimental study of this single-phase flow was reported in (14). Within the bulk liquid in such vessels, the layer adjacent to the stationary outer cylinder separates as the flow approaches the free surface and a standing eddy forms. If the free-surface is between the highest points of both cylinders, the flow is known as partially-reversed (Fig 2). If the inner cylinder interrupts the free-surface, there is a slight carry-over in the thin film travelling with the rotor, but the bulk fluid is forced into "completely-reversed" flow (Fig. 3). The stability of this reversed flow along the concave surface of the outer cylinder is known as the Dean problem (15) and, like the Taylor problem of Couette flow, it is concerned with infringement of Rayleigh's stability criterion for the gradient of angular momentum in rotating flow. The experimental finding (14) that lowering the free-surface, increases the rotation rate for the onset of Taylor-vortices has been theoretically explained (15).

This paper describes the entrainment performance of an annular apparatus with a rotatable inner cylinder, achieved under several different operating conditions at solid concentrations up to 55% by volume. The particles used were 2.5mm diameter calcium alginate pellets in water. The pellets were made from a 0.5% sodium alginate solution and possessed the same still water settling velocity (u=15.8mm s^{-1}) as the largest commonly occurring plant cell clusters, which are typically of linear size, $2 > d_p > 3$mm (12). With the kinematic viscosity of water $v = 1x10^{-6}\ m^2\ s^{-1}$, this gives a particle Reynolds number, $u.d_p/v = 40$. The specific gravity of these pellets (1.014) is also representative of many moist vegetal materials found in industries such as the food processing trade.

The speed of complete suspension was measured to evaluate the entraining capacity of these flows. This bulk process engineering index is the lowest speed at which all the surface area of the particles is available for processing. In the interests of improved basic understanding of the fluid dynamics in mixing equipment, we also report diagnostic observations of the entrainment patterns of these particles and offer practical interpretations of some key physical factors governing their motions in these ring-vortex flows, along with their relationship to processing performance.

Both vertically and horizontally mounted Taylor-Couette systems were examined, in each observing the dependence of the flow, particle patterns and the speed of complete suspension on the solid volume fraction, Ø. Raising Ø provides a first approximation to the evolving biomass in a bioreactor. The effects of varying the liquid volume in the horizontally lain partially filled vessel and the effects of aeration on the three-phase flow in an upright spinning annulus were also recorded.

EQUIPMENT, MATERIALS AND METHODS

Taylor-Couette Apparatus

The main dimensions of our Taylor-Couette rig are listed in Table 2. The annulus was constructed from 5mm thick transparent acrylic plastic tubes. Four holes (5mm bore) were drilled equidistantly around the top and bottom plates, so that they were in the centre of the annulus. An extra 21mm bore hole was drilled in the top plate over the annulus. The holes were rubber bunged. To sparge the annulus, four 21 gauge hyperdermic needles (length 40mm) were inserted through the rubber bungs in the bottom plate, to give distributed aeration in the annulus - *cf.* the rotorfermentor (8). The bungs were removed from the top plate to allow the gas to exit. A variable speed d.c. motor (86 rev. $min.^{-1}$ maximum) drove the inner rotor, via interchangeable pulleys and a drive belt. The rotational speed was measured by an optical tachometer. Air was supplied to the sparged annulus by an air pump (2L $min.^{-1}$ maximum). A water bath was placed around the cylinder to maintain the temperature of the vessel's contents at $19^{\circ}C \pm 1^{\circ}C$. The temperature was checked at the start and finish of each run, with a mercury glass thermometer, either placed inside the vertical annulus or by the method of Hayes and Hutton (16) when the axis of the annulus was horizontal.

Calcium Alginate Pellets

Calcium alginate pellets of 2.5mm ± 0.5mm were formed by dripping 0.5% sodium alginate into a stirred 0.1M calcium chloride solution. To aid visualisation, Congo red dye was added to the sodium alginate solution. Constant pellet size was maintained by filling a beaker to its brim and placing it in an overflow tray. The pellets were dripped from the height at which separate pellets were first formed in the swirling liquid. Hence with 2.5mm diameters they were smaller than the pendant drops. The calcium alginate particles were surface dried with absorbent paper. The volume of the pellets was measured by liquid displacement.

Experimental Procedures

The vessel was filled with known volumes of distilled water and alginate pellets and then mounted in the desired orientation. The speed of the rotor, N, was slowly increased to a required value - *i.e.* it was not impulsively started. The observation speed was maintained for five minutes substantially in excess of the spin-up time $\sqrt{(h^2/v.w_1)}$ - where h is the length of the annulus and w_1 is the angular velocity of the rotor (17) - which is about 100 seconds for N=100 rev. min.$^{-1}$. The speed of complete suspension, N_S, was determined visually by the method of Zweitering (18), namely the speed at which no particles are stationary on the vessel floor for longer than one second. For the tests with the upright rig, the experiment was continued until the speed of complete suspension was reached.

To examine the effects of different liquid levels in the horizontally mounted vessel, it was filled with 0.2L of calcium alginate pellets and 1.8L of water, corresponding to an initial solid volume fraction, Ø = 0.1. Then a sequence of tests were undertaken, each time draining 0.1L of water from the vessel. In each case, the flow patterns were observed while the rotor was turned at N = 100 rev. min.$^{-1}$.

The tests in the sparged vertical-axis vessel were performed with 0.2L of calcium alginate pellets and 1.7L of water, contained within the rig fitted with four spargers. Three rotor speeds were used; 0, 100 and 200 rev. min.$^{-1}$, and two air flow rates, 0.2L min.$^{-1}$ and 1.0L min.$^{-1}$.

RESULTS IN THE HORIZONTAL-AXIS VESSEL

Fully-filled Annulus

The main regime patterns of this slurry-flow are shown in Figure 4 and the speeds of complete suspension are plotted against the volume fraction of pellets in the Figure 5. This criterion, although subjective, was reproducible to within 5%. Even at low rotor speeds embryonic signs of the Taylor vortex flow pattern were observed when the pellets migrated along the bed. The speed at which particles were entrained upwards decreased with rising Ø, in part because the top of the pellet bed approached the bottom of the rotor.

For low solid volume fractions, 0.1≤Ø≤0.3, and intermediate values, 0.35≤Ø≤0.4, the pellets were first entrained into the azimuthal flow near the inner cylinder. With increasing rotor speed and hence particle entrainment around the annulus, a ring pattern of Taylor vortices became apparent. This pattern always consisted of 6 vortex pairs or Taylor-cells, as expected (2) in a vessel of this aspect ratio, h/d = 12.7 - where d is the gap width. At a speed, which depended on Ø, "ditches and ramps" appeared in the sedimentary bed of pellets (Fig. 6). The ditches were under the vortices and the ramps were below the vortex inflows. The pellets were entrained into the rapid outflowing jets at the centres of Taylor vortex pairs and then transported around the annulus by the Couette component of the flow. Deposition occurred preferentially beneath the weaker inflow zones between adjacent vortex pairs. On increasing the rotor speed, the bed pattern eventually mutated from 6 ditches and 7 ramps (*i.e.* 6 vortex pairs) to 5 ditches and 6 ramps (*i.e.* 5 vortex pairs). For rotor speeds between these two states the bed pattern was unsteady, fluctuating between up to 7 ditches with 8 ramps and as few as 4 ditches and 5 ramps. The speed at which the transition was complete, increased with Ø, from N = 39 rev. min.$^{-1}$ for Ø = 0.1 (Fig. 6), to N = 70 rev. min.$^{-1}$ for Ø = 0.3. The transition always occurred below the speed of complete suspension. Above N_S, the flow was more homogeneous and the Taylor vortex patterns were less marked.

At rotor speeds slightly above the onset of entrainment, the systems with higher particle concentrations (≥0.35) were marked by the appearance of a reversed-flow, visualised as pellets not far from the inner cylinder travelling in the opposite direction to the rotor and those pellets carried around with it. For 0.35≤Ø≤0.5, this standing eddy appeared at speeds from N = 10 rev. min.$^{-1}$ down to N = 2 rev. min.$^{-1}$ (see the bottom right of Figure 4). With increasing rotor speed, this eddy moved out to the external wall of the annulus. Then with further increases in speed the reversed-flow was separated axially by the encroaching Taylor vortices, so the regions of annulated reversed-flow were eventually above the ramps. Figure 7 shows how along the inside of the outer cylinder, there were pellets travelling countercurrently in the regions between Taylor-cells. With Ø=0.3, back-flow amongst the pellets on the outer edge of the annulus was observed at N=45 rev. min.$^{-1}$. More detailed examination might well confirm that the regime transition to zone 5 in Figure 4 is continuous, as indicated by the dashed line on Figure 4. For Ø ≤ 0.45, the standing eddy disappeared as the rotor speed was increased towards N_S. However, for Ø = 0.5, it persisted to N = 150 rev. min.-1, which is 30% above N_S. This persistence of the eddy is an indication of the vertical gradient in pellet concentration.

With increasing solid volume fraction, the Taylor vortex pattern first changed from 6 to 5 vortex pairs, but only at Ø = 0.5, did the flow mutate from 5 Taylor-cells to 4 and then 3 vortex pairs. After this the central Taylor-cell shrank in length compared to the two Ekman-driven-cells (Table 3) and the standing eddy eventually disappeared. An imbalance in the end conditions probably accounts for the loss of axial symmetry for N > 105 rev. min.$^{-1}$ (Table 3). Physically, we can think of these changes in Taylor-cell number and Taylor-cell length as being associated with the potential energy, which goes to suspend the particles, resulting in decreased kinetic energy available to drive the vortex motions. Another factor is the increasingly non-Newtonian rheology of these more concentrated slurries. At rotor speeds when the Taylor-cell number of Newtonian flows was invariant, reduced numbers of Taylor-cells have been observed in single-phase non-Newtonian Taylor-Couette flows (*e.g.* with viscoelastic polyacrylamide solutions (19) and with pseudoplastic sodium carboxymethylcellulose solutions (4)).

At the highest solid volume fraction tested (Ø=0.55), the rotor was submerged in the particle bed. A speed of complete suspension was not identifiable under these conditions and there was no evidence of Taylor vortices. Instead we observed a narrow layer of circular Couette-like flow around the inner cylinder (Fig. 8). The thickness of this flowing layer increased with rotation rate. Increasing the rotor speed from zero, we observed the following sequence of events in this choked flow regime (Fig. 4). At 2 rev. min.$^{-1}$ some pellets were entrained into the standing eddy illustrated in Figure 8. At 50 rev. min.$^{-1}$, the entrainment in this eddy filled the annular gap and on the opposite side of the annulus where the rotor was moving upwards there was a particle-free zone of water above stationary pellets. The eddy grew with increasing rotor speed, but never completely encroached into the particle free zone. Even at speeds as high as 100 rev. min.$^{-1}$, there was still a considerable depth of packed bed at the bottom of the annulus. With decreasing values of Ø from this limiting value of 0.55, Figure 5 shows how N_s falls rapidly to become invariant for $Ø < 0.25$.

Partially Filled Annulus

At N=100 rev. min.$^{-1}$, in the fully-filled annulus there were 5 vortex pairs. When the first 0.1L was drained and the flow became partially-reversed (Fig. 2), there were 6 vortex pairs. Six vortex pairs is the number expected at onset of Taylor vortex flow in a fully-filled annulus of this aspect ratio. This number persisted for all partially-filled volumes as low as 1.6L. However the vortex structure displayed significant adjustment with decreasing liquid volume, such that the spacing between the bands of pellets entrained in Taylor-cells and the pellet deposition in these spaces both increased markedly. This decreasing capacity to suspend the pellets is caused by energy being diverted into the standing eddy, leaving less available to sustain the Taylor vortices: effectively the confinement damps the vortices. Because the standing eddy has a lower energy density, it is unable to maintain suspension of particles received from the Taylor vortices and the deposition increases. At 1.7L (*i.e.* Ø=0.12) the pellets were not completely suspended at N=100 rev. min.$^{-1}$ and the resulting "ramps and ditches" of deposited pellets were continually shifting. In comparison, in the fully filled annulus the speed of complete suspension is quite constant at 65 rev. min.$^{-1}$ $0.1 \leq Ø \leq 0.2$ (Fig. 5). The intervortex exchange of particles, first visible with 1.8L of fluid, also increased. This axial transport is probably due to the weakness of the standing eddy relative to the Ekman-pumping from the ends of the vessel (20).

When the fluid volume was reduced to 1.5L particle entrainment in the vortices was no longer visible; instead the pellets were directly entrained into the standing eddy. Although larger than the depth of liquid film on the rotor, the particles were still carried around in the film flows. For partially-filled volumes less than 1.5L. the liquid level was below the top of the rotor and the flow was completely-reversed (Fig. 3). Under these conditions, entrainment decreased and deposition increased dramatically, consistent with expectations since less kinetic energy is available when the contact area with the rotor is reduced. In these tests in the partially-filled vessel where N = 100 rev. min.$^{-1}$, particle dispersion at 1.0L (*i.e.* Ø=0.2) was comparable with that in the fully-filled vessel (same Ø) at a rotor speed of 13 rev. min.$^{-1}$ (*i.e.* just above the entrainment inception speed). At 0.6L the pellet bed was distributed only on the high pressure side of the rotor and there was negligible entrainment into the liquid bulk.

RESULTS IN THE VERTICAL-AXIS VESSEL

Vortex Suspension of Particles

Contrasting with the results from the horizontal-axis vessel, Figure 5 shows that in the vertical-axis vessel, the speed of complete suspension depends almost linearly on the solid volume fraction, herewith $N_s \propto 0.13Ø$. These values of N_s are always greater than those obtained in the horizontal-axis vessel because only the Taylor vortices contribute to particle suspension and their kinetic energy is very much less than that of the Couette-component flow(21). The difference between the two values of N_s, increases with Ø (Fig. 5).

The same equilibrium state was achieved after 60 seconds, when 0.2L of pellets were poured into 1.8L of water already in motion at a rotor speed of 100 rev. min.$^{-1}$, as when the system was impulsively started with a packed bed of pellets. When the rotor speed was increased gradually from zero, the pellets immediately adjacent to it were forced into motion within an otherwise stationary bed, rather like the pattern illustrated in Figure 8. This inner layer expanded in thickness with increasing rotor speed and some pellets from the top of the bed were entrained into a toroidal vortex above it. With further increase in rotor speed, the particle entrainment evolved from this single vortex into the bottom vortex pair, and subsequently the pellets were worked upwards through the column of vortices.

Because the height of the undisturbed packed bed increases proportionately with Ø, it is not surprising that the rotor speed for which the pellets reach the top vortex pair decreases in relation to N_s, as Ø is increased. For example with Ø = 0.15 and Ø = 0.25, the pellets were first entrained within the highest vortex pair roughly by 100 rev. min.$^{-1}$, which is almost N_s for Ø = 0.15, but is 0.6N_s for Ø = 0.25. Even at 200 rev. min.$^{-1}$ for Ø = 0.15, there was still a substantial axial gradient in the pellet concentration. This gradient at speeds above N_s is most accentuated for low Ø values, not just because N_s increases with Ø, but because the potential energy required is larger at low Ø.

Associated with the entrainment of particles was an enlargement of the lowest vortex with a corresponding shrinkage of the vortex above it. At Ø ≤ 0.20 the second vortex eventually grew back to the same size as the other bulk vortices. For solid volume fractions Ø ≥ 0.25, the enlargement was substantial and the number of vortex pairs was reduced from six to five for rotor speeds in excess of N_s. With Ø > 0.3, there were only four Taylor-cells for rotor speeds just less than N_s and the bottom vortex extended up one third of the vessel height. With further increase of N, the expected 12 vortices for h/d=12.7 would eventually develop. However, for Ø > 0.4 only 11 vortices would develop for N < 300 rev. min.$^{-1}$. The existence and stability of such "anomalous modes", with an odd number of vortices, have been studied in single phase flow with h/d as a control parameter (2).

The phenomena reported here are in striking contrast with the observations in the horizontal-axis vessel. Clearly the changes from the "primary mode" (2) pattern (6 vortex pairs) depend on the orientation of the system and the mutations seem to be promoted by higher concentrations of pellets. In the horizontal-axis vessel the pellets are comparatively evenly distributed amongst the bulk vortices, so that changes in the flow patterns are largely axially symmetric. However in the vertical-axis vessel, the vertical density gradient owing to the pellet distribution means that, it is changes occurring in the lower Ekman-connected vortex, which also control the number and form of vortices in the upright annulus.

Sparged Bubbly Flows

When the vessel was sparged with bubbles and the rotor was stationary, the pellets were transported as in a bubble column: *i.e.* entrained upwards by the bubble plumes above the four spargers and descending chaotically between the gas inlets.

With the low gas flow rate of 0.2L $min.^{-1}$, at rotor speeds up to 200 rev. $min.^{-1}$, a weak Taylor vortex pattern was observed. However, we also noticed that pellets were entrained upwards only within a thin layer near the inner cylinder and the bulk of the pellets were hence concentrated at the bottom of the annulus. This is because the bubbles are attracted to the rotor, which is largely - but not solely (22) - due the radial pressure gradient associated with the Couette-component flow. Therefore at the rotor the pellets are carried up with the bubble spirals (Fig. 9) towards the top of the annulus and then descend through the bubble-free fluid at the outside of the annulus, passing through the weak Taylor vortices.

At the higher gas flow rate of 1.0L $min.^{-1}$, and with N=100 rev. $min.^{-1}$ there was no visible evidence of structured Taylor vortices. The descending pellets in the outer flows were transported in an undulating helical paths, and the flow might be described as bubble-column-Couette flow. With the same air flow rate but a faster rotor (N = 200 rev. $min.^{-1}$), the flow exhibited weak Taylor vortices distorted by the waves that were excited by the rising bubble streams. The pellets were much more uniformly distributed than at the lower gassing rate.

So far as practical applications are concerned this upright sparged configuration is capable of adequate dispersion of the particles. Even though the rotor speed tends to disrupt the suspension, the rising bubbles provide an excellent entrainment capacity. On the other hand, the capture and accumulation of bubbles on the rotor surface reduces the torque, which is transmitted to the fluid and hence also the energy available to drive the Taylor vortex structure, so the potential advantages of plug-flow processing must be sacrificed.

CONCLUSIONS

With regard to suspending a settling slurry in Taylor-Couette flow, the system with horizontal-axis is substantially more energy efficient than the upright configuration. In an annulus of radius ratio 0.495 and with solid volume fractions of calcium alginate pellets up to 50%, it has been demonstrated that a state of complete suspension can be achieved at low rotor speeds (<115 rev. $min.^{-1}$) in the unsparged vessel, when its axis is horizontally mounted. When the axis is vertical much faster rotor speeds are required for a complete suspension, owing to the directional distribution of kinetic energy dissipating from the rotating cylinder into the Taylor vortex flow. At high slurry mass loadings there were reductions in the number of toroidal vortices and, because it seems that the concentration of foreign particles affects the length and number of vortices, these effects are dependent on the flow's orientation with respect to gravity. An unattractive feature in the horizontal-axis vessel is the counter-current motion of particles at high solid volume fractions.

For a constant rotor speed, lowering the liquid level in the partially-filled horizontal-axis vessel, causes a "kinematic squashing" effect. This reduces the flows capacity to suspend the pellets and makes the Taylor vortex motions increasingly sluggish. To maximise the ratio of gas-liquid interfacial area to liquid volume in contactor applications of the partially-filled rotating flow annular vessel such as drum-dryers (23) and thin-layer bioreactors (10), one reduces the working volume of liquid. In our vessel, the cost of doing this was a lower degree of fluid agitation, which resulted in unsatisfactory entrainment of the calcium alginate pellets at rotational speeds where the fully-filled vessel would be maintaining a complete suspension.

Since the presence of the interfaces between vortex pairs substantially restricts axial mixing to within Taylor-cells, axial mixing on a scale greater than the annular gap increases with decreasing liquid-solid volume. Because of this and the greater segregation perpendicular to the axial flow, if a low net axial flow were to be superposed at a rotor speed sufficiently high for Taylor vortex formation, the mixing characteristics ought to be less like plug-flow in the completely-reversed flow system, than those in the fully-filled annulus or even the partially-reversed flow system.

With the vertical-axis vessel, we found that whereas bubbles certainly enhance the suspension of particles, their capture and accumulation over the rotor, by impairing the torque transmitted to the fluid, damped the quality of the dissipative pumping motions spanning the annular gap.

This work should be of practical significance to the further development of rotating flow annular equipment, not least because of the fundamental insight on the gravitational factors governing particle motions into and within vortex flows. To build on this we need detailed quantitative experiments, backed by theoretical modelling, of the configurations closest to current industrial exploitations of rotating annular flows, like machines for making carbonless copy paper (24), and to those offering most promise for new applications, such as eukaryote cell suspension bioreactors (12).

ACKNOWLEDGEMENTS

While carrying out this research at Birmingham University, D.A.J. was supported by a S.E.R.C. Studentship award from the Biotechnology Directorate. Gratitude is also due to F.R.E.D. Ltd (R&D Institute, Birmingham University).

REFERENCES

1 Thomas, N.H., Janes, D.A. (1987) *Ann. N. Y. Acad. Sci.* 506, :171
2 Benjamin, T.B., Mullin, T. (1981) *Proc. Roy. Soc. A* 377, :221
3 Kataoka K., Doi, H., Komai, T. (1977) *Int. J. Heat Mass Transf.* 20, :57
4 Sinevic V., Kuboi, R., Nienow, A.W. (1986) *Chem. Eng. Sci.* 41,

:2915
5 Kataoka, K., Takigawa, T. (1981) *AIChE J.* 27, :504
6 Dominguez-Lerma, M.A. Ahlers, G., Cannell, D.S. (1985) *Phys. Fluids* 28, :1204
7 Himmelfarb, P., Thayer, P.S., Martin, H.E. (1969) *Science* 164, :555
8 Margaritis, A., Wilke, C.R. (1978) *Biotechnol. Bioeng.* 20, :709
9 Beck, C., Stiefel, H., Stinnett, T. (1987) *Chem. Eng.* 94, :121
10 Moser A., (1982) *Biotechnol. Lett.* 4, :281
11 Krovak, P., Salvet, M., Sikyta, B. (1984) *Biotechnol. Lett.* 6, :307
12 Janes D.A., Thomas, N.H., Callow, J.A., (1987) *Biotechnol. Tech.* 1, :257
13 Kornegay, B.H., Andrews, J.F. (1968) *J. Water Pollut. Cont. Fed. (Res. Suppl.)* 40, :460
14 Brewster, D.B., Nissan, A.H., (1958) *Chem. Eng. Sci.* 7, :215
15 DiPrima (1959) *J. Fluid Mech.* 9, :621
16 Hayes, J.W., Hutton, J.E., 1972 *Prog. Heat Mass Transf.* 5, :195
17 Snyder, H.A., (1969) *J. Fluid Mech.* 35, :273
18 Zweitering, T.N. (1958) *Chem. Eng. Sci.* 8, :244
19 Beavers, G.S., Joseph, D.D., (1974) *Phys. Fluids* 17, :650
20 Snyder, H.A., (1968) *Phys. Fluids* 11, :728
21 Marcus, P.S. (1984) *J. Fluid Mech.* 146, :65
22 Sene, K.J., Hunt, J..L.R., Thomas, N.H. (1990) *J. Fluid Mech.* (to appear)
23 Kozempel, M.F., Sullivan, J.F., Craig Jr, J.C., Heiland, W.K., (1986) *Lebensm.-u.-Technol.*, 19, :193
24 Nachtergaele, W., Van Nuffel, J. (1989) *Starch/Stärke* 41, :386
25 Janes, D.A. (1988) *Ph.D. Thesis*, University of Birmingham

TABLES

Table 1. Process applications of rotating flow annular devices.

Device	Literature source
Plug-flow reactors	
a) Catalytic	Yacoub & Maron (1984)
b) Polymerisation	(4)
Liquid-liquid extraction column	Thornton & Pratt (1958)
Fast breeder reactor	Bernstein et al. (1973)
Electrochemical reactor	Gabe (1974)
Paper-making machines	(14)
Drum-dryers	(23)
Distillation column	Willingham et al. (1947)
Continuous filter	Tobler (1982)
Bioreactors	
a) Suspension	(12)
b) Biofilm	(13)
Journal bearings	Yamada (1962)
Rotating electrical machinery	Kaye & Elgar (1958)
Artificial lung	Macleod & Reuss (1975)

n.b. Full citations are given in (25)

Table 2. Dimensions of the double cylinder rig.

Inner Cylinder;	Diameter	50mm
	Length	325mm
Outer Cylinder;	Diameter	101mm
	Length	331mm

Table 3. Variations in Taylor-cell (vortex pair) length in the horizontal mode at speeds above 105 rev. min.$^{-1}$.

angular velocity (s^{-1})	left-end cell (mm)	central cell (mm)	right-end cell (mm)
11.0	110	110	110
11.5	120	100	110
12.6	130	90	110
14.7	135	85	110
16.8	135	85	110

FIGURES

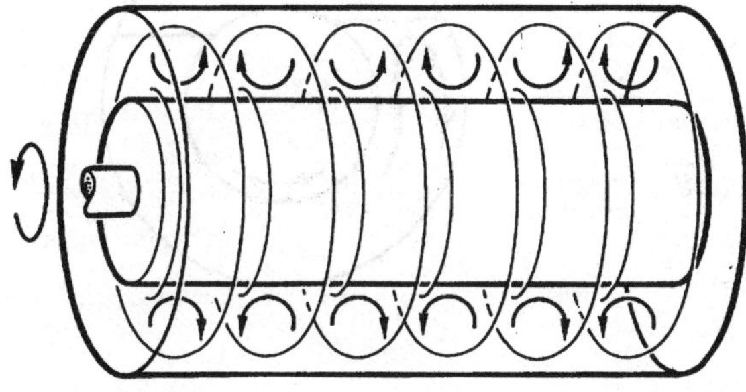

Fig. 1. A schematic view of the Taylor vortex mode of Taylor-Couette flow.

Fig. 2 Partially-reversed flow in the horizontal-axis vessel.

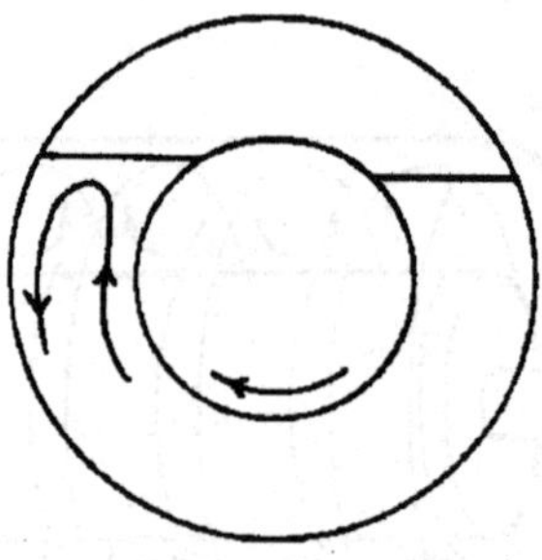

Fig. 3 Completely-reversed flow in the horizontal-axis vessel.

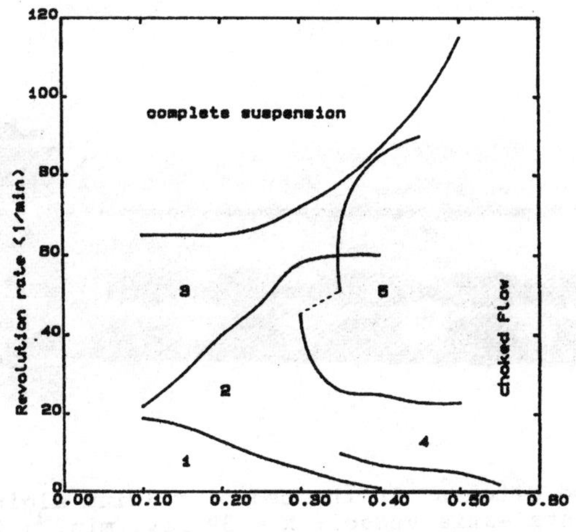

Fig. 4 A schematic of the slurry (suspension) regimes for calcium alginate pellets in the horizontal-axis vessel. 1 = stagnant bed; 2 = particle entrainment; 3 = ramping of the bed; 4 = propagating reverse flow; 5 = reverse flow at outer wall.

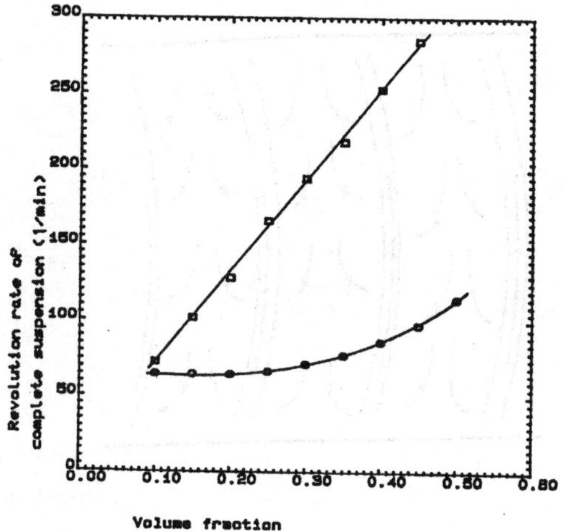

Fig. 5 The effect of vessel orientation on the speed of complete suspension for different solid volume fractions in the Taylor-Couette apparatus. o = horizontal axis; □ = vertical axis.

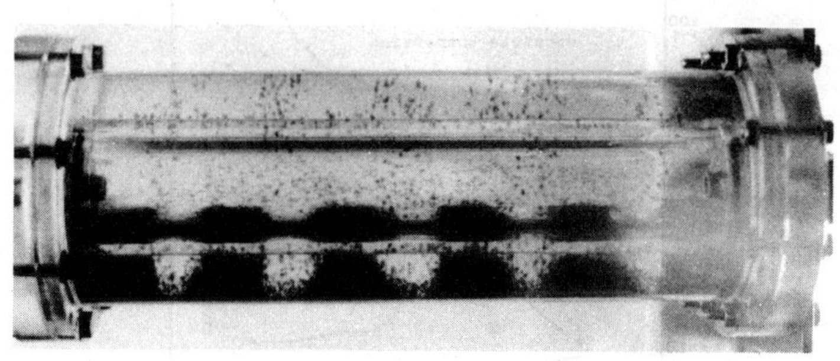

Fig. 6 "Ramps and ditches" in the bed of calcium alginate pellets in the horizontal-axis vessel: N = 39 rev. min.$^{-1}$; Ø = 0.1.

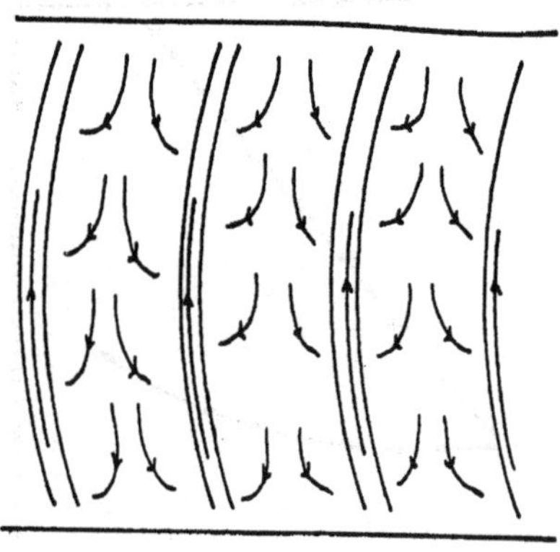

Fig. 7 A schematic of the regions of reverse-flow separating toroidal vortices in the horizontal-axis vessel.

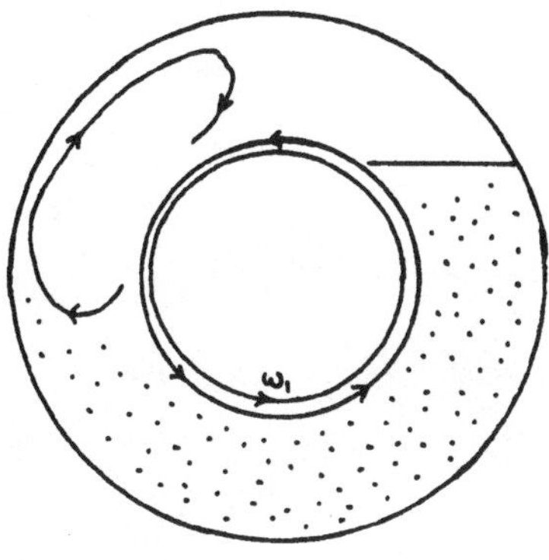

Fig. 8 A schematic of the flow at Ø = 0.55 and N = 65 rev. min.$^{-1}$ in the horizontal axis vessel. Note the standing eddy.

Fig. 9 Bubble spirals in the vertical-axis vessel: N = 200 rev. min.$^{-1}$; gas flow rate 1L min.$^{-1}$.

LATERAL MIGRATION: FOR THE SEPARATION OF DILUTE TWO-PHASE SUSPENSIONS.

G.J. PURDOM † and S.M. RICHARDSON ‡

This report describes the preliminary results of an investigation of the behaviour of drops and particles between two plates. The results of measurement of cross streamline movement, "lateral migration", of neutrally buoyant rigid spheres are presented. Mention is made of how lateral migration could be used to develop this fluid phenomenon for the continuous separation of dilute two-phase suspensions and the technique is shown to have particular merit where the dispersed phase has a low settling velocity or where either phase is shear sensitive.

INTRODUCTION

When bubbles, drops and particles are contained in bounded laminar shear flow, they migrate across stream-lines to an equilibrium position relative to the wall. This peculiar behaviour is known as "lateral migration" and is illustrated by figure 1. Previously this fluid driven artifact has received attention predominantly due to its effect on the viscosity measurement of suspensions (1,2). The purpose of this study is to introduce the use of lateral migration as an engineering tool for the continuous separation of dilute two-phase suspensions. The separator has been conceived as two plates held apart by flat gaskets to form a large aspect ratio rectangular duct. The suspension enters through one end and rapidly acheives fully developed 2-D Poiseuille flow. The initially randomly distributed dispersed phase laterally migrates to an equilibrium streamline. This equilibrium streamline is stripped from the bulk flow by a splitter at the exit resulting in a net concentration of the dispersed phase in the split stream. Recycling of either stream or cascading a number of such separators offers a potential means for further increasing the resolution of the phase separation.

† Harwell Laboratory, AEA Technology, Oxon, OX11 ORA.

‡ Dept. Chem. Eng., Imperial College, London, SW7 2AZ

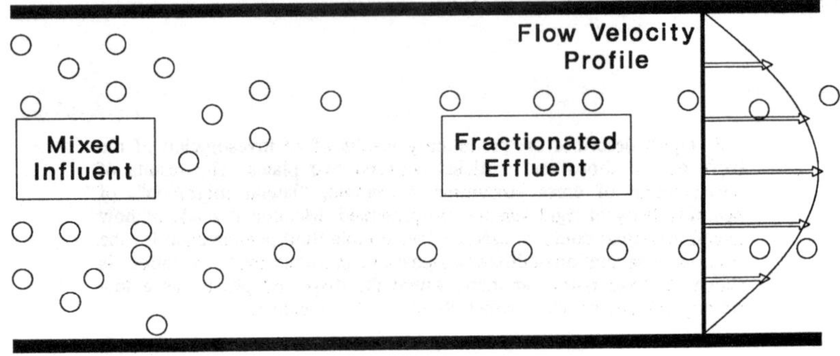

Figure 1. Lateral migration between two plates

The separation of two phase suspensions is often fulfilled by centrifuges or hydrocyclones. However lateral migration has additional advantages over these technologies when the dispersed phase has a very low settling velocity or the product is shear dependent, for the underlying mechanism of lateral migration is not driven by buoyancy forces and since the separation is achieved in laminar flow the shear stress is both small and easily calculated. The design is simple in the extreme and if a number of channels are assembled in a rack in a manner similar to a plate heat exchanger, as shown in figure 2, then the lateral migration separator will both have a small foot print and will be able to be easily disassembled. Further the lack of internal components facilitates on-line cleaning and sterilisation. As a consequence of all these, both capital and running costs are likely to be lower.

This paper reports results of the first part of this programme to quantify the major parameters which affect the rate of lateral migration. The results of lateral migration are easily observed. The migration is less easily predicted from first principles. For this reason only single rigid spheres are considered here.

Droplets with a diameter $<250\,\mu m$ are not susceptible to shear deformation in laminar flow and if derived from an industrial stream tend to accumulate contaminants at the fluid/fluid interface. As a result small droplets will act as rigid spheres. Elastic solids or droplets with greater dimensionless Eotvos and particle Reynolds numbers, as defined

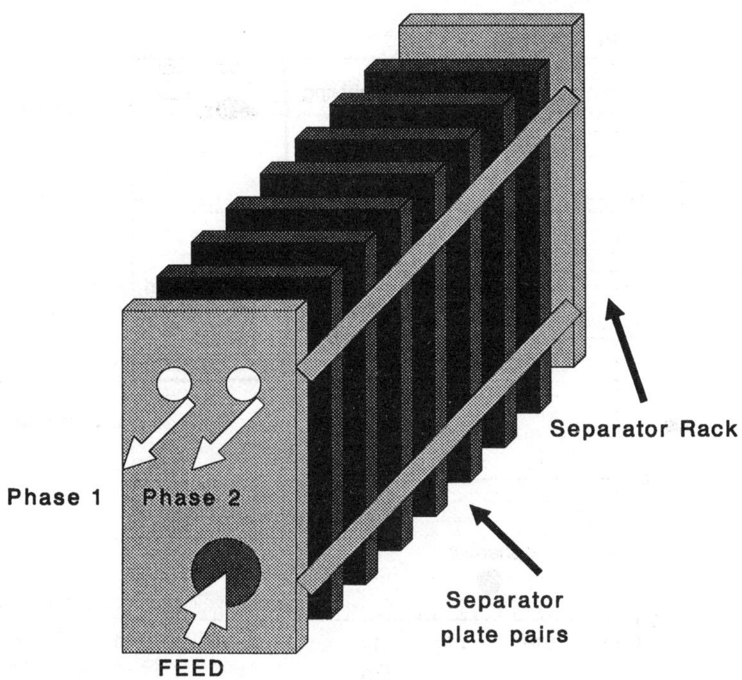

Figure 2. Lateral migration separator rack

below, are more likely to deform to non-spherical volumes according to the approximate operating envelopes shown in figure 3. This shape of droplet is not considered here.

$$\text{Eotvos number, } Eo = \frac{g\,|\rho_p - \rho|\,a^2}{\sigma} \tag{1}$$

$$\text{Particle Reynolds number, } Re_p = \frac{\rho v a}{\mu} \tag{2}$$

where $\rho, \mu, g, \sigma, v, a$ are the density, viscosity, acceleration due to gravity, surface tension, relative fluid/particle velocity and radius. (A variable denoted by a subscript "p" refers to the particle).

However experimental evidence exists (3,4) to suggest that non-spherical particles and especially droplets migrate across streamlines even more rapidly than rigid spheres of an equivalent hydraulic diameter; thus by restricting consideration to rigid spheres this investigation gives a worst case analysis.

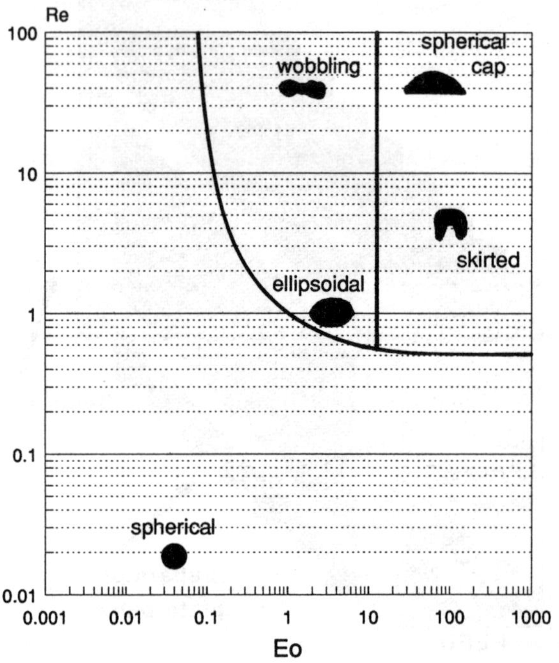

Figure 3. Droplet deformation.

THEORY

Bretherton (5) showed, by an argument of reversibility, that any body of revolution in any Stokes shear flow can not have a velocity component perpendicular to the walls. By implication lateral migration is an inertia driven mechanism. The repercussion is that to study this fluid mechanism the full steady state Navier Stokes momentum equations must be retained if the constitutive equations are to correctly describe the observed behaviour. Because of the geometry (see Fig. 4), boundary conditions and the non-linear nature of the Navier Stokes equations the problem is to a large extent analytically intractable. However considering the case when $Re_p \rightarrow 0$, $a/d \rightarrow 0$, where d is the channel height, and $Re_p \ll (a/d)^2$, Cox and Brenner (6) expressed the particle relative velocity in an asymptotic power series in these components which effectively linearised the problem. They developed expressions for the z-component of the relative particle velocity, v_z, ie the lateral migration velocity, for:

$$\mathbf{v} = \mathbf{v}(v_x, 0, v_z) \tag{3}$$

where for non-neutrally buoyant spheres in 2-D Poiseuille flow it has been shown that:

$$v_x = \tfrac{4}{3} V_{\max} (a/d)^2 - v_\infty + O(a/d)^3 \tag{4}$$

where V_{max}, v_∞ are the centre-line undisturbed fluid velocity and particle settling velocity.

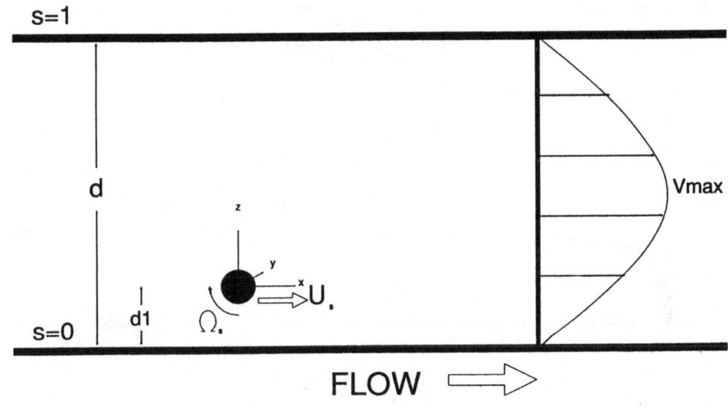

Figure 4. Channel and particle geometry

Ho and Leal (7) evaluated the complex integrals given by Cox and Brenner (6) and later, Vasseur and Cox (8) further improved upon Ho and Leals' solution especially when the particle is "close" to the wall. For neutrally buoyant spheres Ho and Leal found that the channel length required to achieve equilibrium (see Fig. 5) is approximated by:

$$\text{CHANNEL LENGTH} \approx \frac{1.2\, d^4}{Re_T\, a^3} \tag{5}$$

where Tube Reynolds number, $Re_T = \dfrac{\rho V_{\max} d}{\mu}$ (6)

and for neutrally buoyant spheres in solution

$$Re_T = Re_p\, (d/a)^2 \ll 1 \tag{7}$$

Since most applications of industrial interest require $Re_T > 1$ the above theory is not widely applicable; however, comparison to some other workers (9,10) results indicate good agreement outside these mathematically imposed limits.

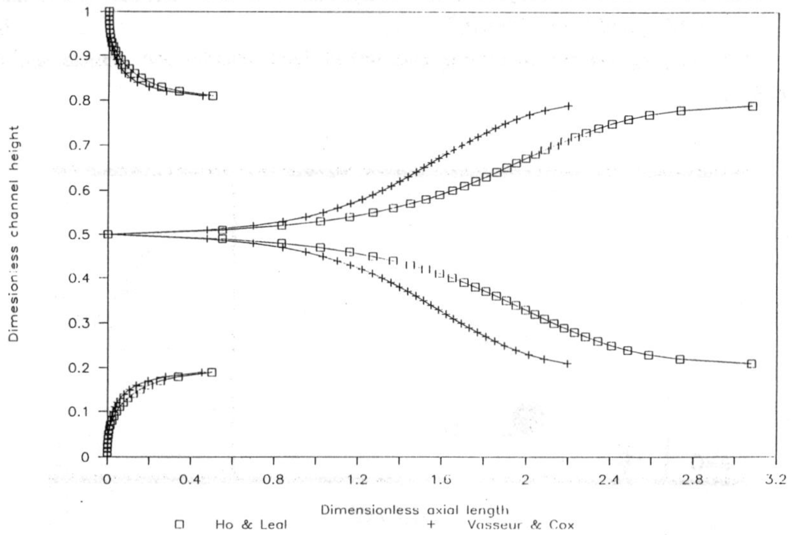

Figure 5. Predicted trajectories of spheres released near centre-line and walls

Numerical methods are not necessarily limited by the same restrictions as the above semi-analytical approach. However to model the flow relative to the moving particle contained in 2-D Poiseuille flow requires a three dimensional grid since the undisturbed fluid flow is asymmetric relative to the centre of the particle. A finite element grid was mapped to the boundaries geometry, including the sphere's surface. Fortunately a plane of symmetry through the sphere centre and vertical to the walls exists which halves the computational load. (This assumes the flow does not separate from the sphere with an asymmetric wakes in the x-z plane. In unbounded quiescent fluid this would require the $Re_p < 5$). Using this grid we hope to evaluate, first, whether the components of the shear stress tensor over the sphere's surface acting to produce a lifting force in the Oz coordinate normal to the wall will give a lateral migration force which is significant when compared to the inherent numerical and machine accuracy. The finite-element package ENTWIFE is being used for the simulation on and run a CRAY-2 super-computer. If successful the results will extend the theoretical understanding both for larger *a/d* ratios, higher Re_p and for when the particle is close to one wall.

EXPERIMENTAL APPARATUS AND METHOD

An experimental rig was designed and constructed to quantify the degree of lateral migration. The channel was scaled-up from the application to facilitate measurement of the displacement due to lateral migration. A large aspect-ratio rectangular channel of variable height, 300*mm* width and 3000*m* length made from upper and lower Delrin plates with clear perspex side-windows was mounted on a rigid frame. Various perspex side-windows from 30 – 15*mm* were used, which gave aspect ratios from 10-20 sufficient to assure 2-D Poiseuille flow at the channels centre.

Figure 6. Horizontally mounted channel with camera trolley

The entire rig supporting the channel and feedbox could be mounted in one of three planes, horizontally, on its side or vertically. This allows experimental control of the influence of buoyancy forces, for example, figure 6 shows the channel mounted hoizontally such that gravity acts in the Oz coordinate. Figure 7 shows the flow arrangement.

Two solutions were used. The first was a binary mixture of ~ 20% vol. glycerol in water. The second was a ternary mixture of ~ 80% vol. glycerol in water with potassium acetate to increase the density. The solution were both verified to be Newtonian over the shear rate

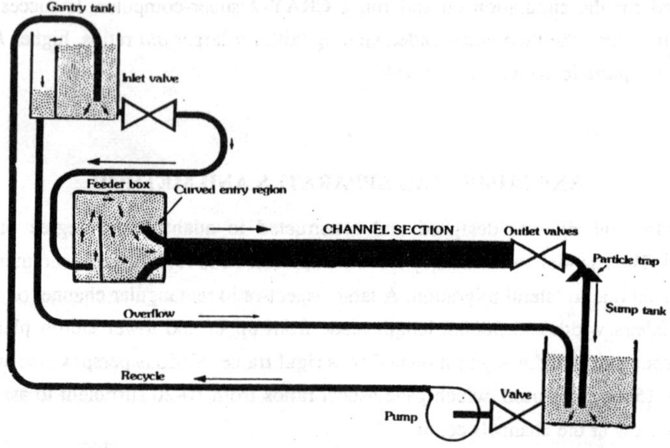

Figure 7. Flow diagram

range used and the exact density was chosen to match the density of the particles used. Two densities of plastic spheres, polystyrene (1050 $kg\,m^{-3}$) and cellulose acetate (1280 $kg\,m^{-3}$), with radii from 3.2 to 0.5*mm*, were used. All sizes and types of sphere were checked by X-ray inspection for voids, which lead to the rejection of a third plastic, nylon. The relative density of the particles was measured by recording the settling velocity in a large quiescent thermostatically controlled water jacketed tank using a CCD video camera and image analysiser. The relative density was calculated from the relationship:

$$\rho_p - \rho = \frac{9 v_\infty \mu}{2 a^2 g} \tag{8}$$

Using this arrangement the steady velocities of a batch of "identical" spheres moving relative to the fixed camera were routinely measured to have a variance in settling velocity of $\sigma_{n-1} < 0.025\ diameter\,s^{-1}$ $(n=20)$.

Each run involved the release of a single sphere from an injection tube 450*mm* downstream of the smoothed inlet into the fully developed Poiseuille flow. A further 50*mm* was allowed for the particle to become iso-kinetic to the entraining fluid both in terms of axial and rotational velocity, then measurement of the particle trajectory was commenced. (It should be noted that 50*mm* is not always a sufficient length to be able to assume that the initial measurements of the particle's trajectory correspond to this iso-kinetic state but including data from the same starting datum makes the comparision of particle trajectories considerably easier). The trajectory of the particle entrained in the flowing fluid within the channel is measured using a moving CCD camera. A low powered collimated light source (~ 4 lux) was shone through one clear perspex side-wall parallel to the upper and lower plates. The particle was imaged by the camera with a f1.4/50*mm* CCTV lens held rigidly relative to the light source on the same optical axis by incorporating into a trolley. The trolley spanned the channel as shown in figure 6. The trolley was propelled parallel to the direction of fluid flow along two linear bearing tracks using a motor and pulley arrangement. The micro-stepper motor with position encoder was controlled by a micro-computer, via an indexer.

The particle image appeared as a silhouette against a white background. Because the particle was highly contrasted it was possible to analysis the image in real-time without any requirement for image-processing. Upto eight 400×400 pixel frames could be analysised per second to determine the centre of gravity of the particle within the field of view. This information was passed to the micro-computer and used in two ways. Firstly, it was used in a self-tuning feed-back controller to control the speed of the trolley such that the particle was made to appear in the centre of the camera's field of view. Secondly, the position of the particle relative to the camera, the motor encoder position and the time were store for post-processing. Once the particle had travelled 1750*mm* and was 500*mm* from the end of the channel the measurements were stopped and the trolley returned to the starting datum position for the next particle. A typical trajectory would involve 150-400 positional measurements.

A number of these trajectories for identical particles released from different initial positions across the channel height build up a trajectory image of the expected behaviour of an initially randomly distributed suspension. Figure 8 shows two instances where lateral migration is and is not significant within the channel axial distance.

RESULTS

The results of this first part of this research programme have shown that lateral migration does indeed exist and is readily observed. Only results for the rig mounted on its side, such that gravity acted in the Oy coordinate, (see Fig. 4) are reported here. Since the buoyancy forces have no component in the x-z plane then the relative particle velocity vector is:

$$\mathbf{v} = \mathbf{v}\left(\tfrac{4}{3}V_{max}(a/d)^2, v_\infty, v_z\right) \tag{9}$$

The buoyancy force is assumed to be small and linearly independent of momentum

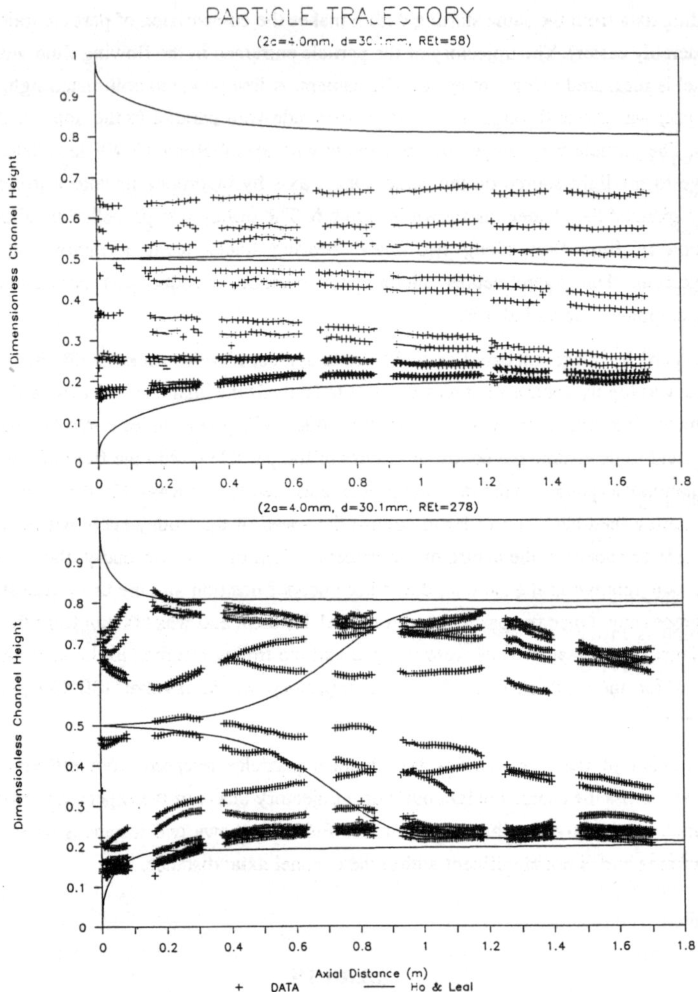

Figure 8. Trajectories of particles within the channel

terms in the x-z plane. (Faxen forces, due to settling in the Oy-direction, are assumed neglible (1)).

Thus since the even small effects of buoyancy forces are removed from consideration by the design, we are able to consider the effects of lateral migration of neutrally buoyant rigid spheres to an accuracy not previously achieved. The experimental array included 65 combinations of material/fluid density, particle size, channel height and tube Reynolds number. Each combination consisted of a replicate of at least eight trajectories. Typical results are shown in figure 8 which also include the predicted migration, according to Ho and Leal (7), of particles released both from the walls and from close to the centre-line. It is apparent that the theoretical results tend to over-estimate the effects of the migration forces for large neutrally buoyant spheres at elevated tube Reynolds number. As a result the axial channel length required for the particle to attain its equilibrium streamline is longer than predicted. Further, the final equilibrium streamline is symmetrically spaced at between ~0.35 and 0.55 of half the channel height from the centre-line, whilst Cox and Brenner (6) predicts 0.6. This result was confirmed when using both 15.0 and 30.0*mm* channel heights. However given these retrictions the results of Ho and Leal as expressed in Eqn. 5 give a reasonable first order of magnitude estimate.

The rig was erected vertically such that gravity acted in the Ox-coordinate. Thus the particle axial relative velocity became:

$$\mathbf{v} = \mathbf{v}\left(\tfrac{4}{3}V_{max}(a/d)^2 + v_\infty, 0, v_z\right) \tag{10}$$

By using buoyancy forces to increase the relative particle velocity the rate of lateral migration is further increased and consequently the channel length required to achieve equilibration is reduced.

CONCLUSION

This paper describes the findings of the most extensive experimental investigation of the behaviour of lateral migration to date. Results for neutrally buoyant rigid spheres are presented. The effects of gravity are eliminated both by careful technique and design. The results conflict with theory which is only rigorous for tube Reynolds numbers less than unity and a small dimensionless particle radius. However in the absence of any other description which predicts the equilibrium migration to a streamline at a channel position at some position midway between the wall and channel centre-line, the theory gives a reasonable first order of magnitude estimate. Attention was focussed on the fact that the lateral migration of a rigid neutrally buoyant sphere represents a worst case analysis and that the channel length required to achieve an equilibrium state for the same ratio of particle hydraulic diameter to channel height is greatly reduced both when the particle has a component of the body force acting parallel to the flow and is non-rigid or non-spherical.

ACKNOWLEDGEMENT

Work described in this report was undertaken as part of the underlying research programme of AEA Technology.

REFERENCES

1. McTIGUE, D.F, GIVLER, R.C. and NUNZIATO, J.W; Rheological effects of non-Newtonian particle distributions in dilute sususpensions. **J. Rheology** 1986, **30** , 5,1053-1076

2. LEAL, L.G; Particle motions in a viscous fluid. **Ann. Rev. Fluid Mech.** 1980, **12** , 435-476

3. GOLDSMITH, H.L. and MASON, S.G; The flow of suspensions through tubes. **J. Colloid Sci.** 1962, **17** , 448-476

4. HILLER, W. and KOWALEWSKI, T.A; An experimental study of the lateral migration of a droplet in a creeping flow **Exp. in Fluids** 1987, **5** , 43-48

5. BRETHERTON, F.P; Slow viscous flow round a cylinder in a simple shear. **J. Fluid Mech.** 1962, **12** , 591-613

6. COX, R.G. and BRENNER, H; The lateral migration of solid particles in Poiseuille flow – I. **Chem. Eng. Sci.** 1968, **23** , 147-173

7. HO, B.P. and LEAL, L.G; Inertial migration of rigid spheres in two-dimensional unidirectional flows. **J. Fluid Mech.** 1974, **65** , 2, 365-400

8. VASSEUR, P. and COX, R.G; The lateral migration of a spherical particle in two-dimensional shear flows. **J. Fluid Mech.** 1976, **78** , 2, 385-413

9. TACHIBANA, M; On the behaviour of a sphere in the laminar tube flows. **Rheol. Acta** 1973, **12** , 58-69

10. VASSEUR, P; The lateral migration of spherical particles in a fluid bounded by parallel walls. PhD. thesis, McGill Univ., Montreal, 1973.

ATTRITION OF POTASSIUM SULHPATE CRYSTALS IN STIRRED VESSELS

P. AYAZI SHAMLOU*, A.G. JONES* AND K. M. DJAMARANI+

Measurements are reported on the attrition of potassium sulphate crystals in a turbulently-agitated liquid of low solubility power. Data from these experiments are discussed in terms of three possible mechanisms of attrition based on purely hydrodynamic conditions in the vessel. The results of this study indicate that crystal attrition may be attributed to turbulent fluid-induced shear forces acting on the surface of the suspended crystals. The scale of turbulent eddies most likely to be responsible for the attrition process fall in the viscous dissipation subrange of the turbulence energy spectrum.

INTRODUCTION

In suspension crystallization, secondary nucleation is the process by which new nuclei (fines) are created and grow from crystals that are already present in suspension. In most stirred-crystallizers, this mode of nucleus generation and growth plays an important role in determining the size distribution of the product crystals and its influence in the design and performance of industrial suspension crystallizers have long been established [Clontz and McCabe 1971; Larson and Bendig 1976; Bauer et al 1974; Jagannathan et al 1980; Wang et al 1981; Ayazi Shamlou et al 1990].

Secondary nucleation is generally considered to involve three stages which in a suspension crystallizer occur concurrently. These are:

Nucleus generation. All sources of secondary nuclei require the presence of stable crystals in suspension. Nucleus generation may the result of one or combination of the following.

(a) Macro-attrition which refers to the mechanical breakage and detachment of relatively large fragments from either the surface of the parent crystals or as a result of their fragmentation.

(b) Micro-attrition which refers to the abrasion of tiny aggregates and asperties of less than, say, 1 μm loosely bound to the surface of the parent crystals, normally being formed at high levels of supersaturation.

* Chemical and Biochemical Engineering, University College London, WC1E 7JE.
+ Present address: Chemical Engineering, Imperial College, Londdon.

c) Disturbance by local fluid mechanical stresses of tiny aggregates (clusters) of solutes undergoing structural changes during their passage through the crystal boundary layer in a moderately supersaturated

solution. It is believed that such changes and disturbances could lead to the formation of new nuclei in the wake of the parent crystals.

Despite the large number of publications in this area the mechanisms of nucleus generation are still not adequately understood, although two possibilities have been identified in the literature. One, which is much more widely reported [Clontz and McCabe 1971; Bauer et al 1977; Larson and Bendig 1976] , concerns contact (or collision) breeding which occurs as a result of impacts of crystals with each other and/or with other solid objects, e.g. vessel walls, baffles or the rotating impeller. These collisions are, in turn, induced by the motion of the fluid. The other mechanism, which in this study is referred to as fluid -induced surface shear mechanism, occurs as a result of the direct interaction between the parent crystals and the surrounding fluid which is normally under turbulent agitation [Jagannathan et al 1980; Wang et al 1981; Ayazi Shamlou et al 1990].

Nucleus transport. The embryonic nuclei are conveyed from the vicinity of the parent crystals to the bulk region of the supersaturated mother liquor. Both fluid turbulence and bulk circulation could contribute to this transportation.

Nucleus survival and growth. For the new nuclei to survive and grow in the bulk region, they must posses a critical minimum size which depends upon the level of supersaturation. Nuclei smaller than this minimum size will dissolve in the surrounding liquid. The higher the level of supersaturation, the smaller will be the size of the nucleus that can survive and grow.

The superimposed effects on secondary nucleation of crystal morphology, level of mother liquor supersaturation and hydrodynamic conditions of the suspension provide major experimental difficulties in any study on secondary nucleation. In particular, interpretation of data on crystal attrition is made difficult by a lack of knowledge on the effects on attrition of phenomena such as surface regeneration time, initial breeding of uncured crystals and rounding-off of sharp edges and corners of crystals by dissolution in locally undersaturated mother liquor.

The aim of this investigation is to determine whether, and if so to what extent, purely hydrodynamics effects in a mechanically agitated suspension crystallizer can induce nucleus generation. Specifically, experimental data are reported for the breakage of crystals of potassium sulphate suspended in a liquid of low solubility power under turbulent agitation. Results from these experiments are discussed and examined on the basis of various hydrodynamic mechanisms of particle breakage.

MATERIALS, EQUIPMENT AND EXPERIMENTS

Experiments were carried out with potassium sulphate crystals suspended in a liquid of low solubility power (methanol) in order to minimize both growth and dissolution of any newly generated nuclei. The parent crystals were grown in a 6L laboratory-scale continuous MSMPR suspension crystallizer

(Jones and Mydlarz 1989). The resulting crystals with a size range between 250 - 1200 μm were sieved to provide the following size fractions: 600-710 μm, 710-850 μm and 850-1000 μm. These fractions were used as the parent crystals in all subsequent nucleus generation experiments.

The crystallizer used in the attrition experiments consisted of a flat-bottom glass vessel with a diameter of 0.128m and a height of 0.18m. the vessel was fitted with four baffles, each having a width equal to 0.013m and a height of 0.18m. The baffles were equally spaced and positioned firmly against the walls of the vessel. The height of the liquid in the vessel was maintained equal to tank diameter in all experiments. Agitation was provided with a 64mm diameter, 45° open turbine with six blades driven by a 1.1 kW infinitely variable speed motor with a range of 0 - 2000 r.p.m. Impeller speed and torque were measured and displayed using a shaft-mounted torque/speed pick-up unit (Ayazi Shamlou and Edwards 1987). Figure 1 shows a sketch of the main apparatus and some of the ancillary equipment used in this work.

Specific experiments were directed towards studying the dependence of the rate of nucleus generation on (i) solids volume fraction which was varied in the range 0.04 to 0.12, (ii) impeller speed in the range 1000 to 1500 r.p.m. and with all speeds above the just suspension values for the crystals and (iii) surface hardness of the impeller which was changed by coating the surface of the stirrer with a layer of a silicone adhesive giving a relatively soft surface finish compared with the uncoated impeller which was fabricated from stainless steel.

An experimental run consisted of pouring a preset quantity of parent crystals of known mean diameter on the surface of the liquid under agitation at a fixed impeller speed. The dependent variable in all tests was the number of fines generated which was assumed to be equivalent to the number of new nuclei. The rate of nucleus generation, i.e. the change in crystal size distribution (CSD) with time, was determined by intermittent sampling of the contents of the vessel. Wet-sieving using methanol as the liquid was used to measure the crystal size distribution with the exception of fractions below the 38 μm, for which a standard Malvern Particle Sizer (model 2200/3300 fitted with a 64mm lens) was used. Further details of equipment, materials, experimental procedures and method of analysis may be found elsewhere (Ayazi Shamlou and Djamarani 1987 and Djamarani 1988).

RESULTS AND DISCUSSION

Experimental results of the change in CSD with time is shown in Figure 2 for parent crystals with an initial mean size of 780 μm, solids volume fraction of 0.12 and impeller speed of 1500 r.p.m. The plots show a continuous drop in the size of the initial crystals which seems to level off after an agitation period of nearly 60 minutes. The decrease in the size of the parent crystals is accompanied by a simultaneous increase in the weight fraction of smaller sizes with the highest increase corresponding to the smallest size fraction, i.e. 1-38 μm.

Additionally, Figure 2 reveals that after a period of about 60 minutes of agitation, the largest crystals, i.e. of size 710-600 μm, also form the largest fraction by weight (nearly 70% of the total) of the crystals in the vessel with the fraction in the size range 1-38 μm making up nearly 26% of the total weight; the other intermediate size fractions make up the remaining

4%. These initial observations suggest that crystal breakage predominately occurs by a process of erosion of fines from the surface of the parent crystals. This conclusion is further supported by the photomicrographs (Fig 3) of a number of crystals before and after an experiment: these clearly show a rounding off of the sharp edges and corners of the crystals after a period of agitation in the liquid.

The effect of impeller speed on the rate of change in the size of the largest fractions and on the rate of generation of fines (the smallest fractions) is shown in Figure 4. The data presented are for parent crystals with a mean initial size of 780 μm and solids volume fraction of 0.04. In Fig 4 three lines are obtained each for one impeller speed indicating that as impeller speed increases the rate of attrition increases.

Figure 5 shows the effect of crystal holdup concentration on the rate of fines generation. The data refer to an experiment made with parent crystals having a initial mean size of 780 μm and an impeller speed of 1500 r.p.m.

The data presented in Figure 6 are for two identical experiments with the exception that in one run, the surfaces of the blades were coated with an even layer of an adhesive (silicone rubber) giving a soft surface finish: the standard uncoated impeller was fabricated from stainless steel. Both experiments were made with parent crystals with a mean initial size of 655 μm, solids volume fraction of 0.04 and an impeller speed of 1500 r.p.m. Evidently, the rate of generation of fines appears to be essentially independent of the surface hardness of the blades of the impeller.

MECHANISM OF CRYSTAL ATTRITION IN STIRRED VESSLES

Several potential mechanisms were previously identified [Ayazi Shamlou and Djamarani 1987; Djamarani 1988; Ayazi Shamlou et al 1990] for breakage of particles suspended in liquids under conditions similar to those used in the present work (Table 1). As far as attrition of crystals is concerned, the most plausible of these mechanisms are crystal-impeller impacts, crystal-crystal collisions and fluid-induced shear forces acting on the surface of crystals as a result of fluid turbulent motion around the crystals.

Full derivation of all the relevant equations may be found elsewhere [Ayazi Shamlou et al, 1990; Djamarani, 1988], but as an example of the basic approach adopted in modelling the attrition process a brief discussion is given below on one of the mechanisms, namely the turbulent fluid-induced surface shear mechanism.

According to the fluid-induced surface shear mechanism of crystal attrition, the relative motion between turbulent eddies and crystals gives rise to shear forces acting on the surface of the crystals and if the magnitude of these forces is higher than crystal surface shear strength, then primary particles (fines) are stripped off the surface of the crystals.

Most crystals produced in an MSMPR crystallizer are expected to contain numerous surface defects, sharp edges, and loosely-bound particles on their surfaces and consequently crystal strength is expected to drop sharply towards the surface. Erosion stops when all the sharp corners and surface defects are stripped off and once the magnitude of the applied shear forces fall below the surface shear strength of the crystals. This suggests that for a fixed set of operating conditions and given parent crystals, the size of the crystals at the end of the erosion process, referred to as the stable crystal size, d_s, is determined by the balance between the applied fluid-dynamic induced shear forces and the crystal surface shear strength.

In many MSMPR crystallizers crystal size ranges between about 10 μm and 1000 μm. Erosion of fines from the surface of these crystals requires turbulent eddies that are equal to or smaller than the size of the crystals: eddies that are significantly larger than crystal size would tend to entrain crystals and thus cause little surface shear. In a turbulently-stirred vessel the eddies that are believed to provide maximum surface shear forces are in the so called universal equilibrium range of the turbulence spectrum. This range is usually subdivided into (i) the inertial convection subrange characterized by energy-containing eddies of scale, l_e, and (ii) the viscous dissipation subrange characterized by eddies of scale, l_d, which is normally expressed in terms of the Kolmogoroff scale of turbulence, l_k. For an stirred vessel operating under fully turbulent conditions, $l_k \approx 50$ μm while $l_e \approx .08D$ and $l_d \approx 5\,l_k$ [Davies, 1972]. This suggests that the lower size limit of crystals most likely to be affected by the turbulent eddies is about 250 μm. In the crystallizer used in this study the upper limit of crystal size likely to be influenced by turbulent eddies of scale upto about l_e is approximately 5000 μm: crystal sizes used in the experiments reported in this work fall well within this range.

From the foregoing discussion therefore, crystal erosion may be attributable to turbulent eddies with a critical scale either in the inertial subrange or in the viscous dissipation subrange of the turbulence spectrum. In either case, the rate of generation of fines may be expressed as follows

$$dn_p/dt = f_b \; N_p \; n_c \qquad (1)$$

where n_c is the number concentration of parent crystals present in suspension, f_b is the frequency of crystal disruption and N_p is the number of particles eroded per disruption.

The frequency of crystal disruption, f_b, depends on the scale of turbulent eddies assumed to be responsible for erosion. Thus

$$f_b \propto (\varepsilon/\nu)^{1/3}\, d_c^{\,2/3} \qquad (2)$$

for turbulent eddies of scale l_e and

$$f_b \propto (\varepsilon/\nu)^{1/3} \qquad (3)$$

for eddies in the viscous dissipation subrange [parker et al 1971].

For monosized parent crystals, which is approximately the case in this study, and assuming a mean size d_p for eroded particles (fines), both N_p and n_c may be expressed as functions of d_c. Thus

$$N_p \propto d_c^{\,2} \qquad (4)$$

$$n_c \propto C_v/d_c^{\,3} \qquad (5)$$

Additionally, the relationship for the stable crystal size, d_s, which is normally expressed in terms of impeller energy input, depends on the scale of the turbulent eddies assumed to be responsible for the erosion process. Thus, assuming turbulence to be homogenous and isotropic, then

$$d_e \propto (\varepsilon/\nu) \qquad (6)$$

if erosion is assumed to be due to turbulent eddies in the inertial subrange of the turbulence spectrum, and

$$d_e \quad [\varepsilon/\nu]^{1/2} \qquad (7)$$

if crystal erosion is assumed to be due to turbulent eddies in the viscous dissipation subrange.

Using Eqns 1 to 7 and assuming that during the initial period of erosion the change in the size of parent crystals is negligible, i.e. $d_p \ll d_c$ and thus $d_c \cong d_e$, yields the following equations for the initial rate of generation of fines

$$dn_p/dt \propto C_v \, (\varepsilon/\nu)^2 \qquad (8)$$

for erosion by turbulent eddies in the inertial subrange of the turbulence spectrum and

$$dn_p/dt \propto C_v \, (\varepsilon/\nu) \qquad (9)$$

for erosion attributable to turbulent eddies in viscous dissipation subrange.

Further considerations of Eqns 8 and 9 and on mathematical equations based on the other two mechanisms (crystal-crystal impacts and crystal-impeller collisions) may be found elsewhere (Djamarani 1988; Ayazi Shamlou and Djamarani, 1988 and Ayazi Shamlou et al 1990]). Table 2 provides a summary of all the equations.

Table 2 together with experimental attrition data reported in this work may be used to assess the suitability of each mechanism in describing the attrition process. For example, according to the information presented in Table 2, attrition of crystals by the turbulent fluid surface shear mechanism with critical eddy size in the inertial convection subrange of the turbulence spectrum predicts that the rate of nucleus generation should have a first-order dependency with respect to crystal concentration and a second-order dependency with respect to power input per unit mass. On the other hand, if eddies in the viscous dissipation subrange of the turbulence spectrum were responsible for erosion, then it would be expected that the rate of nucleus generation would have a first-order dependency with respect to both the power input per unit mass and with the crystal-volume concentration.

Figure 7 presents the experimental data on the effect of crystal volume-concentration on the initial rate of generation of fines. The plot is a straight line with a slope of unity in accordance with both a fluid-induced surface shear mechanism and a crystal-impeller collisional mechanism of attrition. Moreover, the data in Figure 7 in effect rules out crystal-crystal impacts as a possible mechanism since with this mechanism the plot would have resulted in a line of slope equal to two.

The limited data in Figure 8 shows the initial rate of nucleus generation as a function of impeller speed on a log-log plot. A line with a slope of three is obtained suggesting that attrition is more likely to be attributable to a turbulent fluid shear mechanism rather than a crystal-impeller collisional one; with the latter mechanism the plot in Fig 8 would be expected to have given a line with a slope equal to four (Table 2) . It is, of course, desirable that more experimental data are collected to substantiate this conclusion. The results of experiments with the coated and the uncoated impeller shown in Fig 6, however, also appear to rule out crystal-impeller impacts as a source of erosion.

Finally, the limited experimental data obtained in this work plotted in terms of the parameters in the turbulent fluid surface shear mechanism is shown in Figure 9. The data clearly show that the initial rate of nucleus generation has a first-order dependency on $C_v(\varepsilon/\nu)$ as expected.

NOTATION

C_v crystal volume fraction
D impeller diameter
d_p mean size of the nuclei generated
d_c mean size of the parent crystals
d_s size of the stable crystals
f_b frequency of crystal-impeller collisions
l_d scale of energy-containing turbulent eddies
l_e scale of turbulent eddies in the viscous dissipation subrange
l_k Kolmogoroff scale of turbulence
n_c number of parent crystals
N_p number of nuclei generated per disruption
N impeller rotational speed
P impeller power input
P_o impeller power number
R rate of crystal-crystal collisions
t time

Greek letters
ε power input per unit mass
ρ fluid density
ν kinematic viscosity

REFERENCES

Ayazi Shamlou, P. and Djamarani, K.M., 1987, Fluid Mixing III, IChemE Symposium Series, No. 108, Bradford University, pp. 221-234.

Ayazi Shamlou, P. and Edwards, M.F., 1985, Chem. Engng Sci., No. 9, Vol. 40, pp. 1773-1781.

Ayazi Shamlou, P. Jones, A.G. and Djamarani, K., 1990, Chem. Eng. Sci., (in press).

Bauer, L.G., Larson, M.A. and Dallons, V.J., 1974, Chem. Engng Sci., Vol 29, pp. 1253-1261.

Clontz, N.A. and McCabe, W.L., 1971, Chem. Engng Prog. Symposium Series, No. 110, Vol. 67, pp 6-17.

Djamarani, K.M., 1988, PhD thesis, University of London.

Davies, J.T., 1972, Turbulence Phenomena, Academic Press, NY.

Jones, A.G. and Mydlarz, J., 1989, Chem. Eng. Res. Des., 67, pp 283-293.

Jagannathan, R., Sung, C.Y., Youngquist, C.R. and Estrin, J., 1980, AIChE Symposium Series No. 193, Vol. 76, pp. 90-97.

Larson, M.A. and Bendig, L.L., 1976, AIChE Symposium Series No. 153, Vol. 72, pp. 21-27.

Levich, V.G., 1962, "Physicochemical Hydrodynamics", Prentice Hall Inc., NY.

Parker , D.S., Kaufman, J. and Jenkins, D., 1972, Sanit. Eng. Div. ASCE, 98, pp 79-99.

Wang, M.W., Huang, H.T. and Estrin, J., 1981, AIChE No. 21, Vol. 27, pp. 312-315.

MECHANISMS OF PARTICLE BREAKAGE IN STIRRED VESSELS

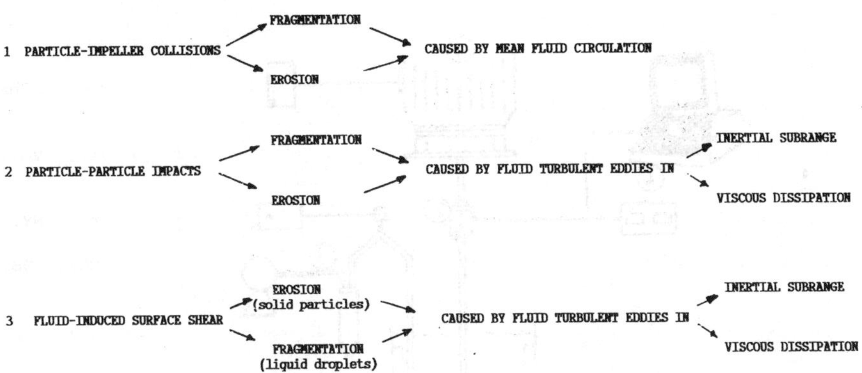

TABLE 1 Particles in suspension in a turbulently stirred liquid can break-up by various hydrodynamic mechanisms

	$dn_p/dt \propto n_c^a (\varepsilon/\nu)^c$ or $\propto n_c^a N^d$		
MECHANISM OF MACRO-ATTRITION OF CRYSTALS	a	c	d
CRYSTAL-CRYSTAL COLLISIONS INDUCED BY FLUID TURBULENCE WITH CRITICAL EDDY SIZE IN THE: (I) INERTIAL CONVECTION; (II) VISCOUS DISSIPATION SUBRANGE OF THE TURBULENT ENERGY SPECTRUM	2 2	① 2	③ 6
CRYSTAL-IMPELLER collisions induced by fluid motion	①	4/3 or ①*	4
CRYSTAL SURFACE SHEAR BY TURBULENT FLUID EDDIES WITH CRITICAL EDDY SIZE IN THE: (I) INERTIAL convection (II) VISCOUS DISSIPATION SUBRANGE OF THE TURBULENT energy spectrum	① ①	2 ①	6 ③

TABLE 2 Dependence of the rate of nuclei generation upon some of the operating variables based on a number of possible mechanisms. Circles indicate present data.

* at constant impeller speed.

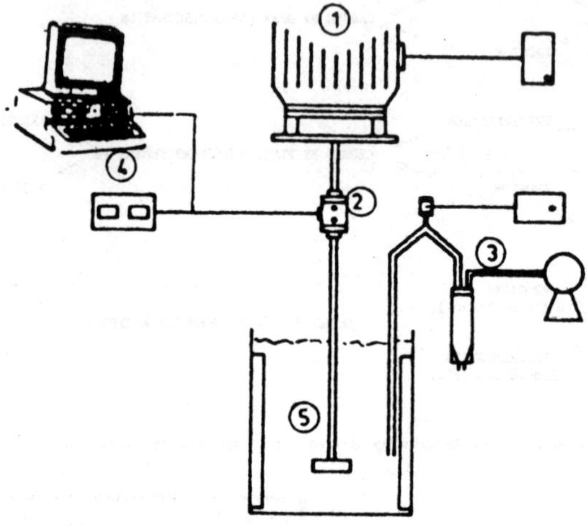

1. motor
2. torque transducer
3. sampling device
4. data recording and display units
5. tank and impeller

FIGURE 1 Experimental set up

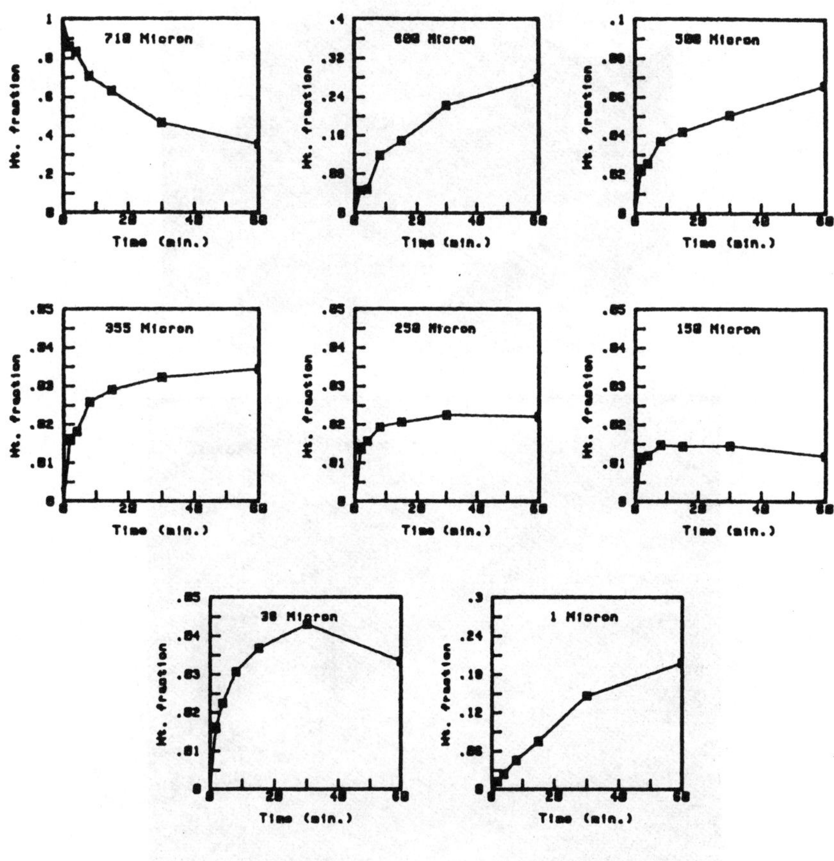

FIGURE 2 Variation in crystal size distribution as a function of time

FIGURE 3 SEM photograph of crystals of potassium sulphate (top) before and (bottom) after an attrition test.

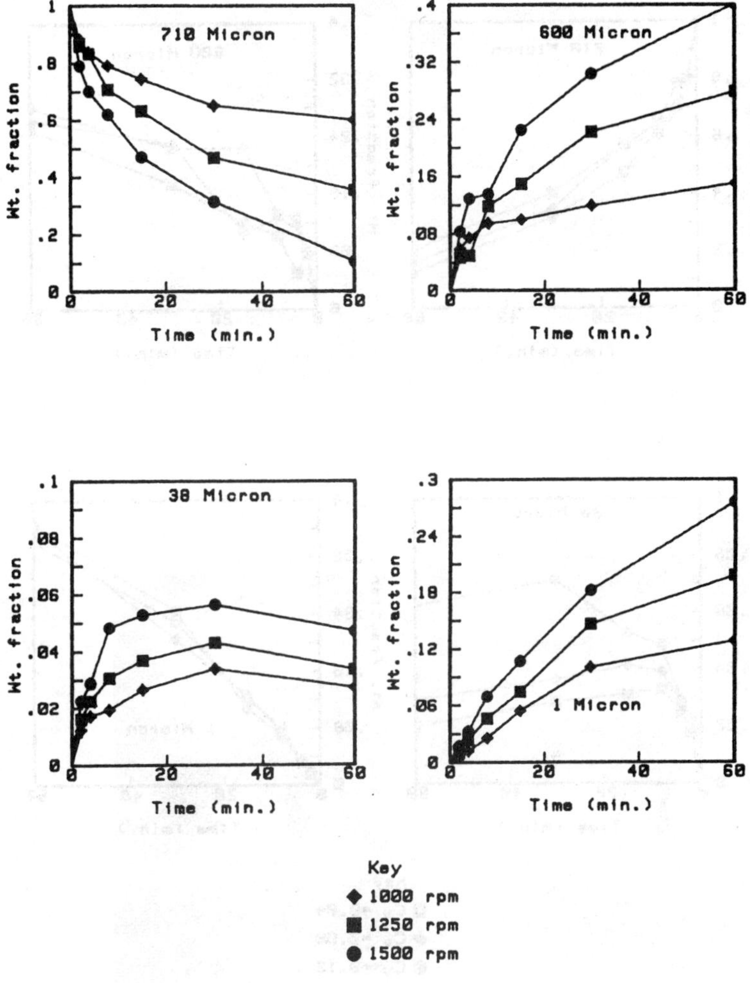

FIGURE 4 Change in crystal size distribution as a function of time (for experimental details see text).

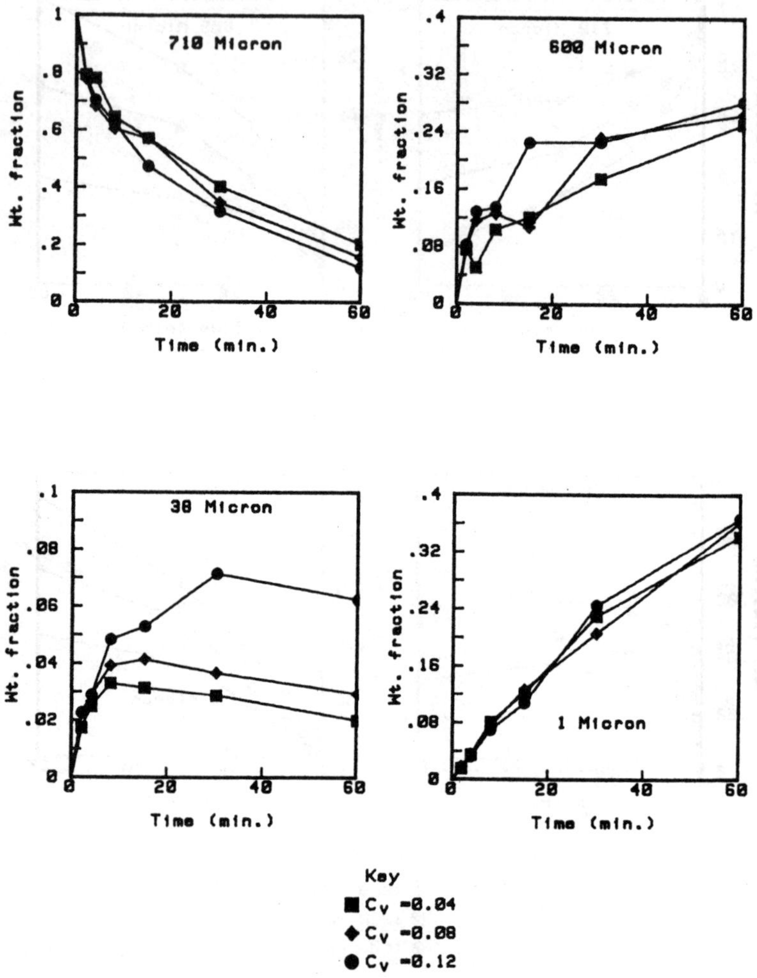

FIGURE 5 Effect of crystal-volume concentration on crystal attrition.

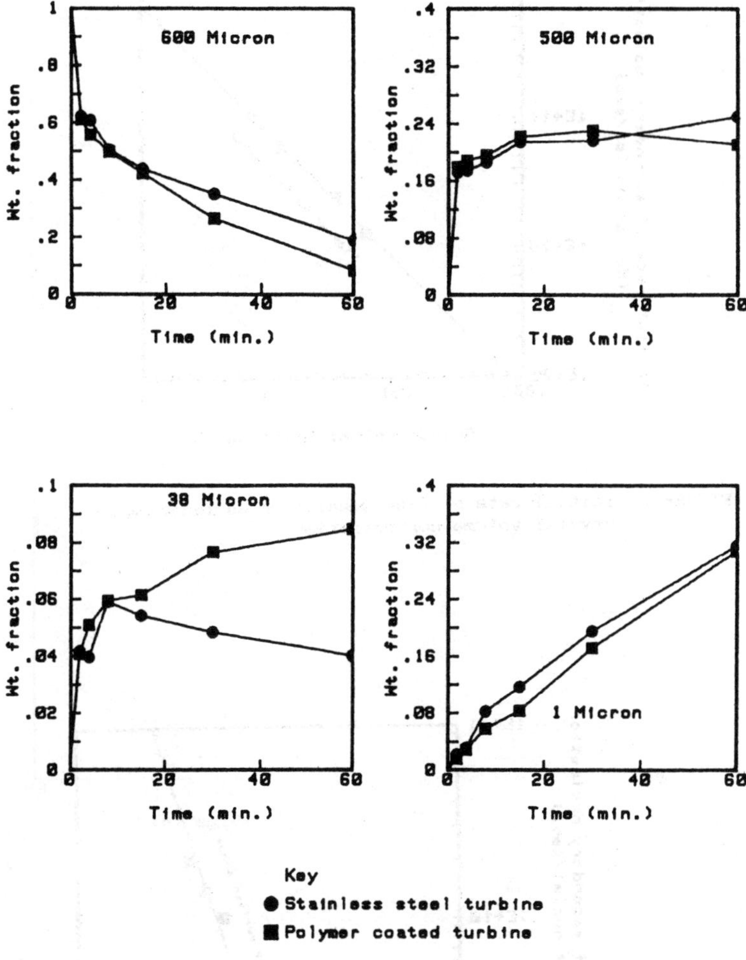

FIGURE 6 Crystal size distribution for coated and uncoated impellers indicate that CSD is practically independent of impeller surface finish.

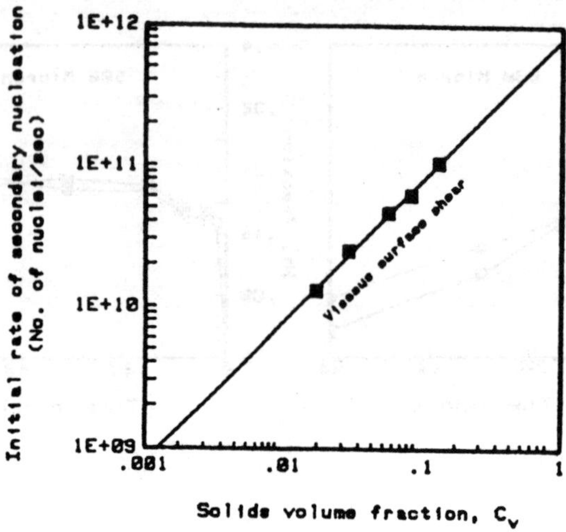

FIGURE 7 Initial rate of fine generation as function of crystal-volume concentration.

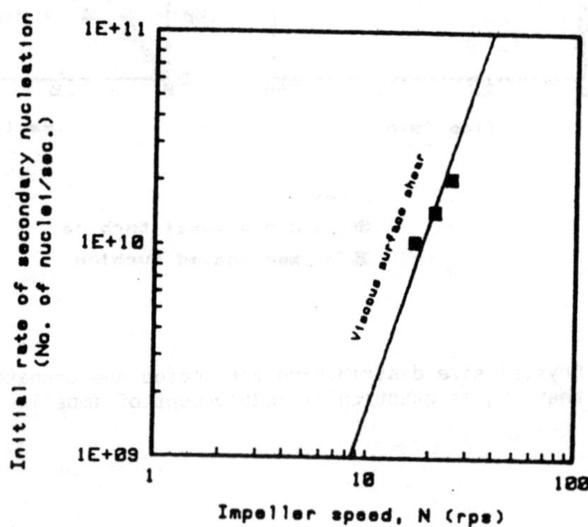

FIGURE 8 Variation of the initial rate of fine generation with changes in impeller speed.

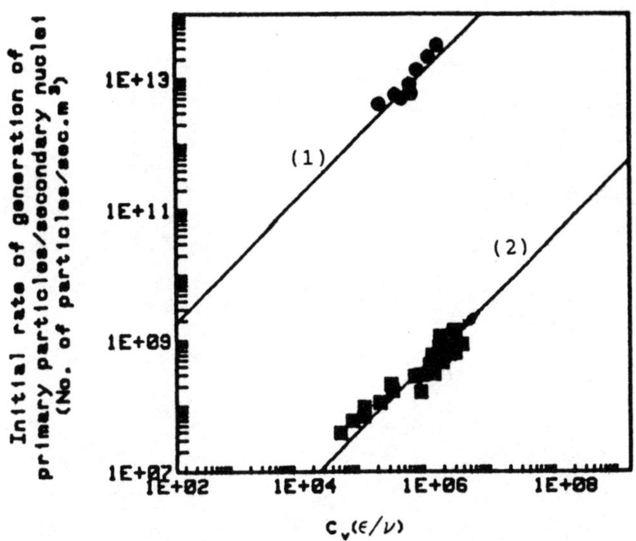

FIGURE 9 Experimental data plotted in terms of model predictions based on the fluid shear mechanism.

(1) data from secondary nucleation tests
(2) data from breakage of model agglomerates (Ayazi Shamlou and Djamarani 1987)

PARTICLE DISSOLUTION KINETICS INSIDE BATCH STIRRED VESSELS

A.Brucato, V.Brucato, L.Rizzuti, M.Sanfilippo*

A simple model for the dissolution of solid particles inside batch stirred tanks is described. The agreement between model predictions and experimental transient concentration data is shown to be very good. The fitting of model predictions to experimental transient concentration data allows the determination of both the mass transfer coefficient and the interfacial concentration, thus providing a way to point out the role played by interfacial resistances in dissolution processes. For the experimental range investigated, these last are found to be negligible. Finally, mass transfer coefficients are found to be almost independent of particle diameter and to depend on the square root of agitation speed.

INTRODUCTION

In the modelling of solid particles dissolution inside mixing tanks, it is usually assumed that the interfacial concentration of solute coincides with the saturation concentration. This last quantity is not always known but it is usually easily experimentally assessed by measuring the concentration attained when the solution is allowed to reach the saturation.

There are cases, however, in which this is not possible, as for instance for solutes whose solutions become supersaturated with respect to some other solid phase, before reaching the saturation with respect to the solute itself. This is the case of a salt known as "Schönite" ($MgSO_4$ K_2SO_4 $6H_2O$) which is industrially relevant as it is involved in some processes aimed at the production of K_2SO_4 . In fact, when dissolved into water, the saturation concentration of the solution with respect to Potassium Sulphate is reached before the saturation with respect to the Schönite itself is attained. Then precipitation of K_2SO_4 is obtained while the dissolution of Schönite goes on. This property, exploited in the final step of K_2SO_4 production, actually inhibits the possibility of experimentally accessing the value of the saturation concentration with respect to the Schönite itself.

In this paper a method for the estimation of the unknown C_i is suggested and tested for the case of Schönite dissolution. The method is based on a simple dissolution model that appears to accurately fit the experimental data. The same approach can be used for solid phases with known saturation concentration in order to test the hypothesis $C_i = C_{sat}$; a test of this kind is also reported and discussed.

* Dip. Ing. Chimica; Univ. of Palermo; Viale delle Scienze; 90128 Palermo (Italy)

Finally, with reference to the particular system studied, the mass transfer coefficient dependence on D_p and N is investigated. This subject has been covered by a number of investigations but agreement on the dependence of k_c on D_p has not been reached yet (1).

DISSOLUTION MODEL

The model adopted for the dissolution of particulate solids in stirred vessels, is based on the following equation:

$$\mathbf{J} = k_c (C_i - C) \tag{1}$$

where **J** is the solute flux leaving the solid interface and C_i and C are respectively the volumetric concentration in the liquid phase at the interface and in the bulk of the solution. Eqn.(1) is actually a definition of the mass transfer coefficient k_c and assumes a simple proportionality between the driving force and mass transfer flux.

The following assumptions are made:

- batch system;
- particle size is the same for all particles but varies with time;
- solute dissolution does not affect significantly the volume of the solution;
- particle shape does not change during the dissolution.

When eqn.(1) is coupled with the mass balance on the dissolved solute, the following equation is obtained:

$$V_t \frac{dC}{dt} = k_c S_t (C_i - C) \tag{2}$$

where S_t is the total solid surface exposed to the liquid.

During the dissolution particle size decreases leading to a decrease in the interfacial area. As it is assumed that the particle shape does not change during the dissolution process, the volume and interfacial surface of one particle can be expressed by the following equations:

$$V_p = \alpha D_p^3 \;\; ; \;\; S_p = \beta D_p^2 \tag{3}$$

where D_p is a suitable characteristic diameter of the particle and α and β are shape factors.

The overall mass balance on the solid particles leads to:

$$D_p = D_{po} (1 - x)^{1/3} \tag{4}$$

where D_{po} is the starting diameter of the particles and x is the dimensionless concentration in the liquid phase defined as:

$$x = \frac{C - C_o}{C^* - C_o} \tag{5}$$

where C_o is the concentration in the liquid phase at time t_o and C^* is the concentration that the liquid phase attains in the hypothesis of complete dissolution of the solid phase ($C^* = M_o / V_t$).

The total solid surface exposed to the liquid at any time during the dissolution process can then be written as:

$$S_t = \frac{(\beta/\alpha)\, M_o}{\rho_p D_{po} (1 - x)^{2/3}} \tag{6}$$

Substituting (5) and (6) into eqn.(2), leads to:

$$\frac{dx}{dt} = A\,(1 - x)^{2/3}\left(x_i - x\right) \tag{7}$$

where:

$$A = k_c\,(\beta/\alpha)\,\frac{C^* - C_o}{\rho_p D_{po}} \quad ; \qquad x_i = \frac{C_i - C_o}{C^* - C_o} \tag{8}$$

With the initial condition x=0 at time $t = t_o$, the analytical solution of eqn. (7) can be written as:

$$(t - t_o) = \frac{1}{A\, y_i^2}\left[\frac{1}{2}\ln\left(\frac{1 - y_i}{y - y_i}\right)^3\left(\frac{y^3 - y_i^3}{1 - y_i^3}\right) + \sqrt{3}\arctan\left(\frac{2y + y_i}{\sqrt{3}\, y_i}\right) - \sqrt{3}\arctan\left(\frac{2 + y_i}{\sqrt{3}\, y_i}\right)\right] \tag{9}$$

where:

$$y = (1-x)^{1/3} = D_p/D_{po} \;\; ; \;\; y_i = (1-x_i)^{1/3} \tag{10}$$

which is similar to the "Cube Root Law" introduced by Hixson and Crowell (2)

It is interesting to note that eqn.(9) is able to deal with all cases of batch dissolution, whether the solid mass initially introduced is bound to be completely dissolved ($y_i<0$), or not ($y_i>0$). The corresponding asymptotic value for y will be respectively 0 or y_i.

In the particular case of $y_i=0$ (i.e. $C^*=C_i$; $x_i=1$) eqn.(9) leads to indeterminate results; in this case however, integration of eqn.(7) is straightforward as two of the factors in R.H.S. of eqn.(7) can be combined into one.

Eqn.(7) can be simplified if the mass of solids initially introduced in the system is small in comparison with the mass that can be dissolved before reaching the saturation: in this case $x \ll x_i$ during the whole run and the analytical solution is much simpler:

$$y = 1 - \left[k_c (\beta/\alpha) \frac{C_i - C_o}{3 \rho_p D_{po}} \right] (t - t_o) \tag{11}$$

which can be put in the form:

$$D_p = D_{po} - B (t - t_o) \tag{12}$$

where:

$$B = A \, D_{po} \frac{x_i}{3} = k_c (\beta/\alpha) \frac{C_i - C_o}{3 \rho_p} \tag{13}$$

Plots of t vs D_p should then appear as straight lines passing from (t_o, D_{po}). If all parameters but k_c are known, experimental determination of the slope in low concentration runs can then be used in order to determine the value of k_c.

EXPERIMENTAL

The experimental runs were conducted in batch: a stirred vessel was filled with a liquid and at time zero a known amount of solid particles was added to the batch while monitoring the dissolved solid concentration.

The stainless steel vessel used was cylindrical, 13 cm in diameter, provided with a dished bottom and four standard baffles. It was stirred by means of a 6.86 cm dia. square pitch propeller, offset from the bottom by 1/3 of the liquid height. The temperature was always maintained at 25°C by means of a thermostatic bath.

All runs were performed using accurately sieved salt particles, so that a narrow size distribution was attained. The particle size was varied from 0.137 mm to 0.9 mm while the agitation speed was varied from 450 rpm to 1300 rpm. Care was taken in order to insure that during all runs all the particles were suspended. As visual inspection inside the vessel was inhibited, the minimum speed required to suspend the particles was estimated by means of Zwietering's relationship (3) for the case of a flat bottom: it was assumed that this value was sufficient to ensure complete suspension in the dished-bottom tank used in this work.

Microscope observation of the particles revealed that they were polycrystalline agglomerates characterized by very irregular shapes. It was then impossible to estimate the value of β/α on the basis of geometrical considerations. The only conclusion that could be drawn by visual inspection was that a value significantly greater than that pertaining to spheres and cubes (β/α=6) had to be expected: in fact irregular shapes exhibit higher surface to volume ratios than compact shapes. In order to estimate the value of β/α, the flow rate of a liquid flowing through a fixed bed of Schönite particles under a known pressure drop was measured. The liquid used was ethylen-glycol as the Schönite is insoluble in it. For greater safety, the glycol was previously saturated with Schönite. The experiments were performed in the laminar range (Re<<10) so that the Blake-Kozeny equation (4) applied. As the physical properties of the liquid phase were known and the bed voidage was easily determined by filling it with known volumes of liquid, the only unknown parameter in the equation was the equivalent particle diameter D_{eq} that could then be easily determined. D_{eq} is related to the "particle specific area" a_v by the simple relationship:

$$a_v = \frac{6}{D_{eq}} = \frac{(\beta/\alpha)}{D_p} \tag{14}$$

from which the unknown ratio (β/α) was estimated. An average value of 11 was found, which is greater than the value 6 exhibited by cubes and spheres, as expected.

The salt concentration in the liquid phase was monitored by means of a conductivity meter (Analytical Controls - mod. 101) whose readings were checked against a non-linear calibration curve previously determined. The time constant of the conductivity measurement apparatus was experimentally found to be smaller than 2 sec.

As liquid phase, a solution 30% water - 70% ethylen-glycol (by weight) was used to suitably slow down the dissolution rate as assessed in preliminary runs. The dissolution time constant was then always much greater than both the time constant of the conductivity meter and the vessel mixing times. The viscosity of the solution was experimentally determined to be 5.84 cp.

RESULTS AND DISCUSSION

The information obtained from experimental runs is in the form of t vs C curves that are easily transformed in the corresponding t vs x curves. Two sets of runs were performed: high final concentration and low final concentration.

High final concentration runs.

In these runs the mass of solute initially added to the batch was such that a final concentration of $8.61 \cdot 10^{-3}$ gr/cm^3 was attained. This value is slightly below $9.0 \cdot 10^{-3}$ gr/cm^3 which is the starting value for Potassium Sulphate precipitation, as experimentally determined in preliminary runs. As stated above, for the system investigated, the saturation concentration with respect to the Schönite was neither known nor experimentally accessible because of the K_2SO_4 precipitation. For this reason a problem arises in the estimation of the interfacial concentration C_i, which is usually assumed to be equal to the saturation concentration. On the other hand, eqn.(9) involves y_i, which is strictly related to C_i (eqns (8) and (10)), as well as the mass transfer coefficient k_c. Eqn.(9) was then utilized, by best fitting procedures, in order to determine both k_c and y_i (i.e. C_i).

A typical result obtained is shown in Fig.1 . The agreement between experimental data and the predicted dissolution curve (curve 1) is very good for the best fitted set of parameters. It shows that the assumptions involved in the derivation of eqn.(9) actually holds within the experimental error. In the same figure two other curves relating to different values of the two parameters are plotted (curves 2 and 3) in order to show the discriminating power of this technique. It can be observed that a 20% change in the value of the two parameters leads to curves that clearly miss the data.

An interesting point is that the short times slope of the predicted curve depends only on the product $k_c\, x_i$, as it can be inferred by eqn.(7) if one considers that in the first instants of the dissolution process the value of x can be neglected in comparison with both 1 and x_i. In fact curve 3 in Fig.1 is almost coincident with curve 1 at short times. Thus the value of the product $k_c\, x_i$ can be firstly assessed by a trial and error fitting on the short times region of the experimental data. Then the value of y_i is assessed by fitting the long times data points. The same results can be put into the form shown in Fig.2 where experimental vs predicted dissolution times (R.H.S of eqn.(9)) are reported.

Two other high concentration experimental runs (runs 2 and 3 in table 1), characterized by different values of D_{po} and/or N, were performed and gave similar results. Moreover the best fit value of the interfacial concentration was always the same (C_i=17.7·10^{-3} gr/cm^3): the unknown interfacial concentration value for the particular system investigated could then be considered as definitely assessed. It is worth pointing out that the accuracy in the prediction of the dissolution curves of the model adopted, which involves the assumption C_i=const, suggests that interfacial mass transfer resistances are negligible in the dissolution rate range. It can be inferred then that the experimentally determined interfacial concentration should equal the saturation concentration.

A second relevant assumption of the model is the independence of k_c of D_p. The soundness of this hypothesis was further confirmed by using directly eqn.(7) in order to compute k_c values at the various particle diameters during the dissolution runs. The result is shown in Fig.3 where particle diameters have been computed by means of eqn.(4). Apart from the scatter of data, dependent on the numerical differentiation, the constancy of k_c over the particle size range investigated in this work is confirmed. The point that during a single run the variation of k_c with D_p may be counterbalanced by an opposite variation of k_c with solid mass concentration, as suggested by Wilhlelm et al. (5) , is not supported by the data of run 3 reported in the same figure.

It is interesting to note that the dynamic determination of interfacial concentration suggested here can be used for systems with known saturation concentration in order to verify the assumption $C_i = C_{sat}$, usually made. Such an experimental run was performed using NaCl particles as dissolving solid (run 18 in Tab.1). More NaCl than needed to attain the saturation was initially added to the system (M_o/V_t = 153.7 ·10^{-3} gr/cm^3 ; C_{sat} = 133.9·10^{-3} gr/cm^3). The result, shown in Fig.3, is a further confirmation of the dissolution model adopted. Moreover, the experimental data are well fitted by the theoretical curve with $C_i = C_{sat}$, showing that in this case also the interfacial resistance to mass transfer is negligible.

Finally, it is interesting to note that eqns (1)-(10) hold also in the case $C_i < C$, i.e. in the case of a crystallizing system, if the system is initially seeded with uniform size seeds and nucleation in the system is inhibited. For this kind of systems the interfacial resistance is usually important. Once the k_c values for such a system have been determined by dissolution tests conducted in similar hydrodynamic conditions, eqn.(7) might be used in order to monitor the interfacial concentration C_i during the run, making it possible in this way the investigation on the functional form of the interfacial resistance in these systems.

Low final concentration runs.

A set of low concentration runs was also performed with the aim of determining the dependence of k_c on the agitation speed N. This kind of experimental runs has the advantage of requiring a smaller amount of sieved solid phase to acquire the information needed. Moreover, in this case the simpler eqn.(12) can be used instead of eqn.(9). Eqn.(4) was used to compute the particle diameter from concentration data and the results are shown in Figs 5 to 8. It can be observed there that straight lines are obtained, as expected. The departure from the predicted straight line of the last data points in each run depends on the fact that the functional form of eqn.(4) amplifies the experimental reading error for points close to the asymptotic concentration.

It is interesting to note that the slopes of these lines are strictly proportional to k_c (see eqn.13), and are almost the same for the runs conducted at the same agitation speed N, confirming that k_c is practically independent of D_p in the size range tested.

The solid straight lines reported in Figs 5-8 , have been obtained by linear regression on data points up to half of the initial diameter. The corresponding k_c values computed by eqn.(13) are reported in table 1 and plotted vs the agitation speed in Fig.9. A dependence of k_c on the agitation speed raised to the power 0.506, can be observed there: this value falls in the middle of the range (0.4 - 0.6) of previously published data for the kind of impeller used in this work, as reported in (1).

Looking at the values of k_c in Table 1, a very week dependence of k_c on the particle diameter seems to appear especially at the higher agitation speeds (1100 rpm) where a dependence on $D_p^{0.05}$ can be observed. The variations are however within the experimental error range so that the conclusion of practical independence of k_c on D_p can be considered as confirmed by these tests.

Recently Armenante (6) reported the following correlation, holding for "microparticles":

$$Sh = 2 + 0.52\, Re_p^{0.52}\, Sc^{1/3} \qquad (15)$$

where

$$Sh = k_c \frac{D_p}{\mathcal{D}_{ab}} \quad ; \quad Re_p = \left(\frac{\varepsilon D_p^4}{\nu^3} \right)^{1/3} \quad ; \quad Sc = \left(\frac{\nu}{\mathcal{D}_{ab}} \right) \qquad (16)$$

The exponent 0.506 here found for N is in close agreement with that of eqn.(15). In fact in the range of impeller Reynolds Numbers utilized in this work, (Re_i = 6500-18700), the exponent of N in the expression of dissipated power is slightly smaller than 3 for impeller-tank configurations similar to the one here utilized (7). On the other hand, eqn.(15) suggests a dependence of k_c on the particle diameter D_p raised to the power -0.3 in the range of Sh>>2 and an even greater dependence is predicted as Sh approaches the value 2. This prediction is not supported by the present data, as discussed above. This different behaviour might depend on the fact that eqn.(15) is claimed to apply to "microparticles" i.e. to particles characterized by D_p smaller than the Kolmogoroff's length scale η (6). On the contrary an estimation of the same scale for the experimental range here utilized (based on an estimated minimum power number ≈ 0.33) showed that in this case D_p was from 1.15 to 11 times greater than η .

For fine particles Asai et al. (8) proposed a correlation similar to eqn.(15) with an exponent for Re_p of 0.58. Many other correlations have been published in the past (9), but none of them seems to be able to combine a dependence on the square root of the agitation speed with the practical independence of particle size found in the experimental range of this work.

CONCLUSIONS

The simple dissolution model resulting in eqn.(9) was tested with two different solid solutes. In each case the model was shown to accurately fit the experimental data confirming its soundness. For the solute for which the saturation concentration was experimentally accessible (NaCl), it was also possible to show that the assumption $C_i = C_{sat}$ actually holds at least in the dissolution rate range examined. For the other solute (Schönite) investigated, it was possible to determine C_i which was found to be constant in each run and independent of D_p and N: this is a strong indication that interfacial resistances are negligible for this solute too, leading to the conclusion that the C_i experimentally determined with these dynamic tests should coincide with the unknown saturation concentration.

In general, it is suggested that eqn.(9) might be successfully applied for:

- modelling and scale-up of agitated batch dissolution apparatuses;
- determination of C_i for modelling and/or design purposes;
- determination C_{sat} for systems in which a steady state determination is inhibited;
- verification of the relevance of interfacial resistances in dissolving and crystallizing systems.

Together with C_i , k_c values are determined provided that the ratio (β/α) is independently estimated, as for instance in the way indicated in the experimental part of this work.

Tests conducted at various agitation speeds and initial particle diameter showed that the mass transfer coefficient k_c is practically independent of D_p while it depends approximately on the square root of the agitation speed.

ACKNOWLEDGMENT

The authors would like to thank the support by "Italkali" S.p.A., Italy to the present investigation.

TABLE 1 - Experimental conditions and evaluated k_c values

run n°	$C^* \cdot 10^3$ [gr/cm^3]	N [rpm]	Dpo [cm]	$k_c \cdot 10^3$ [cm/s]	run n°	$C^* \cdot 10^3$ [gr/cm^3]	N [rpm]	Dpo [cm]	$k_c \cdot 10^3$ [cm/s]
1	8.610	900	0.090	1.50	10	0.861	900	0.045	1.41
2	8.610	900	0.018	1.48	11	0.861	900	0.090	1.45
3	8.610	650	0.045	1.33	12	0.861	900	0.090	1.44
4	0.861	450	0.018	1.02	13	0.861	1100	0.018	1.54
5	0.861	650	0.014	1.16	14	0.861	1100	0.018	1.52
6	0.861	650	0.018	1.22	15	0.861	1100	0.045	1.63
7	0.861	650	0.045	1.23	16	0.861	1100	0.090	1.65
8	0.861	650	0.090	1.21	17	0.861	1300	0.045	1.73
9	0.861	900	0.018	1.39	*18	153.7	1100	0.045	2.72

* Solid solute NaCl; k_c computed assuming $(\beta/\alpha) = 6$

SYMBOLS USED

A = constant in eqn.(7) [s^{-1}]

a_v = particle specific area = part. surf. / part. volume [cm^{-1}]

B = constant in eqn.(12) [cm/s]

C = liquid bulk concentration [gr/cm^3]

C^* = liquid phase concentration for total solid dissolution [gr/cm^3]

C_i = liquid phase concentration at the interface [gr/cm^3]

C_o = liquid phase concentration at time t_o [gr/cm^3]

C_{sat} = saturation concentration [gr/cm^3]

D_{eq} = equivalent particle diameter [cm]

D_p = particle diameter [cm]

D_{po} = particle diameter at time t_o [cm]

J = solute flux at solid-liquid interface [gr cm^{-2} s^{-1}]
k_c = liquid phase mass transfer coefficient [cm/s]
M_o = mass of the solid phase at time t_o [gr]
N = impeller rotational speed [s^{-1}]
Re_i = impeller Reynolds number $\rho D_i^2 N / \mu$
S_p = interfacial surface of a particle [cm^2]
S_t = total solid interfacial surface [cm^2]
t = time [s]
t_o = initial time [s]
V_p = particle volume [cm^3]
V_t = liquid phase volume [cm^3]
x = dimensionless concentration
x_i = interfacial dimensionless concentration

Greek symbols

α = volume shape factor
β = surface shape factor
ε = power dissipation per unit mass [erg s^{-1} gr^{-1}]
η = Kolmogoroff length scale $(\nu^3 / \varepsilon)^{1/4}$ [cm]
ν = kinematic viscosity [cm^2/s]
ρ_p = particle density [gr/cm^3]

REFERENCES

1) Harnby N., Edwards M. F. and Nienow A. W. (Eds), 1985, "Mixing in the Process Industries", Butterworths, London, England.

2) Hixson A. W. and Crowell J. H., 1931, Ind. Eng. Chem. 23, 923.

3) Zwietering Th. N., 1958, Chem. Engng Sci. 8, 244.

4) Bird R. B., Stewart W. E. and Lightfoot E. N., 1960, "Transport Phenomena", Wiley, New York, USA.

5) Wilhelm R. H., Conklin L. H. and Sauer T. C., 1941, Ind. Eng. Chem. 33, 453.

6) Armenante P. M. and Kirwan D. J., 1989, Chem. Engng Sci. 44, 2781.

7) Rushton J. H., Costich E. and Everett H. J., 1950, Chem. Eng. Progr. 41, 395.

8) Asai S., Konishi Y. and Sasaki Y., 1988, J. Chem. Eng. Japan 21, 107.

9) Boon-Long S., Laguerie C. and Couderc J. P., 1978, Chem. Engng Sci. 33, 813.

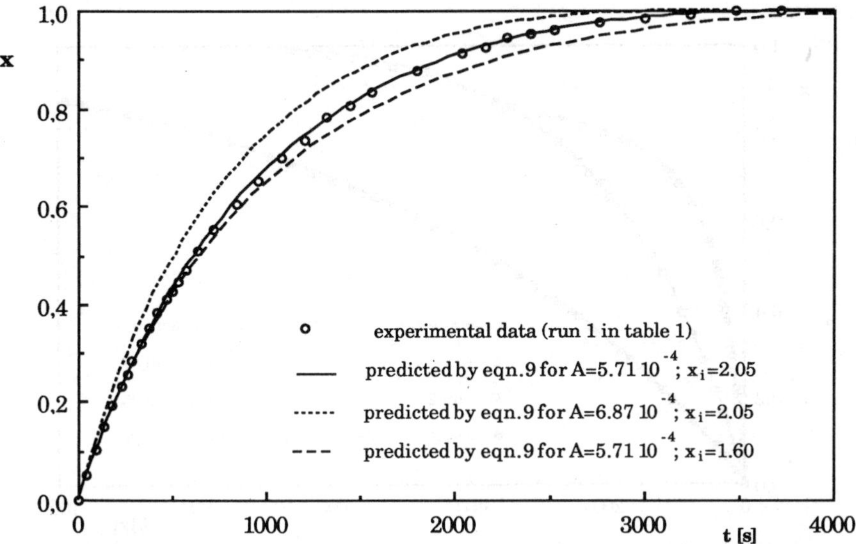

Fig.1: Comparison between experimental data and predicted concentration curve

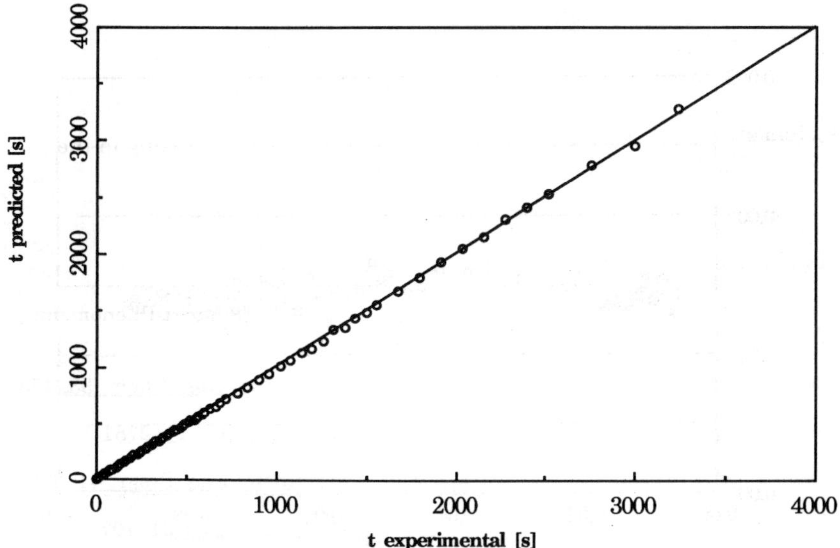

Fig. 2: Times needed for the solution to reach a certain concentration vs times predicted by eqn.9 for $A=5.71\ 10^{-4}$; $x_i=2.05$

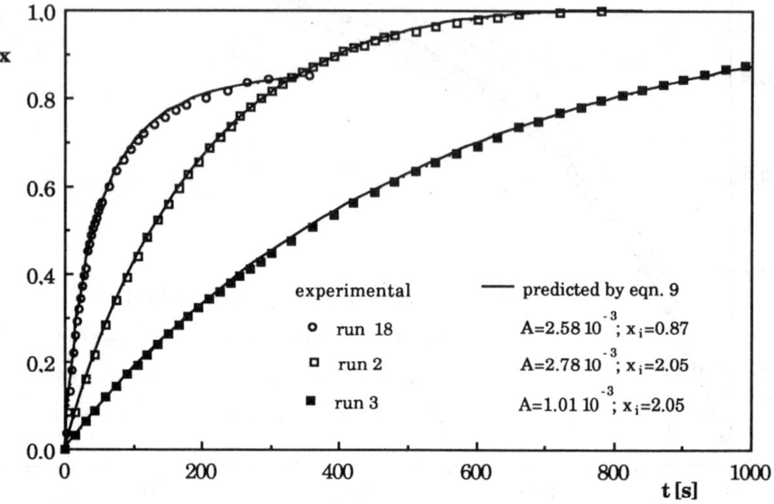

Fig.3: Comparison between experimental data and predicted concentration curve

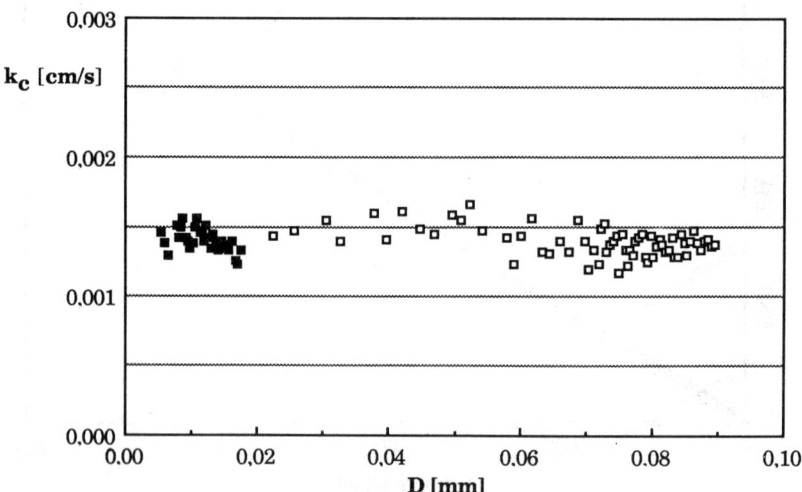

Fig. 4: K_c values (at various diameters) obtained from eqn.7

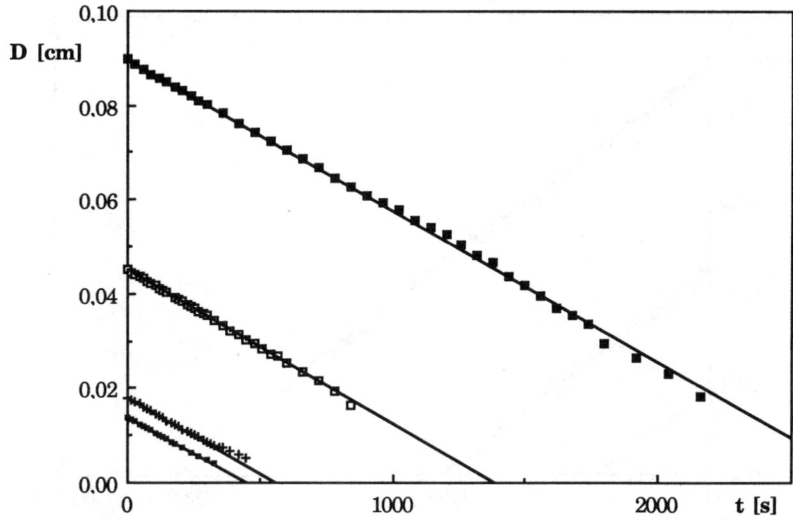

Fig.5: D_p vs time at N=650 rpm: ■ run 8; ❒ run 7; + run 6; ■ run 5

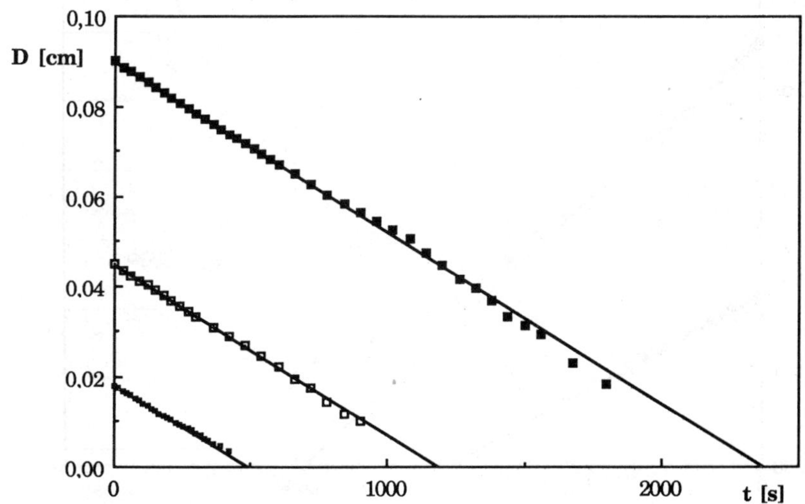

Fig. 6: D_p vs time at N=900 rpm: ■ run 12; □ run 10; ■ run 9

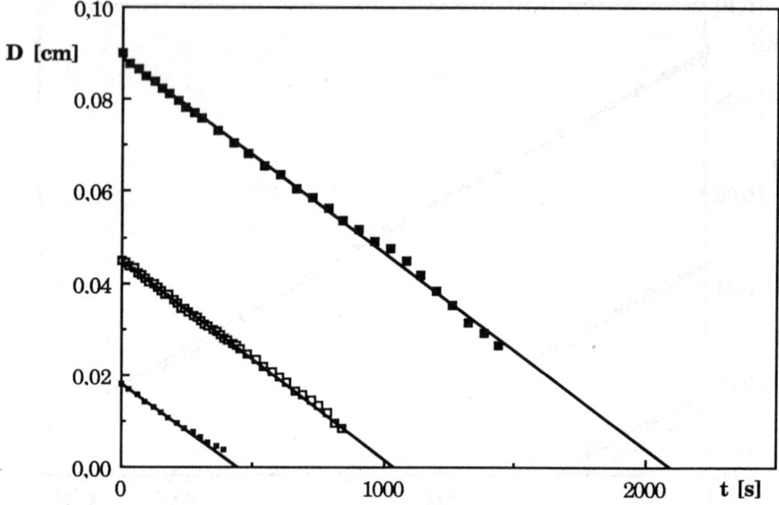

Fig.7: D_p vs time at N=1100 rpm: ■ run 16; □ run 15; ■ run 14

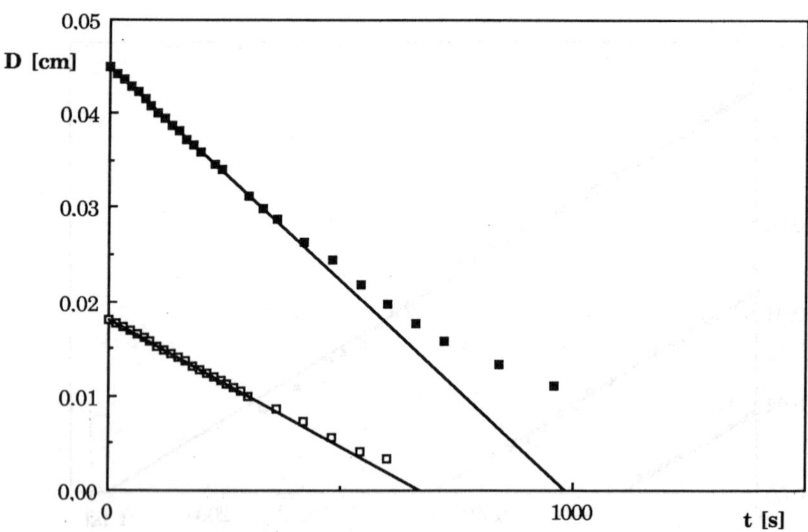

Fig. 8: D_p vs time at N=450 rpm and 1300 rpm: ■ run 4; □ run 17

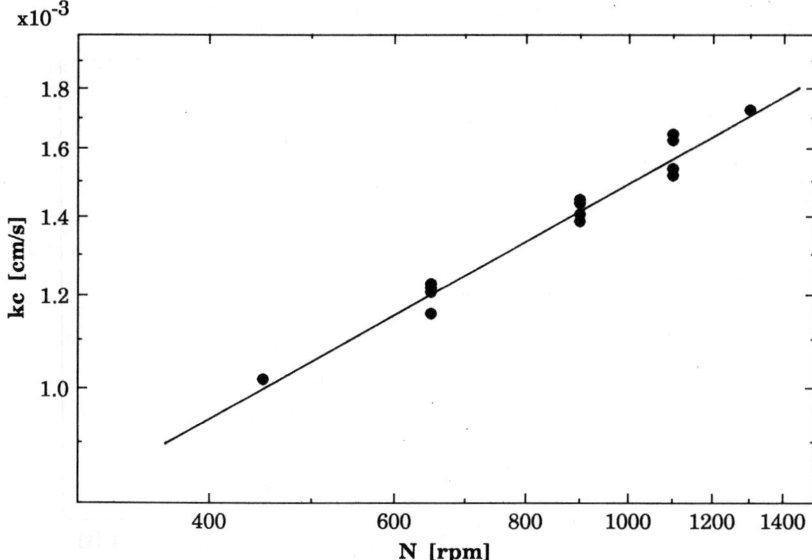

Fig.9: Agitation speed vs mass transfer coefficient (runs 4 - 17)

STIRRING OF CONCENTRATED SLURRIES: A SEMI-EMPIRICAL MODEL FOR COMPLETE SUSPENSION AT HIGH SOLIDS CONCENTRATIONS AND 5 m^3 VERIFICATION EXPERIMENTS

C. Buurman*

Our previous work on complete suspension has been extended to high slurry concentrations, where the interaction of the solids with the turbulent spectrum and the interaction between the solid particles becomes noticeable. Theoretical considerations lead to qualitative relations for complete suspension in which empirical constants may be found from suspension data. A good description of stirrer speed (and power) for complete suspension is obtained for a broad range of conditions and concentrations up to 50 % by volume. Theory, corroborated by the experiments, indicates that, on a larger scale, the specific power required for complete suspension is lower, and that this effect is more pronounced for larger particles.

INTRODUCTION

In the design of stirred slurry reactors, for suspension polymerisation, for instance, or for crystallisers, it is essential for adequate mixer operation that certain important mixer performance criteria are met. In addition to mixing duties related to the homogenisation of the compounds necessary to the process, a straightforward minimum requirement is that the solid particles must be kept sufficiently in motion. In the literature many publications can be found dealing with this problem of "complete suspension"; see, for example, Harnby et al (1). However, a confusing variety of relations for the stirrer speed required for complete suspension still exists.

In a previous paper (Buurman et al (2)) we analysed some experimental data on small- and large-scale suspension vessels with the aim of finding a more reliable interpretation of the effect of the scale of the equipment. Very briefly we came to the conclusion that the mechanism of complete suspension must be explained by the pick-up of particles from the bottom by turbulent eddies. From the turbulence theory we could derive that, in the case where this phenomenon just occurs, the stirrer must dissipate an equal specific power in geometrically similar systems. Deviations from this rule can be explained by the smallness of scale, and of differences in scale, in the experiments and by errors in the scaling of the impellers, such as incorrect scaling of the stirrer blade thickness.

In all this work the effect of the concentration of the solid phase received relatively little attention. Most of the work reported in the literature deals with solids concentrations of less than 20 % by volume. For commercial application of a stirred slurry reactor the concentration is very important because the production capacity of the unit depends directly on it. So

* Koninklijke/Shell-Laboratorium, Amsterdam (Shell Research B.V.), Badhuisweg 3, 1031 CM Amsterdam, The Netherlands

for the design one faces the question of the maximum allowable solids concentration. The solids concentration may affect the mixer design for a reactor via various routes, such as the slurry density and viscosity, the homogeneity of the slurry, but certainly also via the stirrer speed required for complete suspension.

This paper deals with the effect of the slurry concentration on complete suspension. We will derive on the basis of turbulence theory how the presence of the solids affects the stirrer speed and power needed for complete suspension. This leads to a qualitative relation. On the basis of a set of small-scale experiments, with a fixed geometry, but several combinations of solids and liquids, we have verified the theoretical approach and derived empirical constants for the theoretical relation. Finally we report on a series of experiments in a relatively large-scale mixing vessel by which the semi-empirical correlations, and in particular the effect of the scale of the equipment, were verified.

THEORY

In our previous paper (2) the model description for complete suspension was based on the pick-up of particles from the bottom by turbulent eddies. In addition it was assumed that the bottom was sufficiently far from the turbulence generating device, i.e. the impeller, that the turbulent field could be regarded as isotropic. The size of the eddies responsible for the pick-up was assumed to be several times larger than the particles themselves. It was shown that for most suspensions this means that the eddies belong to the range of "energy-containing" eddies. The result was the following relation for the stirrer speed for complete suspension:

$$n_c d^{2/3} = K\, d_p^{1/6} \left[\frac{g\,\Delta\rho}{\rho}\right]^{1/2} F(c) \quad \ldots\ldots (1)$$

It was shown that the observed effect of physical parameters (density of liquid and solid and the particle size) on the stirrer speed for complete suspension was modelled correctly in this manner. In particular this was found to hold for the effect of the scale of the equipment. Therefore we will follow this approach further for the description of the effect of the solids concentration.

In the literature it has long been reported that the presence of solids reduces the turbulence intensity; see, for example, Neesse et al (3). Almost all investigators in this field have observed that, for higher concentrations, a higher stirrer speed is indeed necessary to achieve complete suspension. But due to the relatively low concentrations investigated, the effect is generally marginal. In small-scale experiments we observed that the required stirrer speed for complete suspension could increase significantly at concentrations higher than 30 % by volume (see Figure 1). These experiments also showed that the effect of the concentration could not be modelled by a simple power function, as proposed by, for instance, Zwietering (4).

The physical background to the concentration effect is the increased power dissipation by the presence of solid particles in the turbulent field. These particles can interfere with the turbulent field via two routes:

a. Due to their inertia particles do not follow the motion of the liquid. Consequently, the kinetic energy of a particle changes continuously. Acceleration of a particle occurs at the expense of energy from the turbulent eddies, whereas deceleration leads to energy dissipation in the form of heat.

b. At high solids concentrations in particular, particles will also interfere with each other (collisions and/or hindrance). In a similar way this will lead to increased dissipation of the energy of the turbulent eddies.

Thus, in order for the point of complete suspension to be maintained, a higher solids concentration requires a higher power input to the system.

For this power input we can draw up the following energy balance:

$$\rho \epsilon_t = (1-c)\ \rho \epsilon_1 + c \rho_s \epsilon_s + c \rho_s \epsilon_i \quad \ldots \ldots (2)$$

For the specific energy of eddies with a fluctuating velocity v_e and a wave number k

$$\epsilon_1 \propto v_e^3 k$$

and since we have postulated that the size of the eddies responsible for the pick-up is proportional to the size of the particles, we can write for these eddies

$$\epsilon_1 \propto v_e^3 / d_p$$

For the average specific energy of the solid particles in motion proportionality to the energy of the relevant eddies is assumed; moreover, the time scale of the deceleration/acceleration process is assumed to be proportional to the time scale of the eddies. Hence the power involved, ϵ_s, is proportional to the power in the liquid, ϵ_1, or

$$\epsilon_s = K_1\ \epsilon_1 \quad \ldots \ldots (3)$$

For the power involved in the particle-particle interactions we assume proportionality to the energy of the fluctuating liquid velocity and to the frequency of the interactions:

$$\epsilon_i \propto v_e^2 . f \quad \ldots \ldots (4)$$

This frequency f will depend on

- the cross-sectional area of the particles,
- the effective velocity of the particles,
- the number of particles per unit volume.

The effect of the cross-sectional area is taken as $f \propto d_p^2$, and the effective velocity is again assumed to be proportional to the fluctuating velocity component, so $f \propto v_e$. For this description of the frequency of the interactions we must consider the number of particles per unit volume as a measure of the free path length in three dimensions. For dilute suspensions the size of the particles is negligible compared to this free path length, but in the concentrated slurries considered here we must take it into account.

$$\lambda \propto d_p \left[\frac{c_{max}}{c}\right]^{1/3} - d_p \quad \ldots \ldots (5)$$

c_{max} represents the maximum concentration at which all particles are in contact with one another and we cannot speak of a slurry any more. For the frequency of the interactions it follows that

$$f \propto 1/\lambda^3 \propto \frac{1}{d_p^3 \{(c_{max}/c)^{1/3} - 1\}^3} \quad \ldots \ldots (6)$$

Combining the three effects mentioned above and substituting in equation (4) yields:

$$\epsilon_i \propto \frac{c}{\{(c_{max}/c)^{1/3} - 1\}^3} \cdot \frac{v_e^3}{d_p} \qquad \text{or}$$

$$\epsilon_i \propto \frac{c}{\{(c_{max}/c)^{1/3} - 1\}^3} \cdot \epsilon_1 \quad \ldots\ldots (7)$$

So far we have considered the turbulent spectrum as isotropic. Thus the energy of the eddies is given by

$$E(k,t) \propto v_e^2/k \propto \epsilon^{2/3} k^{-5/3} \quad \ldots\ldots (8)$$

This simply means that for a sequence of decaying eddies the power involved remains constant (until viscous dissipation occurs). However, the presence of the solids causes extra power dissipation, so that in the sequence of decaying eddies the turbulent energy will decrease more rapidly than indicated by equation 8. In other words, in the presence of solids the exponent of the wave number k increases. For such a case the power of eddies at a wave number k becomes

$$v_e^3 k \propto \epsilon (k/k_0)^{-a} \quad \ldots\ldots (9)$$

The exponent $-a$ expresses here the increased steepness of the turbulence spectrum and k_0 is the wave number of the initial eddies generated by the impeller. Again we can now replace k by $1/d_p$ and k_0 by $1/d$ because the size of the generated eddies must be directly proportional to that of the generating device, i.e. the scale of the mixing equipment. This transforms (9) into

$$\epsilon_1 \propto v_e^3 d_p \propto \epsilon (d_p/d)^{-a} \quad \ldots\ldots (10)$$

It will be clear that the value of a depends on the solids concentration and the densities. For low concentrations a must approach zero (the spectrum remains unchanged); for higher concentrations the effect may become comparable to the increased steepness of the spectrum as found in, for instance, the discharge stream of an impeller (Van der Molen and Van Maanen (5)). As a first approximation we have taken for a the value found for the impeller stream (5/4) and applied it only for the high concentration effect, i.e. of the particle- particle interaction (eq. 7)

$$\epsilon_i \propto \frac{c\ (d_p/d)^{5/4}}{\{(c_{max}/c)^{1/3} - 1\}^3} \cdot \epsilon_1 \quad \ldots\ldots (11)$$

(From Figure 1 it can be seen that, in practice, this becomes noticeable at concentrations higher than 25 % by volume.)

For the total concentration effect we can substitute ϵ_s and ϵ_i in the power balance of equation 2, which yields

$$\frac{\epsilon_t}{\epsilon_1} = \frac{n_c^3}{n_{c0}^3} = 1 - c + K_1 c \frac{\rho_s}{\rho} + K_2 \frac{\rho_s}{\rho} \cdot \frac{c^2\ (d_p/d)^{5/4}}{\{(c_{max}/c)^{1/3} - 1\}^3} \quad \ldots\ldots (12)$$

(N.B. for eq.1 $F(c) = (\epsilon_t/\epsilon_1)^{1/3}$)

This relation is rather complicated, certainly compared to earlier proposed correlations. Zwietering (4), for instance, has accounted for the concentration effect by a simple power function

$$F(c) \propto c^{0.13}$$

The important features of equation 12 are that it accounts for effects of

- density of solid and liquid phase
- maximum solids concentration
- particle size
- scale of the equipment

Moreover, in the present approach, F(c) becomes 1 if the concentration approaches zero, as may be expected.

EXPERIMENTAL RESULTS

From small-scale laboratory experiments a series of data was available on the stirring of concentrated slurries. Most of these experiments had been carried out in a 10 l perspex mixing vessel of a common geometry for suspension studies (Figure 2), equipped with an axial turbine impeller and four baffles. The systems investigated were sand-water suspensions with particle sizes of 200, 800 and 2200 μm, coal-water slurry (800 μm), polystyrene beads in octanol and butanol (800 and 2200 μm) and spent cracking catalyst in water (50 μm). From these tests we have calculated the constants in relations 1 and 12. The following values were found

$K = 1.2$ to 1.4
$K_1 = 2.5$
$K_2 = 10.7$

Figure 1 shows a comparison of the experimental results with lines according to the suspension model as described. It can be seen that the model agrees very well with the observed concentration effects. In particular, the sharp increase in the stirrer speed for complete suspension in the concentration range above 20 % and the effects of density and particle size are reflected by the model in a consistent way. Moreover, the exponent for the slope of the turbulence spectrum is found to be a reasonable assumption.

Because all data had been collected from a single vessel the effect of scale could not be verified without an additional experiment on a scale considerably larger than 10 l. Equation 12 suggests that the increase of the suspension speed at high concentrations would become less important in larger equipment. For practical applications this is an important favourable effect and we decided to perform a large-scale verification test to check this. A series of measurements of the stirrer speed for complete suspension was made in the 10 l vessel and for the same concentration range the measurements were repeated for an 80 l and also a 4.5 m^3 mixing tank. The geometry of the mixers was carefully made similar. The system used was a suspension of silica spheres in water with a particle size of 3 mm. Figure 3 shows the results of both series of experiments and the drawn lines are predictions according to equation 12. The scale effect was accurately predicted by the model.

CONCLUSIONS AND FINAL COMMENTS

On the basis of the turbulence theory we have extended our model for complete suspension to high slurry concentrations.

Experiments have shown that the model predictions hold for a variety of particle sizes and densities.

The model includes the maximum solids concentration, where all particles are in contact with one another.

The model also gives realistic values for very low concentrations.

The predicted scale effect has been verified experimentally: for high slurry concentrations in large-scale reactors less specific power (W/kg) is required for complete suspension than in (geometrically similar) small-scale equipment.

SYMBOLS USED

a – exponent for the slope of the turbulence spectrum
c – volume concentration
c_{max} – maximum volume concentration of solids
d – impeller diameter (m)
D – tank diameter (m)
d_p – particle size (m)
$E(k,t)$ – energy per wave number interval (m^3/s^2)
f – frequency
$F(c)$ – concentration effect
g – gravitational acceleration (m/s^2)
k – wave number (m^{-1})
K, K_1, K_2 – constants (–)
k_0 – wave number of initial eddies generated by impeller (m^{-1})
n – stirrer speed (s^{-1})
n_c – stirrer speed for complete suspension (s^{-1})
v_e – fluctuating velocity component (m/s)
ϵ – specific power dissipation (W/kg)
ϵ_i – specific power, interactions (W/kg)
ϵ_l – specific power, liquid (W/kg)
ϵ_s – specific power, solids (W/kg)
ϵ_t – total specific power (W/kg)
λ – free path length of particles (m)
ρ – liquid density (kg/m^3)
ρ_s – solids density (kg/m^3)
$\Delta\rho$ – density difference (kg/m^3)

REFERENCES

1. Harnby,N.,Edwards,M.F.,and Nienow,A.W.,1985,"Mixing in the Process Industries", Butterworth,London,England.

2. Buurman,C.,Resoort,G.,and Plaschkes,A.,1986,Chem. Eng. Sci. 41,2865.

3. Neesse,T.,Schubert,H.,Liepe,F.,and Möckel,H.-O.,1977,Chem. Tech. 10, 544.

4. Zwietering,Th.N.,1958,Chem. Eng. Sci. 8,244.

5. Van der Molen,K.,and Van Maanen,H.R.E.,1978,Chem. Eng. Sci. 33, 1161.

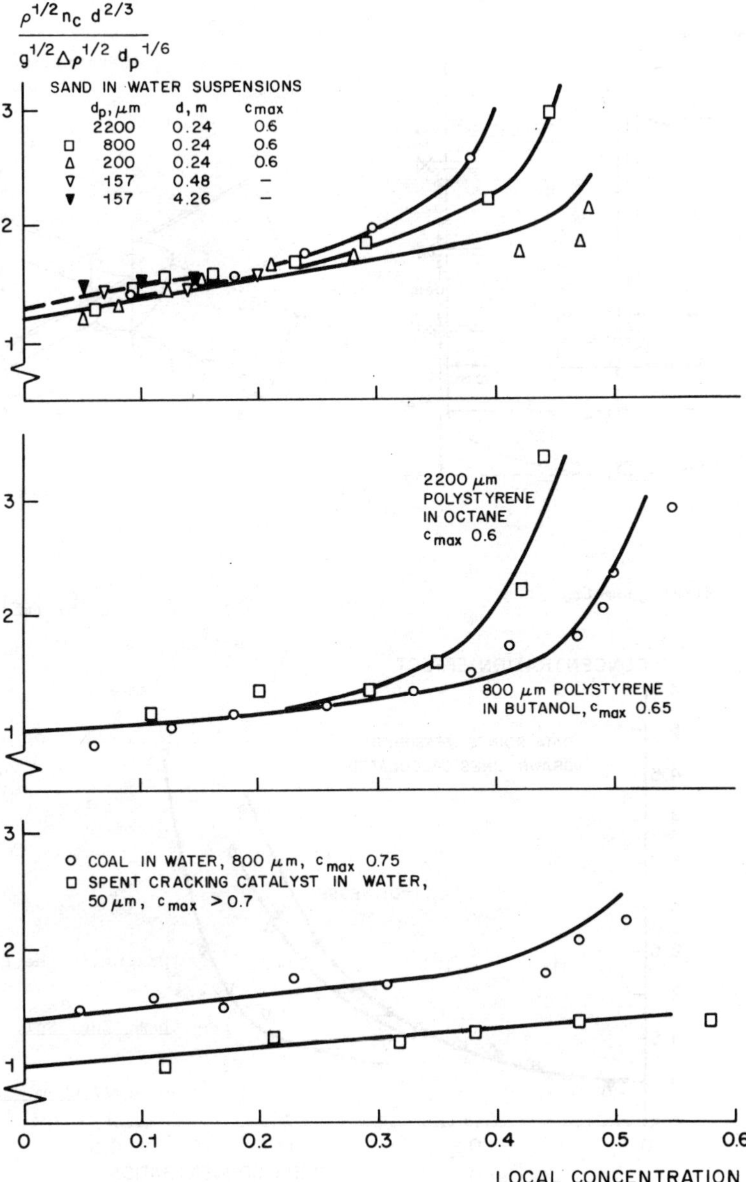

Figure 1 Effect of concentration and particle size on complete suspension

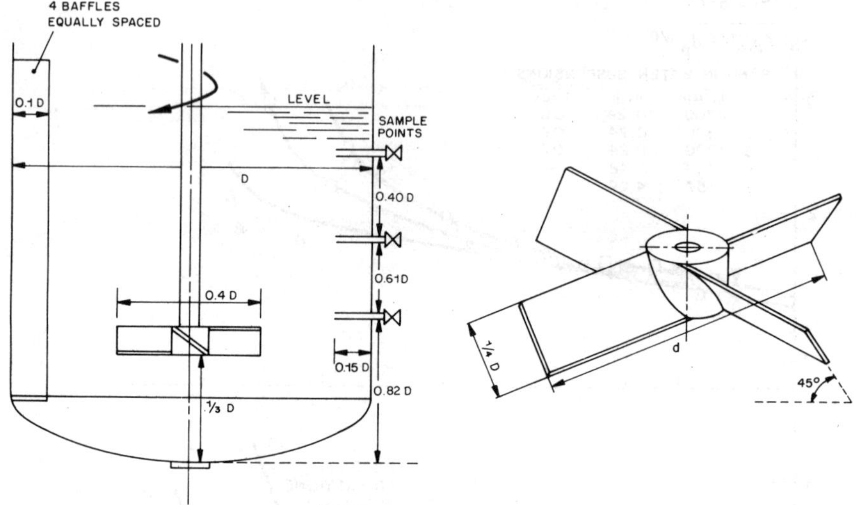

Figure 2 Mixer geometry

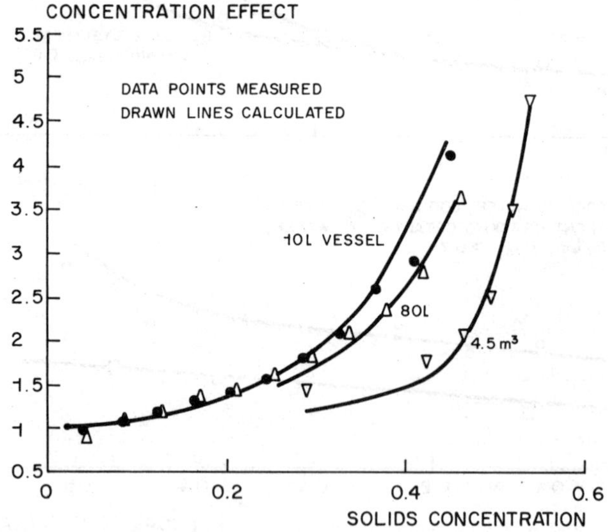

Figure 3 Concentration effect on complete suspension for various scales

A STUDY OF SOLIDS-LIQUID MIXING IN A MODEL MIXER

B.Bland* and A J Matchett**

The mixing of non-absorbent particulate solids with liquids has been investigated in trough mixers, in which a blade rotated. The liquid was added in the range 3-20% giving 3 phase systems. Sands and ballotinis were used as the solids with bimodal size distributions. Water and cmc solutions were added as liquids. Methods of sampling and analysis were developed to allow liquid content and size distribution to be measured. The rate at which material built up on walls was measured and seen to take place in 3 stages. A novel mixing index was developed, based upon the depth of a uniform film of liquid.

INTRODUCTION

This paper deals with the mixing of non-absorbent particulate solids with liquids in systems where there is insufficient liquid to fill the voids. Such systems are 3 phase, consisting of solids, liquid and interstitial air.

Teoh investigated the mixing of Loch Aline sand with 1% carboxymethylcellulose solution (1% cmc) in Z blade and ribbon mixers [1][2][3]. He observed the phenomenon of caking, where material stuck to walls and blades in the mixer. He saw the necessity for sampling the cake separately from the bulk in the body of the mixer and showed that the cake had a higher liquid content and was more densely packed than the bulk. Bland followed this work by demonstrating that the cake also contained a greater proportion of fine particles ($\leq 300\ \mu m$) than the bulk material [3][4]. Caking therefore results in both solids and liquid segregation. Teoh analysed the composition of samples from the bulk and was able to propose a mixing index I, where:

$$I = (x/s)/(\sigma_o/\mu) \qquad (1)$$

x is the mean liquid content of the samples from the bulk,
s is the standard deviation of the liquid content of the bulk,
σ_o is the variance of an unmixed system,
$\sigma_o = \mu(1-\mu)$, μ is the mean liquid content based upon the liquid added.

I was able to allow for the formation of cake and was independent of operating conditions and composition for the Loch Aline sand system used [1][2]. Teoh also identified 3 stages in the mixing process: an initial rapid reduction in I, described by first-order kinetics; a second slower mixing regime and a steady state where I was equal to 0.01.

* Rowntree PLC, York, UK
** Chemical Engineering, School of Science and Technology, Teesside Polytechnic, Middlesbrough, Cleveland. UK

This paper reports some further investigations of the mixing processes in solids-liquid systems. This work was performed in purpose-built mixing cells. This allowed experiments to be performed under much greater control than was possible in the industrial mixers. Mechanisms of mixing, segregation and the structure of the mix were therefore studied in detail and some aspects are reported in this paper [4].

EQUIPMENT AND EXPERIMENTS

Test Cells

Two cells were built which consisted of troughs 140 mm diameter by 120 mm in length. One was fabricated from perspex for visualisation and one was made from stainless steel, with a detachable wall.

Each cell had an axial shaft, onto which was fixed a blade, as shown in Figure 1. The blade ran the length of the cell and the clearance between the blade and the trough could be adjusted by the locating threads which held the blade to the shaft. The perspex cell was hand cranked, while the steel cell was motor driven at 50 rpm. The blade was rotated clockwise in order to mix the material in the cell.

The steel cell had a removable wall on the side where the blade left the trough, as shown in Figure 2. It was shown in previously reported work that the cake could only form on the wall where the blade left the trough [2][3][4]. This was demonstrated both experimentally and theoretically [3]. The wall could be slid out of the cell and the cake weighed and sampled directly.

Materials and Sampling

1. Solids

Synthetic solids systems were made up with a wide size distribution from sands and ballotine.
SANDS-Garside sand (GA) between 710 and 800 μm was mixed with Loch Aline sand (LA) (<300 μm) to produce a bimodal size distribution of a granular type of particle. The amount of fine material was controlled by adjusting the ratio of GA to LA in the initial mix.
BALLOTINI-ballotini grade A (600-800 μm) was mixed with grade AB (<300 μm) to produce a bimodal system of spherical particles equivalent to the sand system. The proportion of fines was changed by changing the ratio of the two components.

2. Liquids

The above solids were mixed with liquids which included:-
Water
Cmc solutions 1%-5% by weight
The liquids were mixed with the solids in the range 3-20% total weight basis. Saturation occured at about 20% liquid dependent upon the solids packing.

3. Sampling and Analysis

The bulk and the cake were sampled separately. Samples were taken from the bulk on a random basis, using a reference grid and random number tables. An insertion sample probe was used to remove a fixed volume of sample from a given location. This is shown in Figure 3 [1][2][3][4]. Sample size was fixed at 0.25 g and 10 or 15 samples were taken at time.

Samples were taken from the cake on an ordered basis. A 4x4 sample grid was placed over the cake and the probe pushed onto the wall to collect the complete depth of sample. The amount of material collected was dependent upon the amount of cake which had formed and there was no cake at some locations for some conditions.

A sample was weighed and then placed in an oven to dry overnight to constant weight, at 86°C. It was then re-weighed. This enabled the liquid content to be determined. If the sample had cmc in it, it was carefully washed in distilled water, filtered and dried to remove the cmc. The sample was gently broken into individual particles and reweighed. The sample was then passed through a 350 µm sieve and the number of oversize particles were determined by counting them. This was related to the weight of largeparticles by multiplying by the average weight per particle determined from previous calibration. Hence, the solids composition of each sample was determined.

4. Experimental Method

The experimental techniques outlined in this paper were the result of much painstaking development and checking [4]. The validity of the choice of batch size, sample size, number of samples, blade clearance and other factors are given elsewhere and are too voluminous to present in this paper [4].

A summary of experimental conditions and the range of factors investigated are given in Table 1. Experiments were performed as follows:-

700 g of solids were first dry-mixed in the cell to ensure a uniform distribution of fines. The blade was then rotated until it was above the solids and the liquid carefully poured onto the surface of the solids avoiding splashing onto the walls or the blade. The blade was then rotated for the required period of time, after which it was stopped. Samples were taken from the bulk, for analysis. In the case of the steel cell, the wall section was removed and weighed to determine the amount of cake. Samples of cake were then taken for analysis. The machine was re-assembled and the blade re-started until the next sample time.

CAKE FORMATION

Cake build-up was only observed in the nip region of the wall where the blade left the trough, as expected from previous work [3][4]. The amount of cake was measured and typical build-up curves are shown in Figures 4 and 5.

The build-up consists of 3 stages:-
Initial caking or sludge formation
Initialisation-liquid mixing in the bulk with little further caking
Ordered caking-caking follows zero-order kinetics.

Initial caking occurs in the first few revolutions of the blade. Liquid is pushed onto the wall by the blade and forms a wet base onto which dry particles stick. The cake is very wet and sludge-like. It is often unstable and can fall back into the bulk. There is little particle segregation associated with this type of adhesion. When the viscous forces, or interfacial tension forces in the liquid are sufficiently large, then agglomerates are formed, which enrobe the liquid in protective particles. The liquid is thus unable to wet the wall and this initial stage of material build-up is either very small or absent.

The initial caking occurs in systems which are unmixed. After this stage, there is a period in which the main action is the mixing of liquid with the solids, in the bulk. There is very little caking in this period. The time taken for this to complete has been called the Time of Initialisation (TOI) [4]. Observations in the perspex cell suggest that the TOI is completed when films of liquid can be observed to have formed on the wall of the trough. It has also been proposed that the end of the TOI marks a reasonably good distribution of liquid on the surfaces of the particulate solids [4].

After the TOI, the amount of cake increases in a linear manner with time. The rate of increase has been termed the rate of deposition (RD). Cake formed in this ordered manner exhibits a higher liquid content and contains a greater proportion of fine particles than the bulk, showing the solids and liquid segregation effects observed in previous work [3][4]. A constant value of RD would be expected if the liquid were well-distributed and the amount of cake formed were a small fraction of the bulk. The blade would then expose the wall to a similar caking environment at each revolution and the amount of material sticking to it would increase by a constant amount per revolution.

The build-up of cake can take one of several patterns as shown in Figure 6 where the curves are differentiated into Types A, B amd C according to the values of TOI and RD. Values of TOI and RD are given in Table 2. A detailed analysis is given elsewhere [4], but two points will be highlighted in the present paper.

It can be seen that as the viscous effects increase with an increase in cmc concentration, then there is a tendency for Type C curves with low values of initial caking. This is due to the formation of agglomerates, which trap the liquid and give very poor overall liquid distributuion. The relatively long TOI values represent the time taken to break the agglomerates down and release the liquid contained within them. In such circumstances, the mixer should be designed as an agglomerate comminutor, for a rapid and successful mixing process (assuming that the maldistributions inherent in the agglormerates is unacceptable).

There is also evidence to suggest that the TOI is a measure of the time at which the liquid is well distributed on the surface of the solids. This suggests that the speed of mixing may be measured by observing the time at which ordered caking commences. This can be done visually and would be a relatively simple task on an industrial plant (compared to taking samples and analysing). However, this hypothesis needs to be tested further, particularly on larger scale mixers.

ANALYSIS OF LIQUID DISTRIBUTION

A method was sought to express the distribution of liquid on the solids in a rational and logical manner, without recourse to the empiricism inherent in Teoh's index I. It was argued that if the liquid were well-distributed, it would exist as a uniform film over all the available solids surfaces. If this simple model were valid, then the ratio of the volume of liquid to surface area of each sample would be constant. The experimental techniques used in these tests allowed the volume of liquid to be calculated from the amount of liquid in each sample (determined by oven drying) and the surface area of the samples of ballotini was calculated from the ratio of large to small particles knowing weight, size and shape of the particles in this case.

A typical plot of liquid volume versus surface area is shown in Figure 7 after 10 miutes mixing time. This shows that the simple model is valid, but also that there is more liquid associated with the cake than the bulk.

It is a graphic illustration that the structure of the cake formed on the wall is quite different to the bulk. Figure 8 illustrates the effect of mixing time and it can be clearly seen that the system is much better mixed after 10 minutes than 1 minute and relatively poorly mixed systems exhibit scatter.

The liquid film depth would appear to offer potential as a physically relevant index of solids-liquid mixing in these 3 phase systems. The liquid depth was determined for each sample of ballotini and from this, the mean depth and standard deviation was computed for each set of experimental conditions. The results are shown in Table 3.

The results show that mixing is improved with time and the standard deviation after 10 minutes is smaller than after 1 minute. The value of the mean depth is related to the amount of liquid added, the more liquid added -the greater the depth. The liquid in the cake is always greater than in the bulk. Table 3 shows that this method is also sensitive to solids composition.

The cake contains more fine particles than the bulk and therefore has a greater surface area per unit mass of solids. Cake would therefore be expected to contain more liquid because of this. However, the liquid depth model demonstrates that there is additional liquid in the cake above that which can be accounted for by the surface area. Analysis suggests that the cake is close to saturation with the liquid (a liquid content of 15-20% weight). Observations suggest that after the TOI, wetted particles contact the trough and liquid spreads from the particles onto the trough. The liquid is observed as a film on the trough wall in the perspex cell. The blade sweeps this liquid into the nip region. This liquid can then enter the voids of the cake by capillary action or by the pressure exerted by the rotation of the blade. Solids segregation has been shown to be due to frictional and adhesion forces acting upon particles at the nip [3][4]. Small particles preferentially adher to the wall.

The simple liquid depth model assumes a uniform film of liquid over all surfaces. However, other workers have noted that a large fraction of liquid is held in the meniscii at points of contact between particles [5]. The model was therefore modified to try to allow for additional liquid at points of contact [4]. This gave a relationship of the form:-

$$X = \rho\lambda\{w_1(\frac{a_1}{m_1} - \frac{a_2}{m_2}) + \frac{a_2}{m_2}\} + \rho\ c\{w_1(\frac{1}{m_1} - \frac{1}{m_2}) + \frac{1}{m_2}\} \quad (2)$$

where

X is the dry weight fraction of liquid in a sample

w is the weight of a size fraction/total dry weight solids

λ is the liquid film depth

ρ is the liquid density

m is the mass of a particle in a particular size fraction

a is the surface area of a particle in a particular size fraction

c is the volume of liquid per particle held in meniscii

subscript 1 refers to the fine particles

2 to the large particles.

The data did not support this more sophisticated model. The data for the bulk and cake did not conform to the same relationship, but the two sets of data clustered around different points. This also supports the simple model in that the structure of the bulk and cake are quite different. However, the range of X and w_l were small and the scatter too great to allow straight lines to be drawn through the data.

CONCLUSION

The mixing of liquids with non-absorbent solids has been investigated at compositions where the systems are 3 phase and multicomponent, using model trough mixers. Methods of sampling and analysis have been devised and applied. The rates of build-up of cake have been measured and shown to consist of 3 stages; an initial stage of sludge build-up; a time ofinitialisation, where mixing takes place in the bulk with little cake formation and ordered caking with zero order kinetics.

The distribution of liquid on the solids has been measured and has been expressed in terms of a mixing index based upon the depth of a uniformly distributed liquid film. The mean and standard deviation of this value have been used to follow the mixing process.

REFERENCES

1. H.A.Teoh, PhD Thesis, Teesside Polytechnic, 1984.
2. H.A.Teoh, A.J.Matchett, Powder Technology, 1985, 44, 27.
3. A.J.Matchett, H.A.Teoh, B.Bland, Fluid Mixing II, I.Chem.E.
4. B.Bland, PhD Thesis, Teesside Polytechnic, 1989.
5. M.Tschapek, S.Falasca, C.Wasowski, Powder Technology, 1985, 42, 175.

Table 1 Experimental Conditions for Mixing Cell

Weight of dry solids	700g
Solids premixing	Dry solids mixed in the cell, prior to liquid additions
Method of liquid addition	Liquid carefully poured onto the solids surface, avoiding contact with walls and blade
Speed of mixing	50 rpm in motor powered cell. Approx 50 rpm in hand cell
Blade clearance	1 mm
Method of sampling bulk	Random-grid and random number tables. Samples taken with probe
Size of sample-bulk	0.25g
Number of samples-bulk	15
Method of sampling cake	Ordered-4*4 sample matrix used. Samples taken with short probe
Size of sample-cake	Dependent upon the amount of cake
Number of sample-cake	16
Sample analysis	All samples analysed for liquid content, amount of large/ small particles - as outlined in text
Solids used	Sands: Garside and Loch Aline Ballotinis
Fines in sand	0, 25, 50, 75, 100%
Liquids	Water cmc solutions 1 to 5%
Range of liquid added	3 - 20% weight basis

Table 2 Time of Initialisation (TOI) and Rate of Deposition (RD) for CMC Systems

System		Sand LA/GA 50/50			Ballotini AB/A 50/50		
Liquid	% added	TOI (min)	RD (g/rev)	System Type	TOI (min)	RD (g/rev)	System Type
1% cmc	4.8%	0.25	0.043	A	0.75	0.034	C
	9.1%	0.25	0.140	A	0.75	0.064	A
	13.0%	0.25	0.422	A	unstable cake		A
	16.7%	unstable cake		A	unstable cake		A
2% cmc	4.7%	2.00	0.037	C	1.00	0.054	C
	9.1%	1.00	0.090	C	1.00	0.072	C
	13.0%	1.00	0.140	C	0.00	0.181	B
	16.7%	1.00	0.203	C	unstable cake		B
3% cmc	4.8%	5.00	0.037	C	2.00	0.049	C
	9.1%	2.00	0.049	C	2.00	0.086	C
	13.0%	2.00	0.069	C	2.50	0.167	C
	16.7%	1.00	0.076	C	0.00	0.215	B
4% cmc	4.8%	6.00	0.015	C	2.50	0.033	C
	9.1%	4.00	0.037	C	3.00	0.062	C
	13.0%	3.00	0.055	C	2.00	0.083	C
	16.7%	1.50	0.081	C	1.50	0.081	C
5% cmc	4.8%	6.00	0.003	C	5.00	0.022	C
	9.1%	4.50	0.010	C	4.00	0.028	C
	13.0%	6.00	0.029	C	4.00	0.031	C
	16.7%	5.00	0.026	C	2.00	0.025	C
			SAND			BALLOTINI	
1% cmc	4.8%	0.00	0.025	B	2.50	0.019	A
50/50	9.1%	0.00	0.087	A	0.50	0.031	A
1% cmc	4.1%	0.00	0.014	A	1.50	0.013	B
75/25	9.1%	0.00	0.032	A	0.50	0.018	A
2% cmc	4.8%	2.50	0.023	C	1.50	0.031	C
50/50	9.1%	2.50	0.021	C	1.50	0.028	A
2% cmc	4.8%	2.50	0.026	C	1.50	0.041	C
75/25	9.1%	2.00	0.044	C	2.00	0.039	C
3% cmc	4.8%	9.00	-	C	3.00	0.023	C
50/50	9.1%	4.50	0.011	C	3.00	0.037	C
3% cmc	4.8%	5.00	0.012	C	3.00	0.021	C
75/25	9.1%	3.50	0.014	C	3.00	0.048	C
4% cmc	4.8%	7.00	0.004	C	5.00	0.021	C
50/50	9.1%	4.00	0.006	C	4.00	0.032	C
4% cmc	4.8%	5.00	0.002	C	5.00	0.017	C
75/25	9.1%	5.00	0.006	C	4.00	0.047	C
5% cmc	4.8%	8.00	-	C	6.00	0.019	C
50/50	9.1%	8.00	-	C	7.00	0.020	C
5% cmc	4.8%	5.00	-	C	6.00	0.016	C
75/25	9.1%	4.50	-	C	6.00	0.036	C

Table 3 Liquid Film Depth in Mixing Systems

System % cmc - % large - % Liquid		Film Depth Mean (μm)	s.d.	Range of Depth (0.05 level)
1% - 50 - 4.7%	1 min	7.83	4.70	5.23 - 10.43
	10 mins	6.67	1.06	6.08 - 7.26
1% - 50 - 9.1%	1 min	14.96	3.41	13.07 - 16.85
	10 mins	14.44	0.91	13.94 - 14.94
Cake	1 min	26.84	2.23	25.60 - 28.08
	10 mins	22.32	1.15	21.68 - 22.96
1% - 25 - 4.7%	1 min	5.48	4.34	3.08 - 7.88
	10 mins	5.85	0.40	5.63 - 6.07
Cake	10 mins	16.20	1.77	15.22 - 17.18
1% - 25 - 9.1%	1 min	12.29	3.90	10.13 - 14.45
	10 mins	11.64	0.53	11.35 - 11.93
Cake	1 min	23.52	2.06	22.37 - 24.66
	10 mins	19.77	1.79	18.78 - 20.76
5% - 50 - 9.1%	1 min	19.23	32.25	1.37 - 37.09
	10 mins	15.06	1.00	14.51 - 15.61
Cake	10 mins	21.82	1.45	21.02 - 22.62
5% - 25 - 9.1%	1 min	31.99	18.73	21.63 - 42.35
	10 mins	12.26	2.33	10.97 - 13.55
Cake	10 mins	17.42	1.83	16.41 - 18.43
5% - 25 - 4.7%	1 min	9.15	16.32	0.11 - 18.19
	10 mins	6.40	0.83	5.94 - 6.86
Cake	10 mins	16.66	2.70	15.16 - 18.16

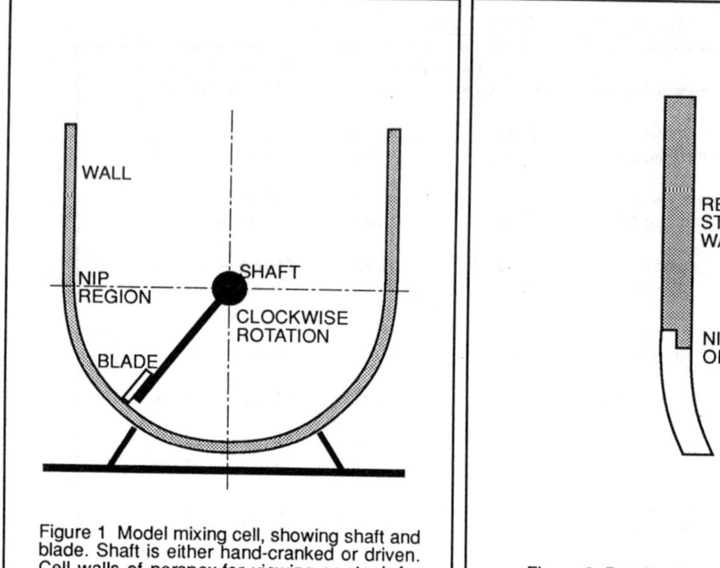

Figure 1 Model mixing cell, showing shaft and blade. Shaft is either hand-cranked or driven. Cell walls of perspex,for viewing or steel, for caking tests

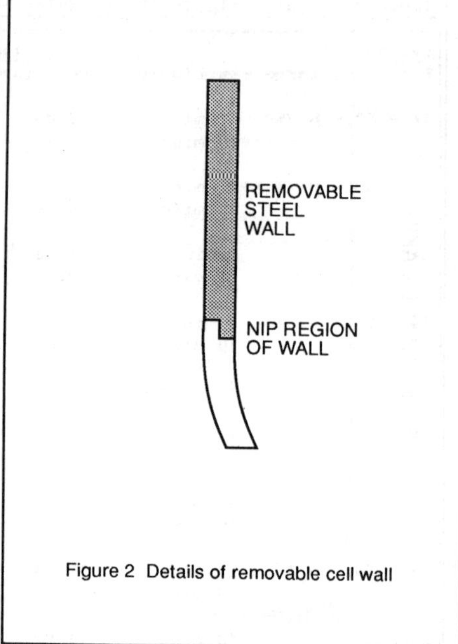

Figure 2 Details of removable cell wall

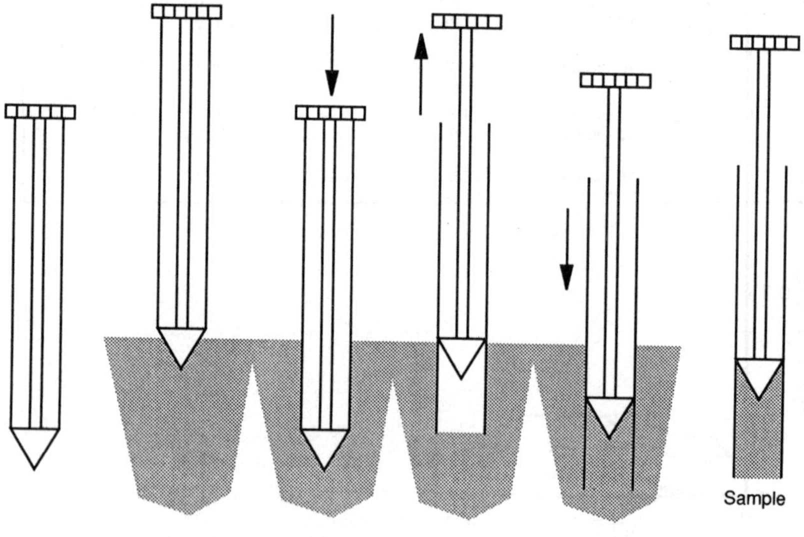

Figure 3 Sample probe and sampling method

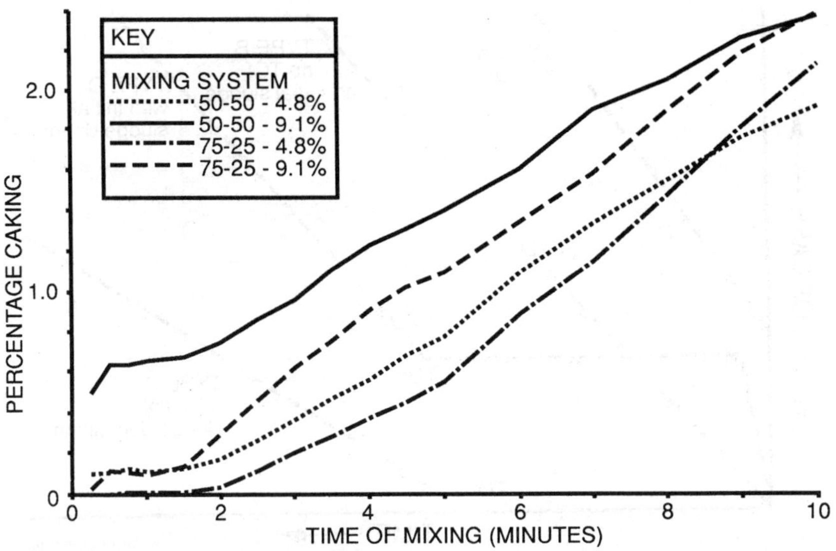

Figure 4 Build-up of cake on cell wall - ballotini + 2% cmc

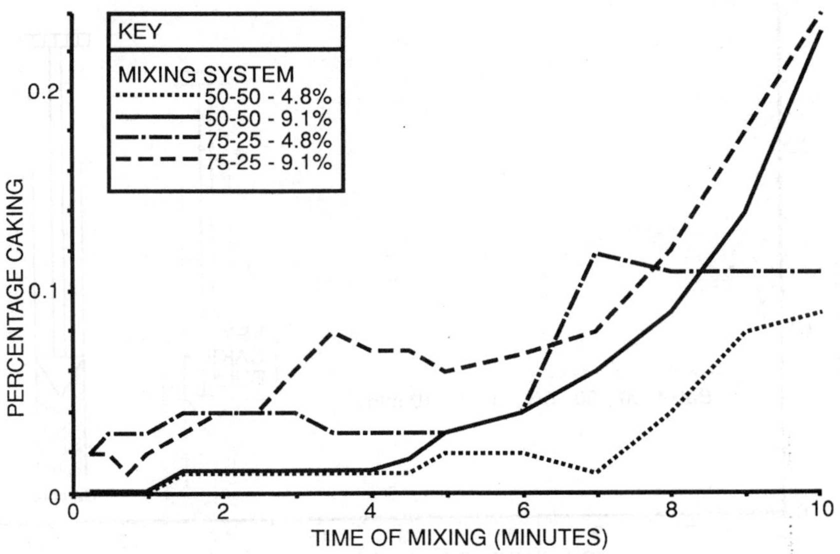

Figure 5 Build-up of cake on cell wall - sand + 4% cmc

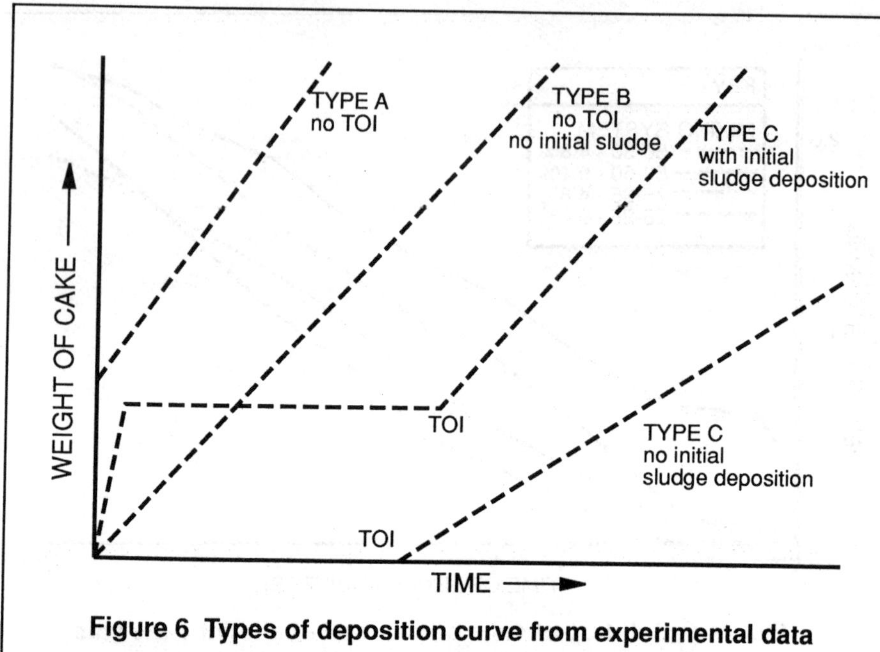

Figure 6 Types of deposition curve from experimental data

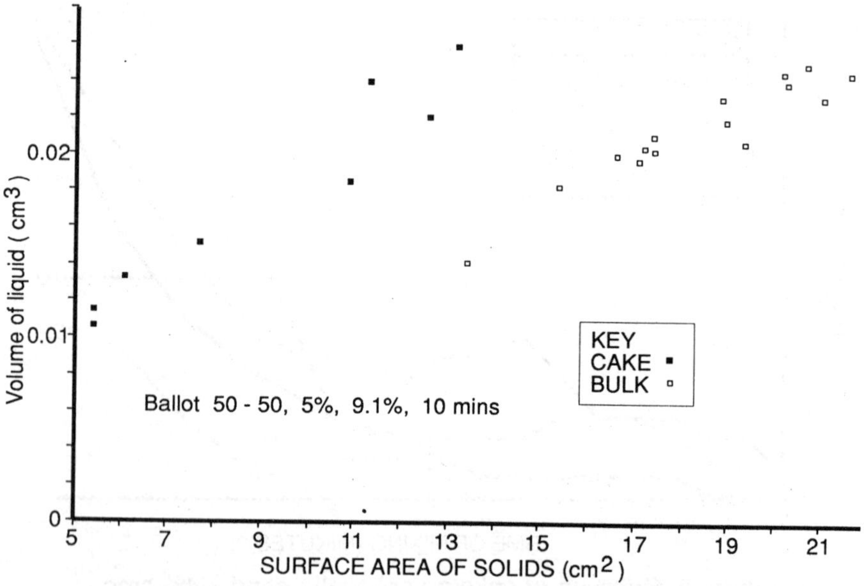

Volume of liquid (cm^3)

0.02

0.01

0

5 7 9 11 13 15 17 19 21

SURFACE AREA OF SOLIDS (cm^2)

Ballot 50 - 50, 5%, 9.1%, 10 mins

KEY
CAKE ▪
BULK ▫

Figure 7 Volume of liquid versus surface area for ballotini samples

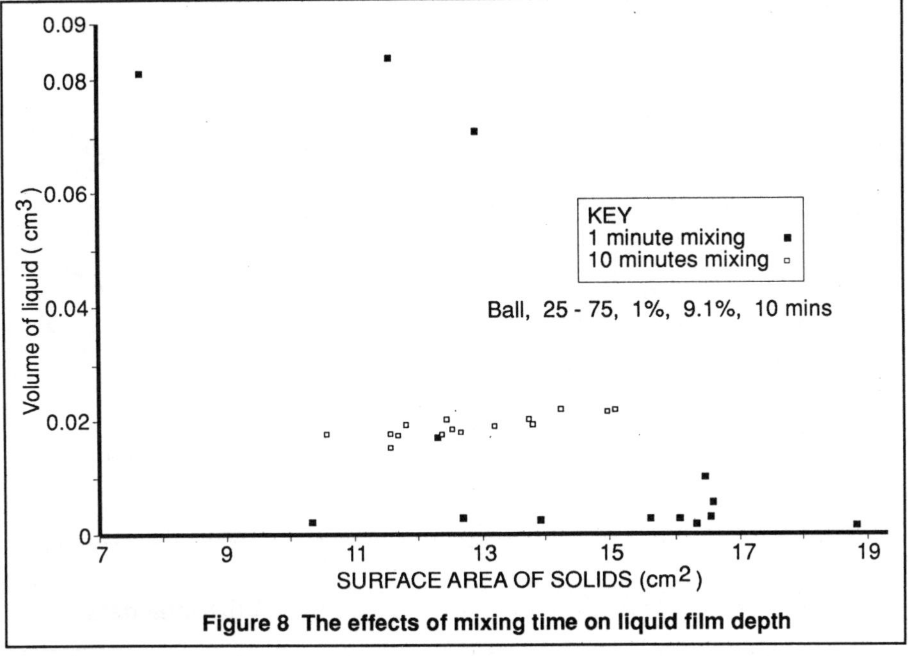

Figure 8 The effects of mixing time on liquid film depth

Poster presentations

SUSPENSION OF SOLIDS IN LIQUID-JET STIRRED VESSELS

P. AYAZI SHAMLOU* AND A. ZOLFAGHARIAN*

Experimental measurements are reported on off-bottom suspension of solid particles in a liquid-jet stirred vessel and a model is provided for the rational interpretation of the data. The model is based on a balance between fluid-dynamic and body forces acting on the particles at the point of off-bottom suspension. Reasonable agreement is obtained between experimental data and predictions from the model.

Solid suspension in a liquid jet-stirred vessel is compared with suspension in a conventional rotary mixer. Similarities between the two systems suggest that the mechanism(s) of suspension in the two mixers are very similar.

INTRODUCTION

Off-bottom suspension of particles in a liquid in a stirred vessel refers to the conditions when the effective weight of the particles becomes equal to the vertical component of the fluid-dynamic forces acting on the particles. When this happens particles begin to lift off and move into the bulk of the liquid in the tank. Attainment of off-bottom (complete) suspension is probably the most important process requirement in any suspension operation involving negatively buoyant particles and its influence on solid-liquid transfer processes in stirred vessels is well established [Harnby et al 1987].

In general, the motion of a liquid in a stirred vessel is induced by external input of energy, and rotary devices, e.g. propellers and turbine impellers, are most widely used for this duty. Consequently, considerable work has been done on rotary mixers in order to establish correlating equations for the estimation of off-bottom suspension for different situations: for mechanically-stirred vessels the so called just-suspension speed of the impeller, N_{js}, is widely used for this purpose [Zwietering 1958; Nienow 1968; Baldi et al 1978; Ayazi Shamlou 1990].

Liquid-jet stirred mixers with their simple design, low installation and operating costs and relatively good mixing performances compared with conventional rotary mixers offer potential applications in a broad range of process engineering operations including suspension of solids in liquids. However, despite a large volume of information on the hydrodynamics of various types of jet [Poreh et al 1967] in solid suspension applications, published studies on jet induced suspension are few in the open literature [Dawson and Trass 1966; Yeckel and Middleman 1987] and no conclusive work has yet emerged as to the mechanism of suspension in this type of mixer.

* Chemical and Biochemical Engineering, University College London, WC1E 7JE.

The aim of this investigation is to examine the ability of a liquid-jet stirred mixer to suspend settling particles in a vessel and to establish a model which provides a rational basis for describing the results.

FLUID-PARTICLE MOTION IN A LIQUID-JET STIRRED VESSEL

Off-bottom suspension of particles by a liquid jet is critically dependent on the choice of the jet and its position in the vessel. Essentially, off-bottom suspension requires the establishment of a radial impinging wall jet at the base of the tank while good bulk up-and-down fluid motion in the vessel is needed to establish an acceptable level of suspension homogeneity.

Figure 1 shows the basic components of a possible liquid-jet stirred mixer which fulfils these requirements. Agitation is provided by recirculating through the vessel a fraction of the liquid using a recirculating pump, a nozzle and a network of pipes. The authors know of no established recommendations for selection of nozzles for different suspension duties and no guidelines appear to exist for the optimum position of the nozzle in the vessel.

For suspension of settling particles in low viscosity fluids, the feeder-nozzle should be fully submerged in the liquid and should be normally positioned with its center coincident with the axis of the vessel and with its head pointing vertically downward, facing the base of the vessel. To minimize pressure losses through the feeder-nozzle any changes from the main recirculating pipe diameter to nozzle outlet diameter should be gradual and smooth. In some applications, specially in cases where suspension homogeneity is not important, line-size nozzles may be used.

The recirculating liquid may be withdrawn either from the top or from the base of the vessel. To minimize the overall pressure drop for flow through the system, the length of the main recirculating pipe to and from the tank should be kept to a minimum by positioning the recirculation pump immediately outside the vessel. The recirculating pump may be used for other duties, e.g. for loading and off-loading the suspension tank and for pumping to and from other vessels. In special applications, fully submersible pumps are also commercially available.

The jet of fluid emerging from the nozzle strikes the center of the base of the tank at 90°, changes direction and spreads radially outwards, sweeping the base of the vessel as it spreads. On hitting the vertical walls of the vessel, the jet changes direction once again and travels upward towards the surface of the liquid.

The motion of the impinging jet of liquid at the base of the tank provides the necessary hydrodynamic conditions for off-bottom suspension of particles while bulk fluid motion in the vessel determines the level of suspension homogeneity.

In practice, attainment of off-bottom suspension and suspension homogeneity are to some extent opposing factors. For example, for otherwise identical conditions reducing the magnitude of the so called just-suspension velocity, U_{js}, (see later) by, say, reducing the distance between the feeder-nozzle and the base of the tank has a detrimental effect on suspension homogeneity. Preliminary experiments carried out at University College London suggest that the feeder-nozzle should be located at about half the height of the liquid from the base of the vessel. This provides both an acceptable level of suspension homogeneity (measured, for example, by the maximum vertical distance travelled by particles in the liquid) and values of U_{js} which are comparable with off-bottom suspension in rotary mixers (measured by, say, the just-suspension speed of the impeller, N_{js}).

EXPERIMENTAL EQUIPMENT AND MEASUREMENTS

Experiments were carried out in order to measure the (off-bottom) liquid-jet velocity, U_{js}, for suspension of a range of particles. In all the experiments, measurements of U_{js} were based on the well-estabilshed 1-2 seconds suspension criterion originally proposed by Zwietering [1958] and subsequently by others [Nienow, 1968 and Harnby et al, 1987] for suspension in rotary mixers.

Figure 2 shows the liquid-jet stirred mixer used in this work. Experiments were carried out with five geometrically similar vessels (Table 1). Each vessel had a flat bottom, fabricated either from perspex or glass in order to facilitate flow visualization of the base of the tank.

A small centrifugal pump with a maximum capacity of 200 gallons per hour was used to recirculate the liquid which in all the experiments reported in this work was water. Additionally, liquid height in the vessel was always maintained equal to vessel diameter.

Three convergent nozzles with outlet diameters equal to 0.005m, 0.007m and 0.009m were used in this work. In addition, experiments were carried out with a line-size nozzle (no converging section) with a diameter equal to the that of the main recirculating pipe which was 0.0127 (Table 2). Preliminary experiments established that the height of the nozzle from the base of the vessel, h, should be approximately half the height of the liquid in the vessel, H. This ratio of h/H resulted in good off-bottom suspension without detrimentally affecting the homogeneity of the suspension. For this reason this ratio of h/H=0.5 was used in all subsequent experiments.

The kinetic power input into a liquid-jet stirred mixer is solely dependent on fluid flow rate through the nozzle and on the pressure drop across it. Fluid flow rate through the nozzle was obtained using an in-line turbine flow-meter with a direct read-out in liters per minute while the pressure drop across it was measured by a differential pressure transducer with a range of 0-1.999 bar.

In the experiments reported in this work, the recirculating liquid was withdrawn from the top of the vessel. To ensure adequate flow to the pump, the suction pipe was expanded at the inlet and had a diameter slightly larger than the outlet pipe diameter which was 0.025m reducing to 0.0127m (nozzle inlet). Preliminary experiments established that, in most cases, at the point of complete suspension there was a layer of liquid at the top of the vessel which was void of particles. The thickness of this clear liquid layer depended on particle size and density: the height of the clear layer generally increased with increase in particle size and/or particle density. Advantage was taken of this phenomenon by positioning the inlet to the pump in this region. This meant that suspension could be achieved with no physical contact between particles and the moving parts of the recirculation device. This method of suspension has obvious advantages in situations where particles are relatively weak and thus prone to breakage. With low density acetal particles and for glass beads with diameter less than about 500 μm, particles were able to travel to the uppermost region of the vessel at the point of complete suspension. To stop these particles leaving the vessel through the suction pipe, the inlet to the pump was covered with a suitable screen mesh with a diameter smaller than the size of the particles.

Table 3 provides information on all the particles used in this work. To measure off-bottom (complete) suspension velocity of liquid jet for a given particle size and concentration, fluid flow rate through the nozzle was gradually increased until the point was reached when no particles remained at the bottom of the tank for more than 1-2 seconds. In case suspension did not occur for a particular nozzle at the maximum possible fluid flow rate through the nozzle, the next smaller diameter nozzle was fitted and the experiment was repeated.

Liquid velocity at the point of complete suspension could be reproduced with an accuracy of 5%.
Further details on equipment, experimental techniques and method of analyses are given elsewhere [Zolfagharian 1990] and a selection of results are presented and discussed below.

MECHANISM OF SOLID SUSPENSION IN A LIQUID-JET STIRRED VESSEL

As fluid flow rate through the nozzle is gradually increased a point is reached when particles in the zone immediately beneath the eye of the nozzle begin to roll radially outwards in the general direction of fluid flow. This produces a central region in the vessel which is void of particles and a ring of particles around the periphery of the base of the vessel. As fluid flow velocity is increased, the diameter of this central region increases as more and more particles accumulate in ring. In this state of motion, particles keep in contact with each other or with the base of the vessel.

As fluid flow velocity is increased further, a point is reached when particles in the ring begin to lift off into suspension. More and more particle become susceptible to this type of motion (lifting/suspension) as fluid velocity is increased.

Once a particle is in the bulk of the liquid, its motion is governed by fluid flow field in the vessel. Inevitably, therefore at some point in its travel the particle will return to the base of the tank. When a particle reaches the base of the tank its motion is then dependent on the motion of the fluid in that region. Often, from its position of arrival at the base of the vessel, the particle rolls towards the ring before being lifted into suspension again.

The motion of liquid and particles at the base of a liquid-jet stirred vessel described above is remarkably similar to solid-liquid motion in a rotary-stirred vessel [Ayazi Shamlou 1990]. One important implication of this observation is that similar mechanism(s) may be responsible for solid suspension in the two types of mixer.

Theoretical considerations of the fluid-dynamic and body forces acting on particles at the base of a vessel [Ayazi Shamlou 1990; Ayazi Shamlou and Zolfagharian 1987] provide the following expression for the critical fluid velocity at the base of the vessel for off-bottom suspension.

$$U_a \propto [d_p (\Delta\rho/\rho_f) g]^{1/2} \quad (1)$$

To lift off a particle, the liquid flowing over the particle has to dissipate energy at a rate which is equal to the product of the vertical component of the hydrodynamic forces acting on the particle and the vertical component of the fluid flow velocity. Assuming that (i) the total fluid energy dissipation rate for off-bottom (complete) suspension is proportional to the number concentration of particles [Oroskar and Turian 1980], and (ii) particles in the vessel are mono-size and spherical so that number concentration may be related to volume concentration, then the total rate of energy dissipation for off-bottom suspension, ε_T, may be expressed by the following equation:

$$\varepsilon_T \propto \rho_f U_a^3 C_v T^3/d_p \quad (2)$$

Equation 2 is independent of the source providing kinetic energy to the liquid in the vessel. In the case of a liquid-jet stirred mixer power input into the liquid in the vessel may be assumed to be equal to total mechanical power delivered to the nozzle provided frictional losses in the nozzle are negligible: this is true when fluid viscosity is low, e.g. water. Thus

$$P_n = 1/2 (Q_n \rho_f U_n^2) = 1/2 (\rho_f U_n^3 \pi d_n^2/4) \quad (3)$$

Additionally, If it is assumed that (i) the energy of the jet at the base of the vessel at the point of complete suspension, ε_b, is proportional to the kinetic jet power delivered by the nozzle, P_n, and (ii) the fluid energy dissipation rate for off-bottom (complete) suspension of particles is proportional to the energy of the jet at the base of the vessel, then Eqns 1 to 3 may be used to give the following expression for U_{js} [zolfagharian 1990]:

$$U_{js} = A \frac{d_p^{1/6} [g \Delta\rho/\rho_f]^{1/2} C_v^{1/3} T}{d_n^{2/3}} \quad (4)$$

The value of the proportionality constant, A, in Eqn 4 includes the superimposed effects of several factors some of which are difficult to quantify at the present stage of development: these include the efficiency of transfer of energy from the liquid to the particles and the relationship between fluid and particle Reynolds numbers and particle drag and lift coefficients at the bottom of the vessel. However, for particle and fluid Reynolds numbers in the turbulent region, the value of the proportionality constant is expected to remain a constant for a given situation. While it is possible to make first-order estimations of the magnitude of the constant A in Eqn 4, it is suggested that its value be obtained experimentally for a given vessel and nozzle configuration.

Eqn 4 suggests that experimental data on just-suspension velocity for various nozzles and vessels may be brought together by plotting the data as U_{js} against the group on the right hand side of Eqn 4. This plot is shown in Fig 4: the solid line is the best line of fit through the data points and has a slope of unity. Evidently, model Eqn 4 adequately describes off-bottom suspension data for liquid-jet stirred vessels.

The only other systematic study on solid suspension with liquid-jet stirred mixers known to the authors is by Racz et al [1977] who provided the following empirical correlation for the estimation of U_{js}

$$Re_n = K\ Ar^{x1}\ [\Delta\rho/\rho_f]^{x2}\ [d_p/T]^{x3}\ [h/T]^{x4}\ [d_n/T]^{x5}\ C_w^{x6} \qquad (5)$$

where Re_n is the critical jet Reynolds number at the point of complete suspension and Ar is the Archimedes number. The magnitude of the exponents x1 to x6 were found experimentally by Racz et al for a range of vessel configurations and for various particles (Table 4).

Substituting the values of K and x1 to x6 in Enq 5 and rearranging the resulting expression gives the following equation for U_{js} for solid suspension in flat-bottom vessels

$$U_{js} \simeq 2\ \frac{\nu^{0.16}\ [\Delta\rho/\rho_f]^{0.28}\ g^{0.42}\ T^{1.16}\ d_p^{\ 0.1}\ C_w^{\ 0.24}}{d_n} \qquad (6)$$

Noting the empirical nature of Eqn 6, the agreement between model Eqn 4 and Eqn 6 is considered to be remarkably good.

One point that merits further discussion is the difference between the value of the exponent of the $[\Delta\rho/\rho_f]$ term in Eqns 4 and 6: evidently Eqn 4 predicts a value of 0.5 for the exponent of this term while Eqn 6 suggests that it should be 0.28. Our own experimental data plotted in Fig 5 suggest that the exponent of the $[\Delta\rho/\rho_f]$ is about 0.4. It should be noted that in all the experiments carried out as part of this work, density difference, $\Delta\rho$, was varied by changing particle density: liquid density remained constant in all the experiments. Moreover, a decrease in $\Delta\rho$ results in a corresponding decrease in U_{js}. With decreasing values of $\Delta\rho$ and U_{js} measurements become increasingly more sensitive to experimental errors, instrument calibration and uncertainties: this is specially relevant when considering the experimental data on U_{js} provided by Racz et al, Eqn 6, [1977] since some of their particles had relatively low densities.

Another point of interest is the effect of fluid viscosity, μ, on U_{js}. Eqn 6 shows a slight dependence of U_{js} on μ (exponent of 0.16) while Eqn 4 suggests no such effect. Eqn 4 was based on the assumption that both fluid and particle Reynolds numbers in the vessel are in the turbulent regions. Under these conditions, the effect of viscosity on particle drag and lift coefficients and on kinetic power input into the liquid is negligible. Consequently, the value of the coefficient A in Eqn 4 remains a constant, and U_{js} is unaffected by viscosity. But, as fluid viscosity increases, and fluid flow in the vessel becomes more and more laminar, the influence of viscosity on U_{js} becomes increasingly more important. An examination of experimental data on U_{js} reported by Racz et al [1977] reveals that some of their data may have been in the transitional region: if so, then a slight effect of viscosity on U_{js} may be expected.

It is also interesting to note from Table 4 that with the exception of the value of exponent x2 for dish-bottom vessels, changing vessel configuration simply changes the magnitude of the proportionality constant K in Eqn 6: the values of the exponents (x1 to x6) remain practically unaffected, suggesting that the same mechanism might be responsible for particle suspension in the various configurations studied by Racz et al.

Finally, it is worth noting the remarkable similarities between Eqns 4 and 6 and the equivalent expressions for the just suspension speed of rotary impellers, N_{js}, as given for example by the Zwietering's empirical expression reproduced below:

$$N_{js} = 2.5 \frac{\nu^{0.1} C_w^{0.13} d_p^{0.2} [g \Delta\rho / \rho_f]^{0.45} T^{0.85}}{D^{1.7}} \qquad (7)$$

for propeller impellers with a downward pumping action, $D/T<0.5$ and $P_o=0.5$ or alternatively by the following semi-theoretical expression developed by Ayazi Shamlou [1990]:

$$N_{js} = 3.5 \frac{[\Delta\rho/\rho_f]^{1/2} g^{1/2} d^{1/6} C_v^{1/3} T}{P_o^{1/3} D^{5/3}} \qquad (8)$$

for propellers and pitched turbine impellers with downward pumping action and $D/T=0.5$.

The similarities between Eqns 4 and 6 on the one hand and Eqns 7 and 8 on the other clearly indicate that similar mechanisms are responsible for suspension in rotary and in liquid-jet stirred mixers.

NOTATION

A constant in Eqn 4
Ar Archimedes number ($g \rho_f \Delta\rho d_p^3/\mu^2$)
C_v solids concentration (volume fraction)
C_w solids concentration (% wt. or wt. fraction)
D impeller diameter and pipe diameter
d_p particle diameter
dn nozzle diameter
g acceleration due to gravity
H Height of liquid in the vessel
h clearance between nozzle and base of tank
h_n height of nozzle
N impeller speed
N_{js} critical impeller speed for suspension in tanks
P_n Jet power input
P_o impeller power number
Q_n liquid flow rate through the nozzle
Re_n jet Reynolds number ($\rho_f U_{js} d_n/\mu$)
T vessel diameter
U_a axial component of local liquid velocity near the base of the direction at the point of off-bottom suspension
U_n liquid velocity at the nozzle
U_{js} just-suspension velocity of liquid jet (at the nozzle)

Greek letters
ε_T total energy dissipation for complete suspension
ε_b mean liquid energy near the base of vessel
μ Newtonian viscosity

ν kinematic viscosity of liquid
ρ_f liquid density
ρ_p particle density
$\Delta\rho$ particle-liquid density difference

REFERENCES

Ayazi Shamlou P., 1990, Fluid Mixing IV, Bradford , this issue.
Ayazi Shamlou P. and Zolfagharian A., 1987, Fluid Mixing III, Bradford, UK: 221-233.
Baldi G., Conti R. and Alaria E., 1978, Chem.Eng.Sci., 33: 21-25.
Dawson A.D. and Trass O., 1966 (June) , Can. J. Chem.Eng.:121-129.
Harnby N., Edwards M.F., and Nienow A.W., (Eds), 1985, "Mixing in Process Industries",(Butterworths).
Nienow A. W., 1968, Chem.Eng.Sci.,14: 1453-1459.
Oroskar A.R. and Turian R.M., 1980, AIChE J., 26:550-558.
Poreh M., Tsuei Y.G. and Cermak J.E., 1967(june), J. Applied Mechnics: 457-463.
Racz I.G., Dees P.J. and Wassink J.G., 1977, Chemie-Ing.-Techn., 49, 10: 841-857: (in German).
Yeckel A. and Middleman S., 1987, Chem.Eng.Comm., 50:165-175.
Zolfagharian A., 1990, PhD thesis, University of London (in preparation).
Zwietering T.N., 1958, Chem.Eng.Sci., 8: 244-253.

Vessel diameter, T(m)	0.153	0.190	0.215	0.240	0.297

TABLE 1 Experiments were carried out in five geometrically similar vessels

Nozzle diameter d_n (mm)	Angle (°) α	Nozzle height h_n (mm)
5	85	42.5
7	86	39.5
9	86	18.0
12.7	line-size	-

TABLE 2 Details of the nozzles used in this work.

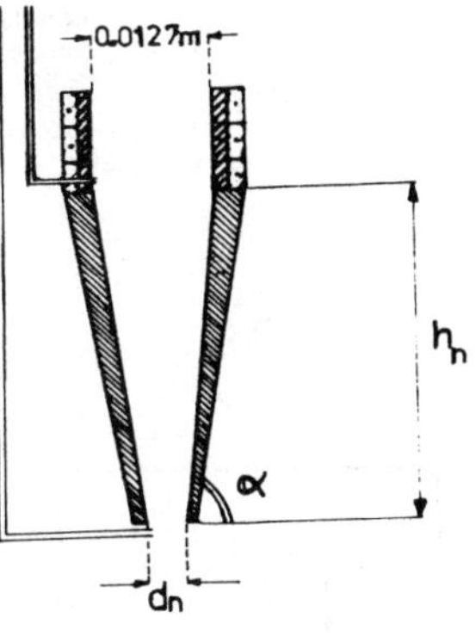

Particles	Particle diameter or particle range (μm)	Particle density ρ_p (kg/m³)
Acetal	2000, 4000, 5000	1300
Lead glass	75-90, 106-125, 150-180, 212-250, 300-420, 420-500 600-720, 1000-1200, 2000, 4000, 6000	2900
Soda glass	1700-2000	2540
PTFE	3000	2140
Zirconium oxide	1680-2000	3800

TABLE 3 Physical properties of solid particles used in the experiments.

Vessel shape	K	x1	x2	x3	x4	x5	x6
Flat-bottom	2.00	0.42	-0.14	-1.16	-0.05	0.06	0.24
Cone-blunt	0.59	0.46	-0.17	-1.34	-0.04	0.09	0.30
Round-bottom	0.15	0.51	-0.17	-1.34	-0.05	0.06	0.33
Dish-bottom	0.42	0.39	-0.07	-1.46	-0.04	0.06	0.29

TABLE 4 Values of the exponents and the magnitude of the constant of proportionality in Eqn 5 for several vessel configuration (data from Racz et al 1977).

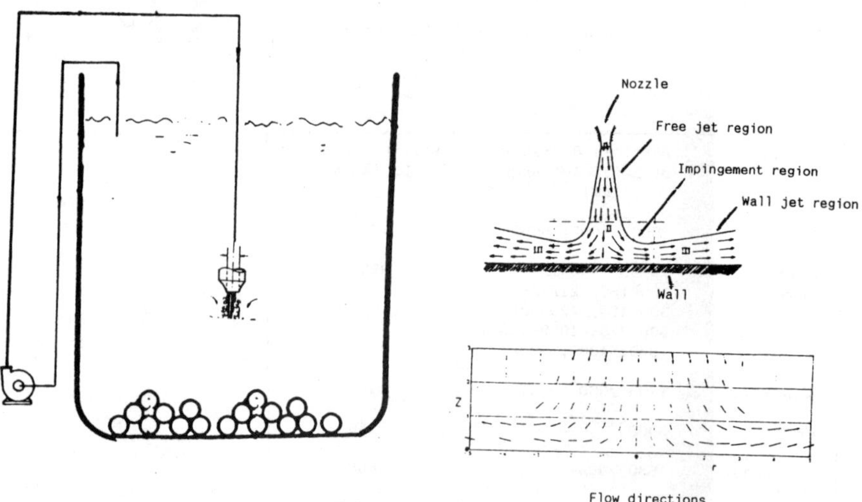

FIGURE 1 Basic components of a liquid-jet stirred vessel for off-bottom suspension of solid particles.

FIGURE 2 Particle motion at the base of a liquid-jet stirred vessel before the point of off-bottom suspension is relatively orderly with particles moving in the in the general direction of the fluid flow.

FIGURE 3 Photograph of the experimental set up used in this study in order to obtain data on off-bottom suspension velocity of liquid jets.

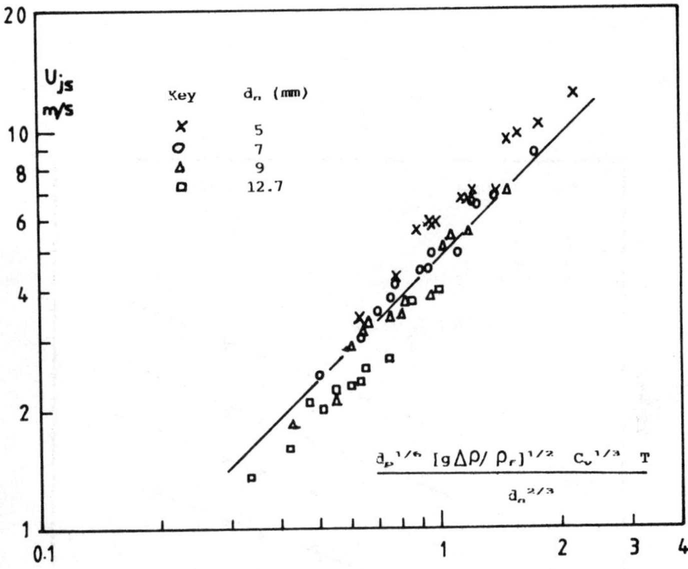

Vessels: 0.153, 0.19, 0.215, 0.24, 0.297m diameter;
Particles: Glass beads (d_p=0.46mm, ρ_p=2900 kg/m^3);
acetal particles (d_p= 2mm, ρ_p=1300 kg/m^3)

FIGURE 4 Experimental data on U_{js} plotted against the group on the right hand side of model Eqn 4 suggests a general applicability of the proposed mechanism of suspension in a liquid-jet stirred vessel.

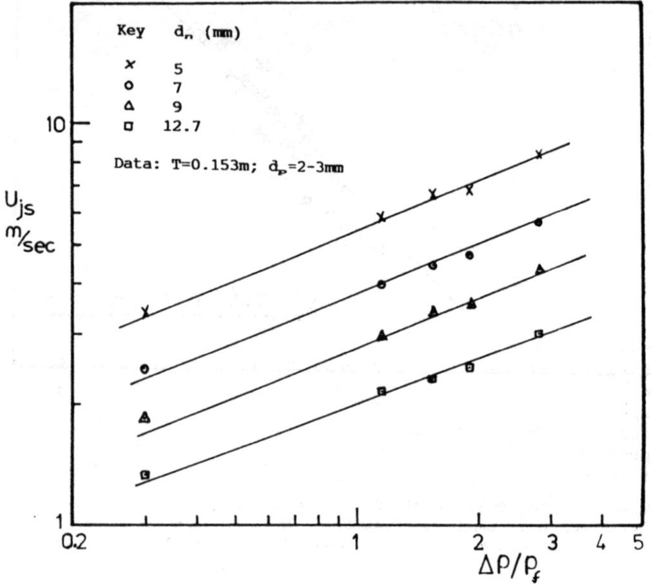

FIGURE 5 Data shows the effect on U_{js} of liquid-particle density difference (see text for discussion).

SCALING-UP OF SOLIDS DISTRIBUTION IN STIRRED VESSELS

A. T. C. Mak and S. W. Ruszkowski*

A conductivity probe was developed in the Fluid Mixing Processes group (FMP) of BHR Group Ltd to measure the local solids concentration in stirred vessels. This instrument offers a convenient means of measuring local solids concentration in large-scale stirred vessels.

As part of the FMP generic research programme experiments have been conducted covering the measurement of power, solids suspension and solids distribution in 0.61 and 1.83 m diameter geometrically similar vessels. Each system consisted of a downward pumping 45° 4-bladed pitch bladed turbine mounted at T/4 clearance. Mixtures of sand and water at concentrations up to 30% by weight were used as test media.

Results indicate that at least equal power per unit volume is required to achieve the same degree of homogeneity for solids distribution on scale-up, in agreement with experimental results reported by other workers. The frequently reported theoretical criterion of constant tip speed underestimates the impeller speed and power required at large scale.

INTRODUCTION

Good distribution of solid particles in a stirred vessel is very often required in solid-liquid processing. For example, a uniform dispersion of solid particles throughout the mixing vessel is necessary to ensure uniform exposure to the process conditions. Despite having numerous applications, the basic mechanisms are still poorly understood and design criteria are rather limited, especially the influence of geometrical configurations and scale-up effects on distribution. This is partly due to the difficulties of developing the necessary measurement techniques and acquiring relevant experimental data, and partly due to insufficient communication between the end users (industrial sectors) and those who are developing the theories (university researchers).

This paper describes the development and application of a conductivity probe which was developed in BHRG to measure local solids concentration in stirred vessels, the measurement of solids distribution in two geometrically similar vessels (Tank diameter, T = 0.61 and 1.83 m) and the interpretation of results.

* BHR Group Ltd., Cranfield, Bedford, MK43 0AJ, UK

TECHNIQUES FOR SOLIDS CONCENTRATION MEASUREMENT

There are many different techniques adopted by researchers in the measurement of solids concentration. Visual observation was employed in the early days. Tracer and streak photography were very often used in conjunction with this technique. It has provided very useful understanding of flow patterns produced by different types of impellers and configurations but this technique is limited to qualitative observations at very low solids concentration.

Predicting solids homogeneity from slurry interface (Fig 7) in the vessel could be misleading. A high liquid/solid-liquid interface does not necessarily guarantee a good distribution within the bulk. In fact, experience within BHRG suggests that a hydrofoil impeller can produce a far better solids distribution with a low level of slurry interface, compared to a T/2 diameter pitched blade turbine.

Because of their non-intrusive nature and the capability of measuring solids concentration in localised regions, optical methods are very popular and much useful work has been produced (eg Magelli et al (2), Shamlou and Koutsakos (3)). However, this technique is often limited to a low solids concentration, depending on the transparency of the test media, and to small scale vessels. Re-calibration is very often required for different test media.

Sample removal is another widely used technique (eg Barresi and Baldi (1)). It is simple and convenient and has much potential. For example, it can be applied irrespective of the nature of slurry content. Also, it is the only technique by which local particle size distribution can be determined. However, this technique is labour intensive and there is still uncertainty over how well such a technique can measure the true local concentration. Further research on sampling techniques is still required.

The conductivity method does offer many advantages in measuring solids distribution. The equipment is relatively cheap and easy to construct. It can measure a solids concentration range from under 5% up to 50% by volume and can reliably detect a 1% change in concentration. Of course, like all the other techniques, conductivity method has its shortfalls; it is instrusive and therefore certain degree of flow disturbance should be allowed for. Any solids likely to cause a rapid change in base conductivity (conductivity of fluid) are undesirable. The conductivity technique is the one which has been adopted by BHRG in measuring solids distribution.

THE SOLIDS CONCENTRATION PROBE

The solids concentration probe was developed by BHRG from a design used at the Warren Springs Laboratory.

The solids concentration probe consists of a "Y" shaped epoxy resin body and a stainless steel tube handle (Fig 3). This is so designed to be robust yet cause a minimum of flow disturbance. Two 10 x 10 mm platinum electrodes are mounted 10 mm apart, at both ends of the epoxy body, to give a measuring volume of 1 cm^3. The conductivity between the two electrodes is measured and transmitted to a computer via a solids concentration meter (Fig 1). The local temperature is measured simultaneously by using a platinum resistance thermometer, enabling correction for the effect of temperature upon conductivity measurement.

This technique is based on a principle that the conductivity of a liquid will be modified by the presence of foreign bodies. Thus, when solid particles pass between the two electrodes of the probe, a change of conductivity will be registered, from which effect the concentration can be calculated, given that the relationship between change in conductivity and volume fraction of solids is known.

Prior to the measurement of solids distribution, the probe was calibrated in a 200 mm diameter liquid fluidised bed (Fig 6). Round grain silica sand up to 55% volume was used. The sand was sieved to 500-700 µm and acid washed to remove impurities. The conductivity measured by the probe, after correcting for temperature, was found to be inversely proportional to the volume fraction of solids.

There was some scatter in the results of these calibrations (Fig 6). This was due to the difficulty of maintaining perfectly homogeneous and steady conditions in the fluidised bed.

Further calibrations were carried out in a small stirred vessel of 150 mm diameter using sand in the size range 90 to 150 µm . The suspension was agitated intensely to maintain a homogeneous distribution of solids in the vessel. The homogeneity of the distribution was verified by making measurements at several points and heights in the vessel. The results of the stirred vessel calibrations were limited to 10% solids concentration by volume, since at higher concentration a homogeneous solids distribution could not be maintained without drawing air down into the suspension. A typical result for a stirred vessel calibration is shown in Figure 5. There is much less scatter, and this is the case for all stirred vessel calibrations carried out with a number of different types of solids.

All the solids tested, with particle diameters ranging from 20 µm up to 700 µm , gave the same calibration:

$$K_C = K_{OR}\ (1 - A\varepsilon_S) \qquad \text{Eqn (1)}$$

the calibration constant, A, was 1.4 in all cases.

Machon et al (5) obtained the same functional relationship between conductivity and solids concentration, working with a radically different probe geometry. Unfortunately the value of the calibration constant was not reported.

The measured conductivity, K_M, and reference conductivity, K_{OR}, were corrected for variations in temperature with the following equation:

$$K_C = K_M \frac{K_{OR}}{K_{OM}} \qquad \text{Eqn (2)}$$

Prior to the start of experiments tests were carried out to check the effect of probe orientation on the reproduciblity of results. For tests with standard baffled vessels, the effect of probe orientation was not unduly significant. In the worst case there was a difference of 10% between the highest and lowest value recorded. The maximum response was observed when the open side of the probe faced the impeller leading edge (Fig 3) this orientation was used for all the tests. The solids distribution results measured in stirred vessels show excellent reproducibility, especially those with a concentration greater than 5% wt.

A conductivity probe for the measurement of solids concentration has also been developed by Musil and Vlk (7), although with a radically different geometry from the BHR Group Ltd probe described here. The Musil probe has been used principally for the measurement of off-bottom suspension speed, N_{JS}, but also for solids distribution measurement (eg Rieger et al (6)).

EXPERIMENTATION

Test Facility

Tests were conducted in two geometrically similar torispherical based stirred vessels with diameters of 0.61 m (T_{61}) and 1.83 m (T_{183}). T_{61} was constructed entirely from perspex and T_{183} was made from mild steel but equipped with four perspex windows to allow observation of flows.

Four vertical strip baffles were spaced equally around the circumference of the vessel. Each baffle was T/12 wide and spaced at T/60 from the vessel wall, giving a total distance of T/10 from inner baffle edge to vessel wall. The vessel geometries are shown in Figure 2.

Two geometrically similar downwards pumping 45° T/2 4-bladed pitched blade turbines were used. The impellers have an overall diameter of 0.31 and 0.93 m and they have an exact scale-up factor of 3. The impeller shaft was centrally mounted in the vessel, with an impeller clearance of T/4, measured from the lowest point on the vessel base to the impeller centreline. The slurry content height, H, was equal to the tank diameter, T, for all experiments giving total slurry volume of 0.165 m^3 and 4.5 m^3 in T_{61} and T_{183} respectively.

Mixtures of mains water and BIS Chelford 95 sand (up to 30% wt) were used. The sand was granular with rounded particles and sieved to a particle size range of 150-210 μm.

Sampling Positions

The measurement of solids concentration in both vessels covered five axial positions: T/6.1, T/3.1, T/2.0, T/1.5 and T/1.2. Another four radial positions (T/60, T/12, T/6 and T/4 from the vessel wall) were measured in T_{183} to check the radial solids distribution profiles. Details of sampling positions are shown in Figure 4.

In all cases, the probe was placed midway between two baffles and kept parallel to the first baffle in an anti-clockwise direction, viewing from the top of the vessel (Fig 3). This introduces the least disturbance to solids passing through the probe sampling volume.

RESULTS AND DISCUSSION

Flow Patterns

At low impeller speeds, the impeller pushes the solids sideways and digs a pit in the solid bed. As the speed increases, the solid bed is partially fluidised and this produces three distinct zones; a clear liquid layer at the top, an unsuspended solid layer on the vessel base and a region of fluidised mixture in between (Fig 7).

As the impeller speed increases further the fluidised region expands, while the amount of unsuspended solids decreases and eventually disappears. With a further increase in impeller speed, the clear liquid layer in the upper region gradually disappears and the solid-liquid mixture passes through a most homogeneous point before the solids redistribute themselves axially and radially.

Thus, if the probe is mounted above a certain height in the vessel, it will measure a gradual increase in solids concentration until the contents reach maximum homogeneity (eg Fig 8).

On the other hand, if the probe is mounted below a certain height in the vessel, it will first measure a very high solids concentration which gradually decreases to the maximum homogeneity level (eg Fig 9).

Relative Standard Deviation

Perfect homogeneous-solid distribution is impossible to achieve (Fig 11-13). Relative standard diviation (RSD) is used to quantify the solids distribution quality so that scaling effects between T_{61} and T_{183} can readily be compared. It is a measure of the deviation of solids concentration of the individual sampling position from the average.

$$RSD_j = \frac{1}{C_M} \left\{ \frac{1}{(n-1)} \sum_{i=1}^{n} (C_{ij} - C_M)^2 \right\}^{\frac{1}{2}}$$

n = 5 for 5 sampling positions
C_{ij} is the solids concentration measured at i^{th} position and j^{th} speed
C_M is the mean bulk solids concentration calculated from the amount of solids added to the vessel at the start of the experiment.

Radial Concentration Profile

Radial concentration profiles with 30% wt solids in T_{183} were measured and some of the results are presented as plots of volume fraction against probe radial position at constant axial position (Fig 8-10). The radial concentration profiles were flat, excepting when the probe was at the clear liquid/solid-liquid interface (eg Fig 10). This normally happened when the probe was mounted towards the top of the vessel, with the impeller running at relatively low speed. The flatness of radial profiles has already been reported by other authors (1,3) but their results were established on vessels of limited sizes (T = 0.39 and 0.225 m). It is encouraging to have the results once again confirmed at a much larger scale. The use of a one-dimensional steady state model by a number of researchers (e.g. 1 and 3) is justified. Thus, to focus on one radial position (ie T/6 from vessel wall) for scaling comparison in this work is acceptable.

Influence of Solids Concentration

An initial study on the effect of solids concentration on distribution was carried out in T_{183}. Two concentrations (15 & 30% wt) were examined and their relative standard deviations were plotted again $N/X^{0.13}$ (Fig 14). This idea originated from Zwietering's correlation (4) for solids suspension $N_{JS} \propto X^{0.13}$, and is recommended by Barresi (1) for solids distribution.

However, this result is published out of interest only because the concentration range utilized was not wide enough to draw a definite conclusion.

Scale-up

The influence of scale on distribution was investigated by comparing the results obtained from the 0.61 and 1.83 m vessels at the same probe radial clearances (ie T/6 from the vessel wall). They are presented as plots of relative standard deviation against tip speed (N D), power per unit volume (N^3 D^2) and Froude number (N^2D) (see Figs 15-17). The results show that the use of constant tip speed as the scale-up criterion for solids distribution under-estimated the power requirement. To achieve the same degree of homogeneity in the 1.83 m vessel, at least equal power input per unit volume is required. This is established by comparing the relative positions of two RSD curves and their minima.

Most other workers who report correlations for the effect of scale on solids distribution (Refs 1, 2, 3 and 8) have either carried out measurements at only one scale or at a very limited vessel size range, and fitted their results to a theoretical model. Although the models differ in their theoretical justifications, they all predict that impeller tip speed should be kept constant to maintain solids distribution on scale-up. The results presented here show clearly that this is not the case.

Buurman et al (9) have correlated their solids distribution results, which were obtained in vessels ranged from 0.24 m to 4.26 m in diameter, in terms of height of homogeneous zone. The results indiated that $N^2D^{1.55}$ is constant for homogeneous suspension. Rieger (6) also investigated the dependency of solids distribution on scale. Rieger's results are inconclusive, showing that neither tip speed or specific power input, used alone, could correlate the data well.

There is clearly a need for more extensive data on solids distribution in stirred tanks, taken at large scales. If the results of further experimental investigation confirm that solids distribution does not scale with impeller tip speed then further work will also be necessary to develop models which adequately represent the mechanisms of solids distributions in stirred vessels.

The speed required to just suspend the solids (N_{JS}) was also measured during experimentation at both scales. Fig. 12 and 13 shows N_{JS} compared with the solids distribution profiles and indicate that at both scales the solids distribution is homogeneous at an impeller speed below N_{JS}. The reason for this is that a very small fraction of the solids remain stationary for longer than one to two seconds and hence fail the Zwietering criterion (4) for just suspension speed even through the vast majority of solid particles are well distributed throughout the vessel. It is likely that other impeller types will achieve N_{JS} and at that speed give poor solids distribution. The relationship between these two factors will be governed by the impeller type and size and vessel base geometry. This is clearly an area requiring further work.

CONCLUSIONS

Experience at BHR Group Ltd has shown that a conductivity probe is a reliable, convenient and accurate way of measuring local solids concentration in stirred vessels. The probe can be used in both large scale and small scale experimental vessels, thus overcoming problems of the interpretation of results when different techniques are employed in different experiments.

The use of one dimensional steady-state models to characterise solids suspension is justified by the data presented here from the 1.83 m vessel.

However, the use of constant impeller tip speed, proposed by many workers for scale-up of solids distribution, is not supported by the results from the 0.61 m and 1.83 m vessels. There is a clear need for further experimental and theoretical work on the effect of scale on solids distribution.

ACKNOWLEDGEMENT

The authors wish to thank the members of FMP (Fluid Mixing Processes) for their financial and technical support and their permission to publish part of the generic research results.

SYMBOLS USED

A = Calibration constant (-)

C_{ij} = Solids Concentration at i^{th} position and j^{th} speed, expressed as volume fraction (m^3 m^{-3})

C_M = Mean solids concentration from calculation, expressed as volume fraction (m^3 m^{-3})

D = Impeller diameter (m)

H = Slurry content height (m)

K_C = Corrected conductivity (microsiemens)

K_M = Measured conductivity (microsiemens)

K_{OM} = Conducivity of water at reference temperature (microsiemens)

K_{OR} = Conductivity of water at measured temperature (microsiemens)

n = Sampling points (-)

N = Impeller speed (rev s^{-1})

N_{JS} = Jus suspension speed (rev s^{-1})

RSD = Relative standard deviation of solids concentration (-)

T = Vessel diameter (m)

X = Mass ratio of solids to liquid in suspension (-)

ε_s = Average volume fraction of solids in vessel (-)

REFERENCES

1. Barresi, A. and Baldi, G., 1987, "Solids Dispersion in an agitated vessel". Chem Eng Sci 42, 2949.

2. Magelli, F. et al, 1990, "Solid Distributions in vessels stirred with multiple impellers". Chem Eng Sci 45, 615.

3. Shamlou, P. A. and Koutsakos, E.,1989, "Solids suspension and distribution in liquids under turbulent agitation ". Chem Eng Sci 44, 529.

4. Zwietering, Th. N., 1958, "Suspending of solids particles in liquid by agitators". Chem Eng Sci 8, 244.

5. Machon, V., Fort, I., Skrivanek, J., 1982, "Local solids distribution in the space of a stirred vessel". 4th European Conference on Mixing, Noordwijkerhout, BHRA.

6. Rieger, F., Ditl, P., Havelkova, O., 1978, "Suspension of solid particles - concentration profiles and particle layer on the vessel bottom". 6th European Conference on Mixing, Pavia, BHRA.

7. Musil, L. and Vlk, J., 1978, "Suspending solid particles in an agitated conical-bottom tank". Chem Eng Sci 33, 1123.

8. Tojo, K. and Miyanami, K., 1982, "Solids suspension in mixing tanks". Ind Eng Chem Fundam 21, 214.

9. Buurman, C., Resoort, G. and Plaschkes, A., 1985. "Scaling-up rules for solids suspension in stirred vessels". 5th European Conference on Mixing, Wurzburg, BHRA.

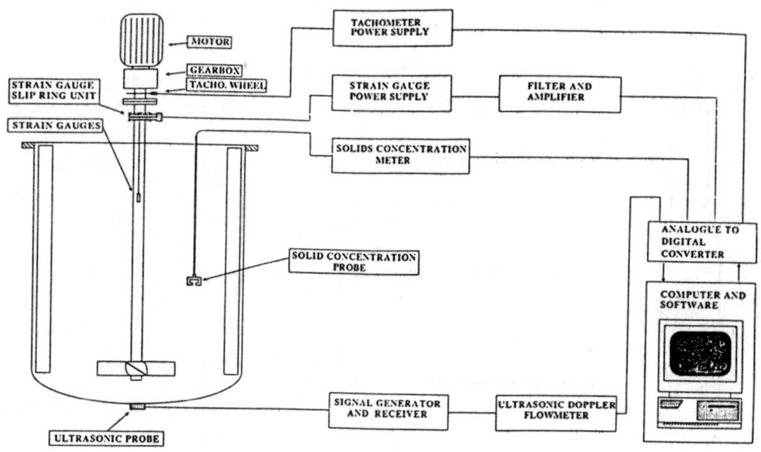

Fig 1 Test Facility

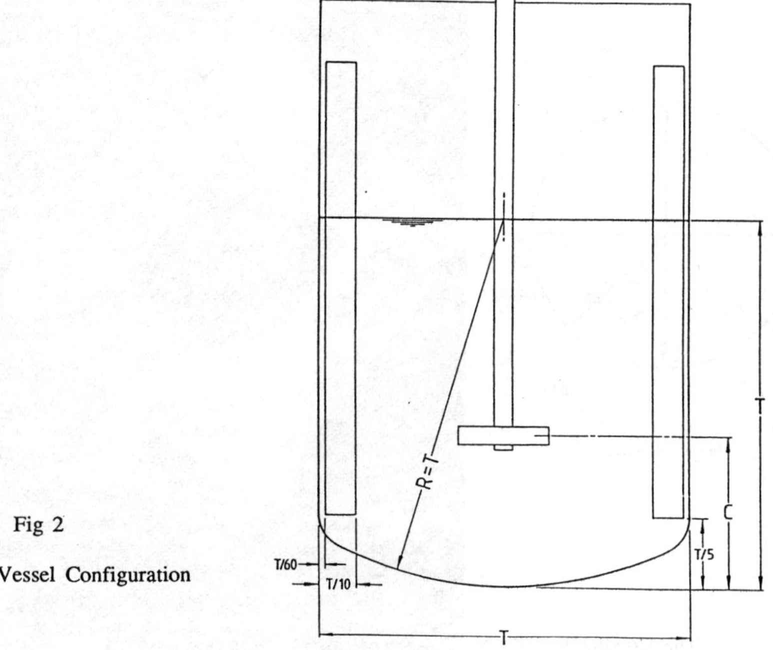

Fig 2

Vessel Configuration

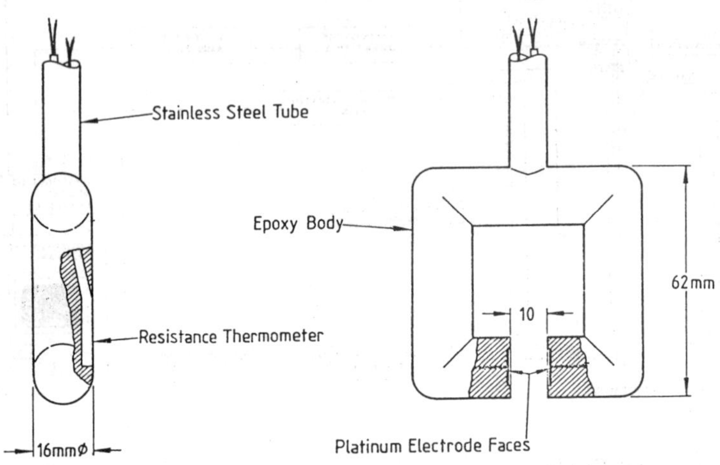

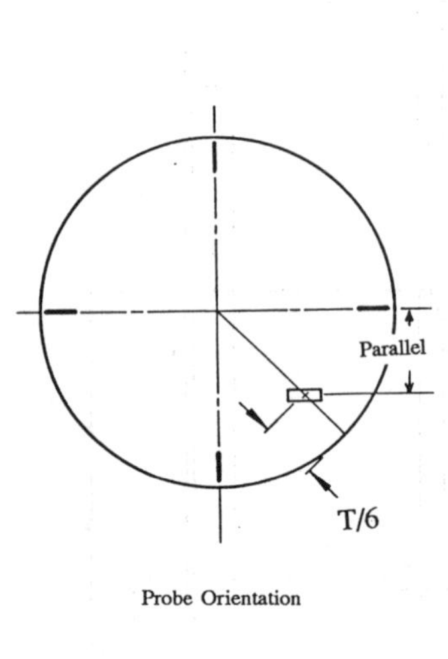

Probe Orientation

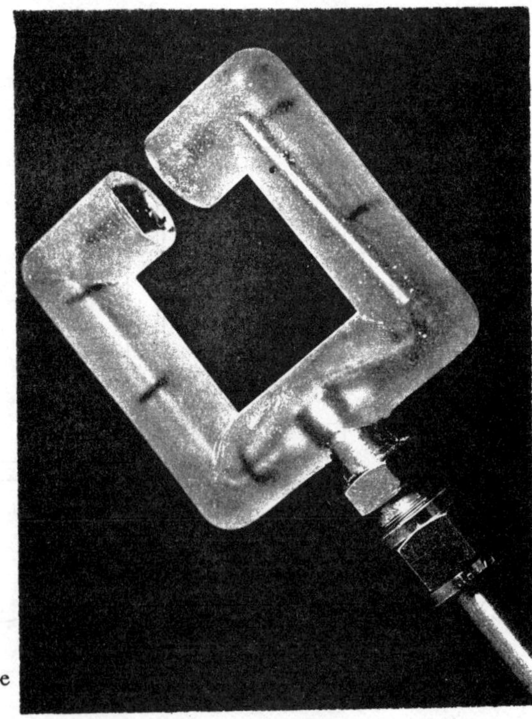

F: The Conductivity Probe

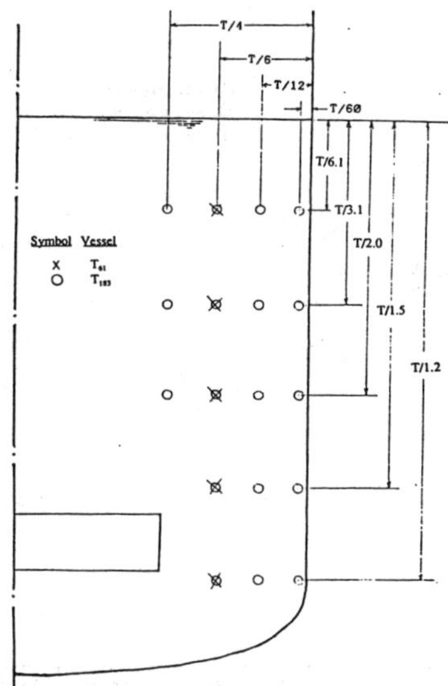

Fig 4 Sampling Positions

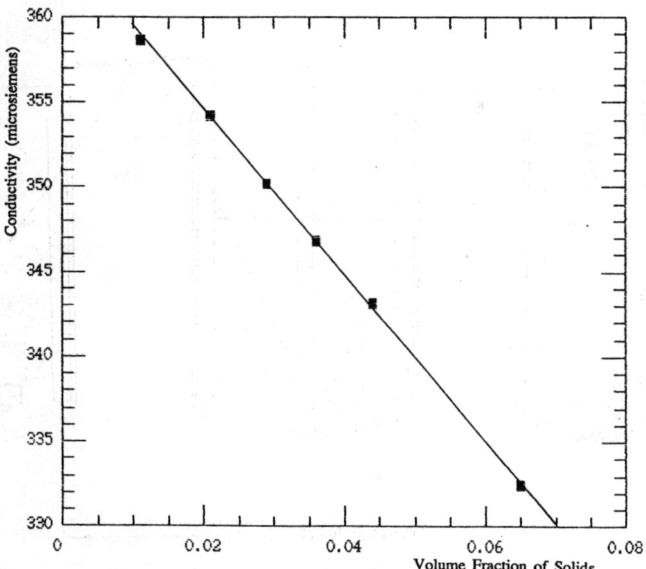

Fig 5 Calibration Confirmation : Conductivity Test in a Well-agitatied 150 mm Diameter Beaker

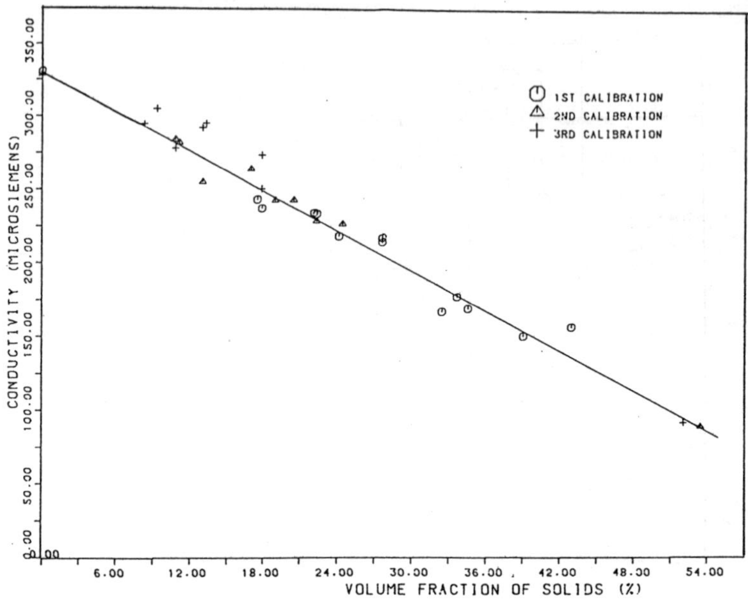

Fig 6 Solids Concentration Probe Calibrated in Fluidised Bed

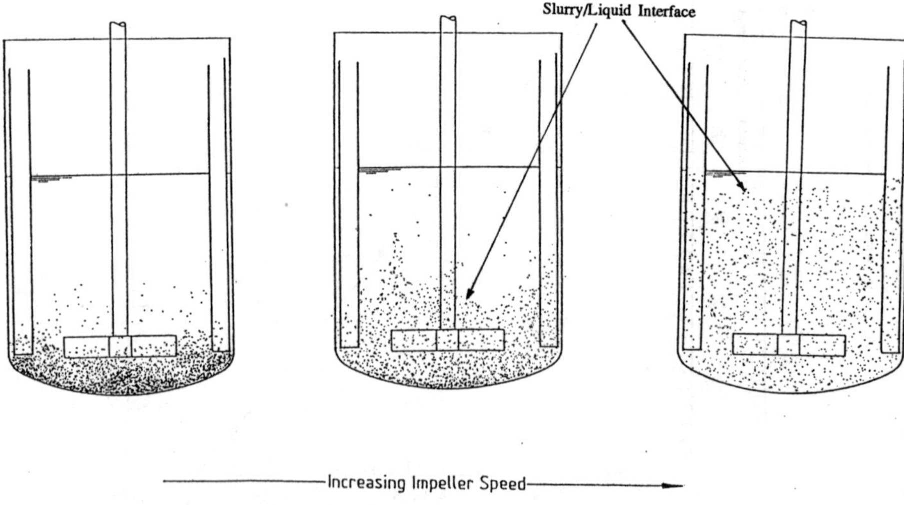

Fig 7 The Distribution of Solid Particles in the Vessel

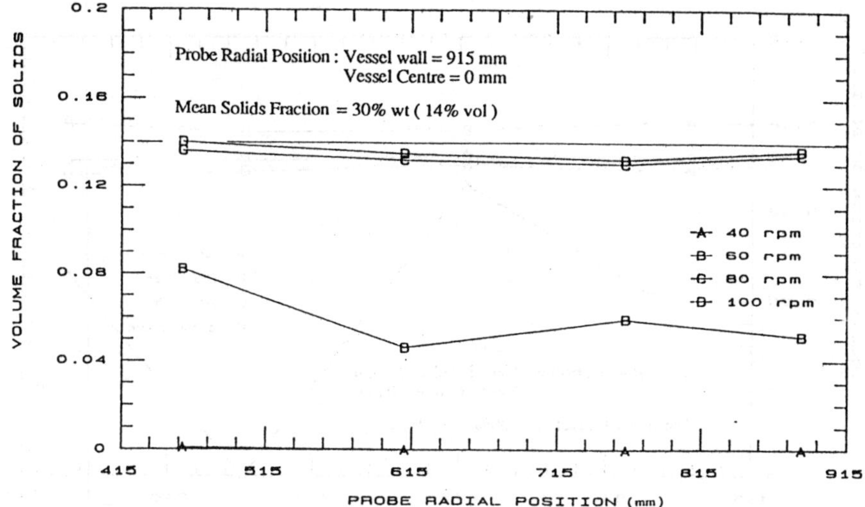

Fig 8 Radial Solids Distribution in 1.83 m Vessel (300 mm below fluid surface)

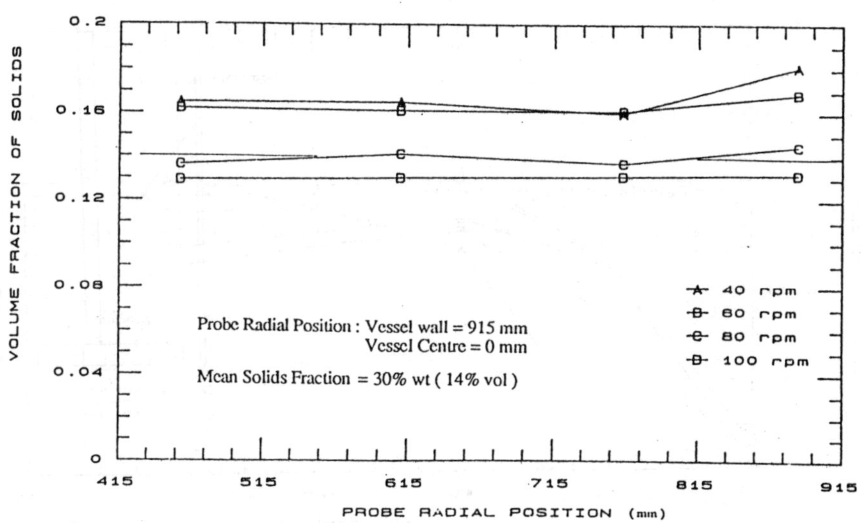

Fig 9 Radial Solids Distribution in 1.83 m Vessel (900 mm below fluid surface)

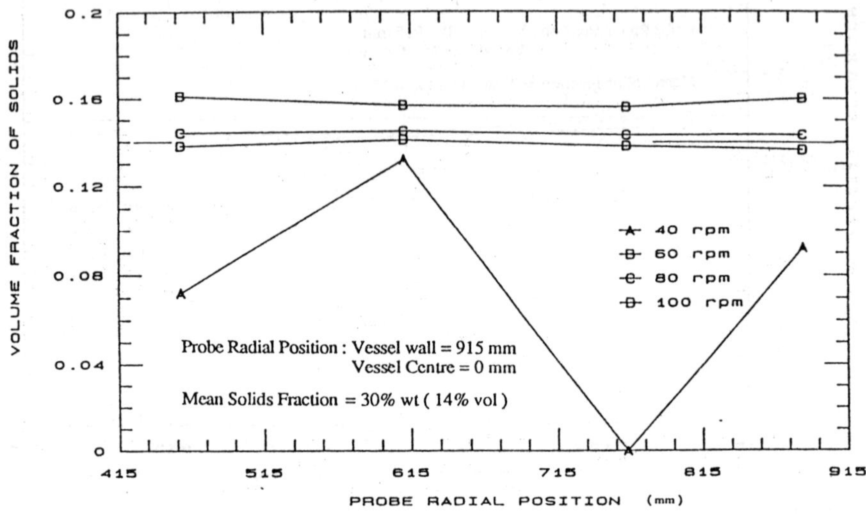

Fig 10 Radial Solids Distribution in 1.83 m Vessel (600 mm below fluid surface)

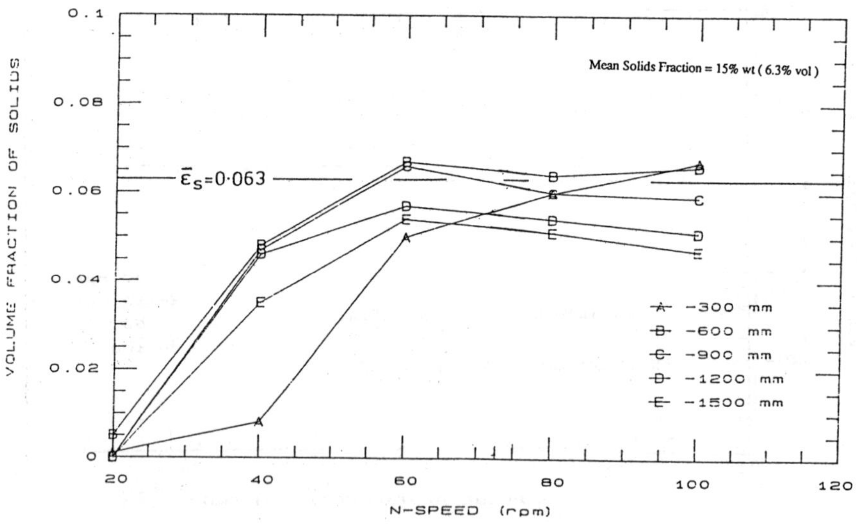

Fig 11 Solids Distribution in 1.83 m Vessel at T/6 Side Clearance

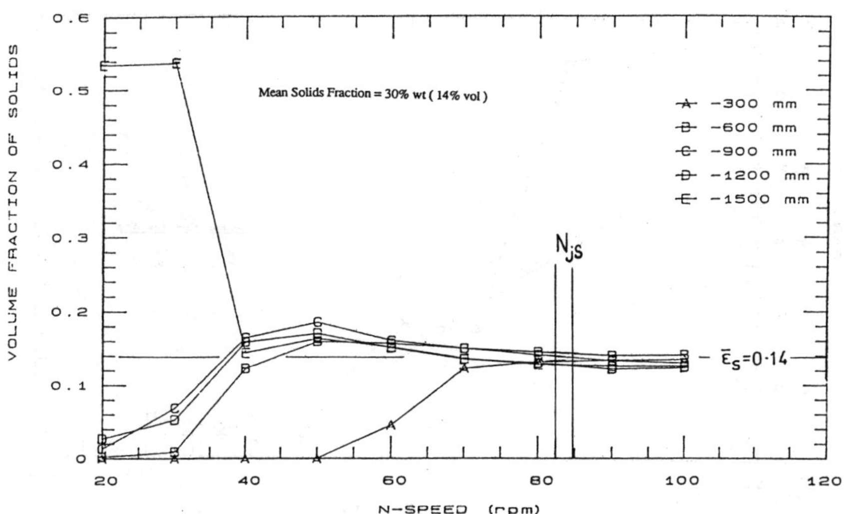

Fig 12 Solids Distribution in 1.83 m Vessel at T/6 Side Clearance

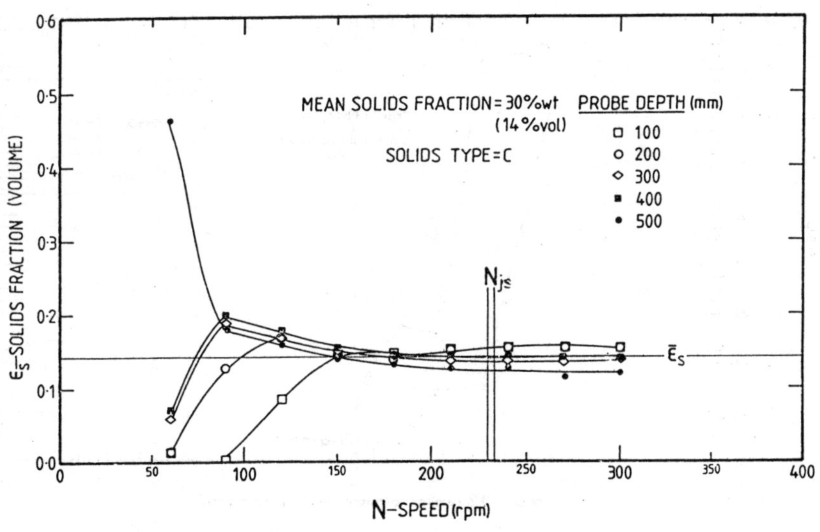

Fig 13 Solids Distribution in 0.61 m Vessel at T/6 Side Clearance

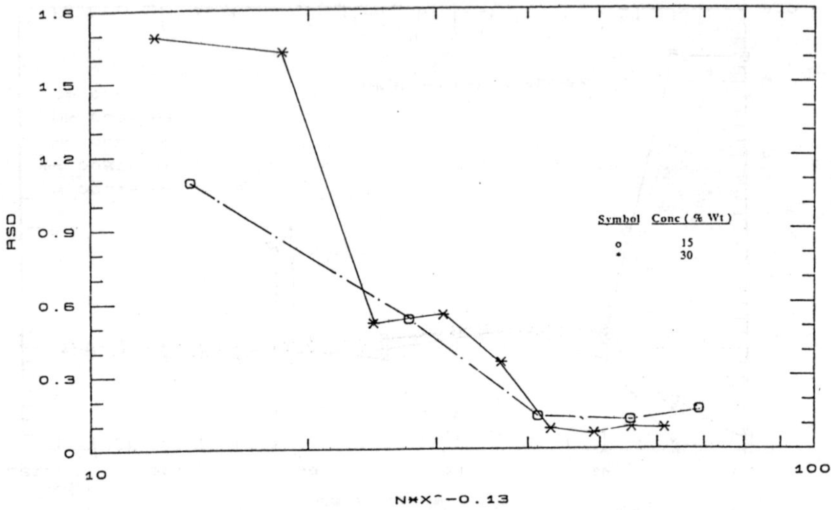

Fig 14 Effect of Solids Concentration on Homogeneity (T_{183})

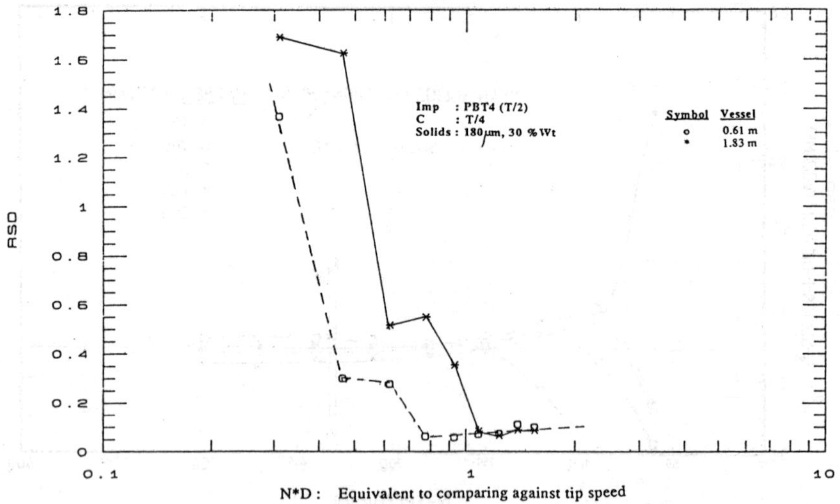

Fig 15 Effect of Geometric Scale on Homogeneity (correlated by tip speed)

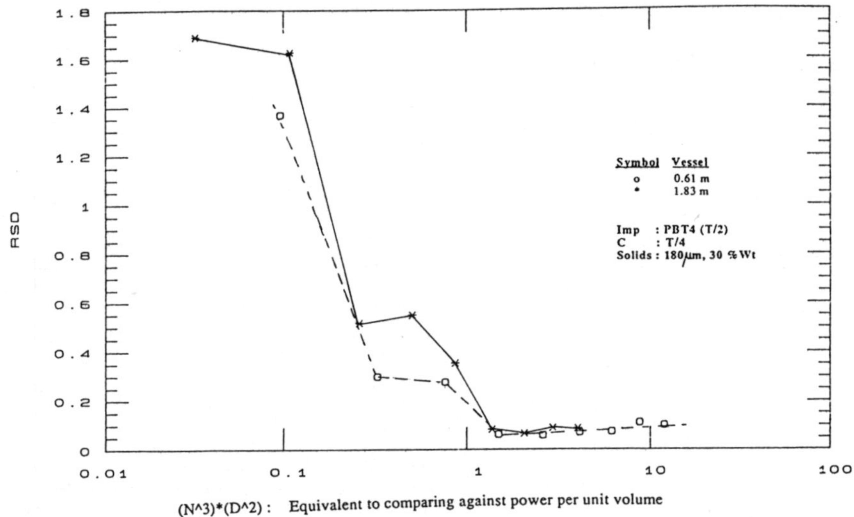

Fig 16 Effect of Geometric Scale on Homogeneity (correlated by power per unit volume)

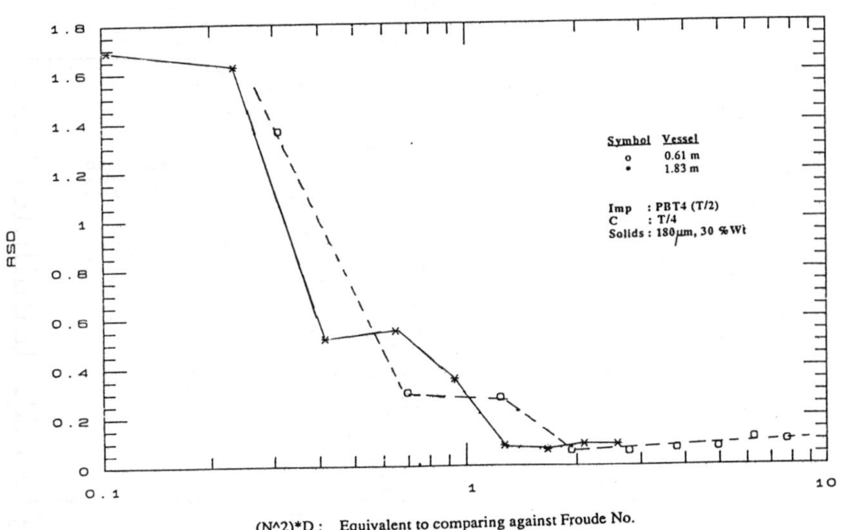

Fig 17 Effect of Geometric Scale on Homogeneity (correlated by Froude Number)

MECHANISM OF SUSPENSION OF COARSE PARTICLES IN LIQUIDS IN STIRRED VESSELS

P. AYAZI SHAMLOU*

A discussion is provided on fluid-particle motion and suspension in a mechanically-stirred vessel and a new model is presented for the estimation of the minimum impeller speed for complete suspension based on fluid-dynamic and body forces acting on particles on the floor of the vessel.

Experimental data are reported on just-suspension speed for a range of vessels, impellers and particles. Data from these experiments interpreted in terms of the model equation and also compared with previously published correlations indicate a general applicability of the proposed mechanism of suspension.

INTRODUCTION

The minimum speed of impeller required to suspend particles in liquids in a mechanically-stirred vessel has been a subject of increasing interest to researchers [Zwitering, 1958; Nienow, 1968; Ayazi Shamlou and Koutsakos, 1989] because of its industrial significance [Harnby et al, 1985]. Most treatments of the subject however have been based on some arbitrary and often ill-defined state of particle motion at/or close to the base of the tank and experimental data on minimum impeller speed are normally presented in the form of empirical correlations. While each correlation is useful in describing experimental data over the range of conditions for which it was obtained, confident extrapolation of data requires a knowledge of the possible mechanism(s) responsible for suspension. The need for a better understanding of the ways in which particle motion and suspension occur in stirred vessels provided the motivation for this work. A possible mechanism is provided to describe the initiation of particle motion and the subsequent particle suspension based on the mean fluid circulation at the base of the vessel. For coarse particles, the model equation provides the correct form of the functional relationship between minimum impeller speed for suspension and some of the important parameters affecting it. Suggestions are made as to how the basic model equation may be corrected to provide numerical values of the minimum impeller speed for various states of suspension.

FLUID AND PARTICLE MOTION IN STIRRED VESSELS

A single particle initially at rest on a surface, e.g. on the floor of a vessel or on a bed of particles, gradually passes through various states of "motion" with increasing liquid velocity.

Initially, when liquid velocity is low, particles remain in their stable rest positions and in contact with each other and/or with the base of the vessel.

* Chemical and Biochemical Enigneering, University College London, WC1E 7JE.

As liquid velocity increases, local instability results and individual particles begin to roll out of their stable rest position and slide in the general direction of fluid flow. During this rolling and sliding state, individual particles maintain their physical contact with each other and/or with the floor of the vessel. Progressively more particles become unstable and therefore susceptible to this type of motion, as the flow velocity is increased. Depending on the mean direction of fluid flow, the rolling and sliding of particles at the base of the vessel result in the formation of either a dune at the center of the tank or a ring at the periphery of the vessel.

Increasing liquid velocity further results in a third state of motion when the first particle, usually at the top of the dune, begins to lose contact with its neighbouring particles and lifts off into the bulk of the liquid. In this state of motion, the particle is truly in suspension. Once in suspension, the particle follows the general direction of liquid motion and circulation which inevitably means that at some point in time the particle will return to the bottom of the vessel. The state of motion of a particle following its arrival at the bottom of the tank depends on whether it physically touches the base and comes to a momentary rest. If so, then the particle will follow the basic three states of no-motion, rolling/sliding and lift-off/suspension. Increasingly more particles follow the three basic modes of motion as liquid velocity is increased.

Evidently, a number of transitions in particle motion are possible and the numerical value of the so called minimum impeller speed depends largely on which transition is used to define the mode of suspension. Unfortunately, most researchers in this field have used critical conditions at some arbitrary point in their experiments. This has inevitably led to variation in their results. The problem is further complicated because the transition from one state of motion to the next is gradual and takes place over a range of liquid velocity. In other words, there is no truly critical condition for which particle motion and/or suspension begins suddenly as that condition is reached.

FORCES ACTING ON A PARTICLE

Consider a liquid flowing over a bed of mono-size, loose, non-cohesive spherical particles resting on the floor of a vessel, Fig 1. The forces acting on a particle resting on the surface of the bed are many and varied in nature. In the discussion that follows the liquid flow field near the base of the vessel is considered to be uniform and steady and the forces acting on the particles resting on the surface of the bed of material are gravity forces of weight and buoyancy, hydrodynamic lift normal to the direction of the fluid flow and the drag force parallel to flow direction. Most treatments of hydrodynamic forces on a particle on a surface consider only drag: lift does not appear explicitly. But because the constants in the resulting theoretical equations are determined experimentally and because lift is normally assumed to depend on the same variables as drag, the effect of lift, regardless of its importance and origin, is automatically considered. In the present discussion the two hydrodynamic forces are separated in order to describe, and to provide some justification for, the various modes of particle motion, from the initial rolling and sliding to subsequent lift-off and suspension.

Lift force on a particle

The lift force may arise as a result of one or combination of several sources, some of which, although recognized as important, are not as yet amenable to rigorous examination in real situation.

A sphere in a flowing liquid with a velocity gradient experiences a lift force due to the accompanying pressure difference. Saffman [1965] solved the equation of motion for the flow of a fluid over a particle close to a rigid wall in the viscous regime and proposed the following equation for the lift force acting on the particle in the Stokes' region

$$F_L = \frac{6.4 \mu U_f d_p^2 (du_f/dy)^{1/2}}{4 \nu^{1/2}} \qquad (1)$$

where u_f is the velocity of fluid at vertical distance y from the wall (Fig 1). The direction of this force is such that it causes the particle to move away from the wall. Studies in which this lift force created the radial motion of the particles in pipe flow showed that experimental measurements were too high by a factor of 5.0 compared with estimated values based on Eqn 1. This discrepency is partly explained by the fact that for particles in motion close to a solid boundary additional sources of lift exist which are not accounted for in the derivation of Eqn 1. One obvious source is the lift associated with circulation due to spin of the particle in a fluid. Spin is often imparted on a particle as it is dislodged by rotation from its rest position. A second source of lift which is thought to be present even if the particle does not spin at all is due to displacement towards the solid boundary of the stagnation point of the fluid striking on the particle and to displacement away from the boundary of the axis of the wake, for particle Reynolds number greater than about 0.2. It is perhaps important to note that all these sources of lift are associated with liquid circulation in the vessel.

<u>Drag force on a particle</u>
For a particle close to a rigid wall, the drag force, F_D, imparted on the particle by the flow of a fluid over it is given by [O'Neill 1968]

$$F_D = 5.1\pi \mu d_p U_a \qquad (2)$$

where U_a is taken to be the velocity of the liquid at mid-point of the particle. Eqn 2, like Enq 1, was derived for laminar flow over particles in the Stokes' region. No comparable expressions are available for the case where particle Reyanolds number is in the turbulent region. In such cases, an engineering approach is to assume that both these hydrodynamic forces are functions of local particle Reynolds number and may be expressed in terms of a lift and drag coefficient respectively. Thus

$$C_L = \frac{F_L}{{}^1/_2 \rho_f U_m^2 A_p} \qquad (3)$$

and

$$C_D = \frac{F_D}{{}^1/_2 \rho_f U_m^2 A_p} \qquad (4)$$

where A_p is the projected area of the particle and U_m is the mean characteristic velocity of the liquid close to the particle. Studies on minimum suspension velocity in horizontal pipelines suggest that, for particle reynolds number greater than about 0.2, a probable value of the lift coefficient is between 0.5 and 1.0 and that approximately $C_L/C_D = 1/4$ (Thomas 1962).

Effective weight of a particle
The force due to the effective weight of a spherical particle of diameter d_p and with a density ρ_p in a liquid of density ρ_f may be expressed as follows

$$F_w - F_B = F_E = \frac{d_p^3 (\rho_p - \rho_f) g}{6} \quad (5)$$

MECHANISM OF INCIPIENT MOTION OF PARTICLES

When the mean local velocity of the liquid flowing over a particle at rest on a bed of loose non-cohesive particles is increased sufficiently, the particle becomes unstable and begins to roll and slide along the base of the vessel. At the base of the vessel, the local velocity of the liquid is assumed to have two components: the axial component, U_a, and the radial component, U_r. Based on what has been said so far about the two hydrodynamic forces acting on a particle close to a wall, it is assumed that the radial component of liquid flow velocity (the component of velocity which is parallel with the base of the vessel) gives rise to both a lift force and a drag force. The axial component of liquid flow velocity (the component which is perpendicular to the base of the tank) on the other hand results in only a drag force. Furthermore, since little information is available on the liquid velocity profiles close to the base of a tank under conditions of interest to this work, for the purpose of this discussion it is assumed that $U_a \propto U_r \propto U_m$.

Referring to Fig 1, at the point of dislodgment and assuming that the various forces act through the centre of mass of the particle, the moment of all the forces about point O must be zero [Ayazi Shamlou and Zolfagharian, 1987]. Thus

$$F_{Lr}X + F_{Da}X + F_{Dr}Y = F_E X \quad (6)$$

Dividing Eqn 6 by X and noting that for a spherical particle initially at rest in its most stable position the ratio of Y/X is close to unity [Ayazi Shamlou and Zolfagharian, 1987] gives the following equation for the balance of forces at the point of dislodgment of the particle:

$$F_{Lr} + F_{Da} + F_{Dr} = F_E \quad (7)$$

Substituting the equation for the lift force (Eqn 3), the effective weight of the particle (Eqn 5) and for the drag force (Eqn 4) in Eqn 7 and writing the resulting equation in terms of the minimum liquid velocity, U_{r1} (or U_{a1}) gives the following expression for the local liquid velocity at the point of incipient motion of the particle.

$$U_{r1} = K \, [g d_p \Delta\rho / \rho_f]^{1/2} \quad (8)$$

The constant of proportionality , K, in Eqn 8 contains all the numerical constants involved in its derivation as well as the two hydrodynamic coefficients of lift and drag. The magnitude of these two coefficients are expected to change from point to point at the base of the tank because of the spatial variation in the velocity of the liquid. However, for a given position at the bottom of the tank, the values of C_D and C_L are assumed to remain constant provided that local particle Reynolds number is in the turbulent region. Therefore, while the numerical value of U_{r1} depends on K, the functional relationship suggested by Eqn 8 remains unaffected by the absolute value of K. The spatial variation of C_D and C_L also means that particle movement and particle suspension (see later) at the bottom of the tank occur over a range of liquid velocity as is often observed experimentally.

The net effect of the action of the various forces acting on a particle is that eventually the particle is dislodged and begins to move in the general direction of fluid flow velocity. Progressively more particles take part in this mode of motion as the mean liquid velocity is increased. If, for example, the mean direction of liquid velocity is towards the center of the vessel, the motion of particles results in the formation of a dune at the center. Increasing the mean liquid velocity further causes particles from the periphery of the dune to roll up the inclined surface of the dune. Gradually, the height of the dune increases as more and more particles reach its apex. Throughout this period of relative motion between the particles and between the particles and the base of the vessel individual particles maintain their contact with each other and/or with the base of the tank.

MECHANISM OF INCIPIENT SUSPENSION OF PARTICLES

When liquid velocity is increased sufficiently, the particle at the apex of the dune lifts off and loses all contacts with its neighbours. The particle is now considered to be truly in suspension and the liquid velocity at this point corresponds to the initiation of suspension.

At the point of incipient suspension, the relevant forces acting on the particle are those that act perpendicular to the surface (Fig 1). The force balance on the particle at this point is:

$$F_{Lr} + F_{Da} = F_B \qquad (9)$$

Substituting for the various forces (Eqns 3 to 5) in Eqn 9 and writing the resulting expression in terms of the axial component of local fluid velocity, U_{a1}, leads to the same functional relationship as that given in Eqn 8, the only difference being in the magnitude of the coefficient K. One important implication of this observation is that essentially the same mechanism is responsible for the various modes of particle motion. This is in accord with visual observations of the base of the vessel during experiments involving particle suspension. These observations provide unequivocal evidence that the transition in particle motion from one mode to the next is relatively smooth.

Minimum impeller speed for particle suspension in stirred vessels

From the foregoing discussion, it may be concluded that the numerical value of the mean liquid flow velocity for initiation of particle motion and for subsequent particle suspension depends on the definition of the state of particle motion. However, the dependency on system parameters (particle and fluid properties and geometrical configuration) of the various minimum impeller speeds appears to be unaffected by the mode of particle motion. This means that the establishment of the basic expression for minimum impeller speed for particle suspension may be based on the conditions at, say, the point of incipient suspension. The resulting expression provides the functional relationship between the minimum stirrer speed and some of the important parmeters affecting it. The numerical value of the constant of proportionality of this equation then depends on the criterion used to define the state of suspension and may be obtained experimentally by plotting the data on N_{js} in the form suggested by the functional relationship. This approach is used below to obtain an expression for the so called just-suspension speed of the impeller.

At the point of incipient suspension of a particle, it is assumed that the rate of dissipation of fluid energy for particle lift-off is due to and is given by the total fluid flow forces (drag and lift) acting on the particle [Oroskar and Turian 1980]. Thus

$$\varepsilon_p = (F_{Lr} + F_{Da})U_{a1} \qquad (10)$$

If it is assumed that the total rate of energy dissipation, ε_T, required to entrain all the particles is proportional to the total number of particles, n_p, in the liquid, then

$$\varepsilon_T \propto \rho_f U_{a1}^3 \, d_p^2 \, n_p \qquad (11)$$

For mono-size spherical particles, Eqn 11 may be written in terms of the volumetric concentration of solids, C_v, as follows

$$\varepsilon_T \propto \rho_f U_{a1}^3 C_v \, T^3/d_p \qquad (12)$$

where the height of liquid in the vessel is assumed to be equal to vessel diameter, T.

To proceed further with the development of this model, it is necessary to relate the energy needed for particle suspension to the kinetic energy of the liquid. The kinetic energy of the liquid depends on the geometry of the equipment and on the source providing the motion. For mechanically stirred vessels the external source providing the motion of the liquid is a rotating impeller for which the energy input is usually expressed in the following form

$$P_o = P/\rho_f N^3 D^5 = \text{constant} \qquad (13)$$

under turbulent flow conditions.

Now, the energy imparted on the liquid by the impeller decreases continuously with distance from the impeller. In the absence of any experimental and/or theoretical information on the form of this energy decay, it is assumed that the energy of the liquid near the base of the vessel may be expressed by the following relationship

$$\varepsilon_b = K_1 P = K_1 P_o \, _f N^3 D^5 \qquad (14)$$

where the magnitude of the constant K_1 depends on several factors including the efficiency of transfer of kinetic energy of impeller to the surrounding liquid and, in turn, from the liquid to a particle on the verge of suspension. Few authors have measured local velocities and energy dissipations for the range of conditions of interest to this work. In the absent of this information it is assumed that for a given set of operating conditions the value of K_1 in Eqn 14 remains a constant.

Substituting in Eqn 12 for ε_T using Eqn 14 and for U_{a1} using Eqn 8 gives, after some rearrangement, the following expression for impeller speed, N_{is}, at the point of incipient suspension of particles:

$$N_{js} = A \frac{(\Delta\rho/\rho_f)^{1/2}\, g^{1/2}\, d_p^{1/6} C_v^{1/3} T}{P_o^{1/3}\, D^{5/3}} \qquad (15)$$

The value of the coefficient A in Eqn 15 is obtained experimentally and contains all the relevant numerical and all the proportionality constants involved in its derivation. Additionally, in obtaining the magnitude of the constant of proportionality A from experimental measurements, any suspension criterion may be used. For example, if the 1-2 seconds criterion is used to define the state of suspension, then, in Eqn 15, A will have a value corresponding to this state and the corresponding impeller speed is referred to as the just-suspension speed, N_{js}: the form of the expression given by Eqn 15 remains unaffected by the definition of the state of suspension.

EXPERIMENTAL VERIFICATION OF THE JUST-SUSPENSION CORRELATION

Experimental data on N_{js} were obtained for various glass beads (ρ_p= 2540 kg/m^3) and for acetal particles (ρ_p =1300 kg/m^3) suspended in water in four flat-bottom glass vessels (T= 30 cm, 24 cm, 19 cm and 15 cm). In all experiments reported in this work particle concentration was equal to 1% of the slurry weight in the vessel.

Each vessel was equipped with four baffles (W/T=0.1) equally spaced around the vessel, firmly positioned against the vertical walls and fixed slightly above the base in order to avoid particles getting trapped behind the baffles at the bottom of the tank.

Four propeller and two 4-bladed, 45° open, turbine impellers (w/D=1/4) were used. In all experiments the height of the impeller from the base of the vessel was maintained at a constant ratio of h/T=0.25 and liquid height in the vessel was approximately equal to vessel dimater. The impellers were driven by a 1.1 kW motor with an infinitely variable speed control in the range 0-2000 r.p.m.

Measurements of N_{js} were carried out by visual observations of the base of the vessel, using the well-established 1-2 seconds criterion first proposed by Zwietering [1958]. Power input to the impeller and impeller speed were measured simultaneously using a shaft mounted torque transducer [Ayazi Shamlou and Edwards 1985].

Full details of experimental procedure, equipment and materials are given elsewhere [Zolfagharian 1990]. Figure 2 provides a sketch of the main rig.

A detailed discussion on the verification of Eqn 15 using experimental data obtained specifically for this purpose may be found elsewhere [Zolfagharian 1990]. Figure 3 shows some typical data on N_{js} plotted against the group on the right hand side of model Eqn 15. The solid line which has a slope of unity represents the best line of fit through the data points. The plot also suggests that coefficient A in Eqn 15 has a value of 3.5. An important feature of the plot in Figure 3 is that, within the range of experimental conditions covered, the data on N_{js} for the two group of impellers are practically indistinguishable. In other words with A=3.5, model Eqn 15 may be used to estimate N_{js} for propellers and for turbine impellers.

Comparison with previous work

Most previous researchers in this field have expressed their data on N_{js}

with empirical correlations of the following form:

$$N_{js} \propto d_p^{a1} \; C_v^{a2} \; (\Delta\rho/\rho_f g)^{a3} \; D^{a4} \; (\nu)^{a5} \qquad (16)$$

with the constant of proportionality depending upon the geometry of the system. A review of the available information seems to indicate that

$0.14 < a1 < 0.67$	for	$100 < d_p < 1000$ μm
$0.10 < a2 < 0.3$	for	$0.0008 < C_v < 0.36$
$0.40 < a3 < 0.5$	for	$0.05 < \Delta\rho/\rho_f < 10.1$
$-0.50 < a4 < -1.0$	for	$0.17 < D/T < 0.7$
$0.00 < a5 < 0.17$	for	$1 < \mu < 20$ cP

TABLE 1 RANGE OF EXPONENTS IN EQN 16

For example, a good empirical correlation that is often recommended for the estimation of N_{js} is as follows [Zwietering 1958]

$$\frac{N_{js} D^{0.85}}{\nu^{0.1} C_{sw}^{0.13} d_p^{0.2} (g\Delta\rho/\rho_f)^{0.45}} = k[^T/_D]^{\delta} \qquad (17)$$

where the magnitudes of the constant k and exponent δ depend upon the size of the impeller and its clearance from the base of the tank (Table 2).

IMPELLER	δ	k
2-bladed paddle	1.4	1.8
propeller	0.85	2 - 3
6-bdaded disc turbine	1.5	1.45

TABLE 2 Values of the constants in eqn 17

It is interesting to note that the value of the group on the right hand side of Eqn 17 is only weakly dependent upon the type of impeller: typically, for $^D/_T = {}^1/_3$ and $^h/_H = {}^1/_4$, the group on the right hand side of Eqn 17 has a value in the range 6 to 8 for propeller, vaned disc, turbine disc, 2-bladed paddle ($^D/_w = 2$ and $^D/_w = 4$) impellers. This is in accord with the observation made in this investigation regarding the value of the constant A in Eqn 15.

Moreover, for propellers with k=2.5 and δ =0.85 Eqn 17 reduces to

$$N_{js} = 2.5 \frac{(\Delta\rho/\rho_f \; g)^{0.45} \quad ^{0.1} \; d_p^{0.2} \; C_{sw}^{0.13} \; T^{0.85}}{D^{1.70}} \qquad (18)$$

The model Eqn 15 which is based on the mean fluid circulation over the particles at the bottom of the vessel gives exponents of d_p, $\Delta\rho/\rho_f$, g, C_v, D and T which are reasonably close to the corresponding values given in Eqn 18. The exponents of all the variables in Eqn 15 are also well within the range of values reported in Table 1.

There is some experimental difficulty in obtaining reproducible data on N_{js}. One reason for this stems form the difficulty encountered in consistently defining the critical conditions based on the 1-2 seconds criterion. Moreover the visual observation of the base of the vessel normally used for

N_{js} measurements is highly subjective and thus prone to experimental uncertainties. Despite the nebulous definition of the just-suspension condition there seems to be a reasonable agreement between model Eqn 15 and published results from several previous workers (see Table 1).

Additionally, some recent measurements of N_{js} for suspension of settling particles in viscous liquids [Panesar 1988,] have shown that while both Eqns 15 and 18 describe well the dependency of N_{js} on the parameters shown in the two equations, the absolute values of N_{js} estimated by Eqn 18 are too high by as much as a factor of about 2.5 compared with measurements of N_{js}. This can be explained, at least qualitatively at this stage, by examining Eqn 15 and noting that as impeller Reynolds number decreases ($Re = \rho_f ND^2/\mu$), flow becomes laminar and thus P_o is expected to increase. In a similar way, an increase in liquid viscosity will lead to a decrease in particle Reynolds number giving higher values of C_D and, probably, C_L. As a result of these changes the magnitude of the proportionality constant, A, in Eqn 15 is expected to decrease indicating that, for otherwise identical conditions, increasing liquid viscosity should lead to a decrease in the minimum suspension speed, which is perhaps not intuitively obvious at first.

Another point that deserves further investigation is the effect on N_{js} of particle diameter. Equation 15 developed in this work and most correlations presented in the literature are applicable to situations where particle diameter is relatively large and/or where fluid viscosity is relatively low, i.e. to cases where particle diameter is larger than the thickness of the boundary layer at the base of the vessel. In these cases, the hydrodynamic forces (lift and drag) acting on the particles are due to the mean circulation of the fluid near the base of the vessel. Some recent experimental data obtained by Zolfagharian [1990] shown in Fig 4 and reported previously by Baldi et al [1987] suggest that as particle size decreases its effect upon the minimum suspension speed increases, i.e. the exponent of d_p in Eqn 15 is expected to be larger than 1/6. This suggests that suspension of fine particles might occur by a different mechanism to that which was used to derive Eqn 15. However, it is questionable whether particles that are so small in diameter move into suspension as single grains or as clusters. If particles move as clusters, then the size and the effective density of these clusters are important from the point of view of suspension, and not the properties of the primary particles forming the cluster. It is therefore not possible at this stage to draw definite conclusions from these observations regarding suspension of fine particles.

A full discussion on differences between the model presented in this work and previously published models will be the subject of a separate paper but, since in most experiments involving solid suspension fluid flow (based on impeller Reynolds number) is in the turbulent region, it is appropriate to comment briefly on the role of turbulence in particle suspension. A number of models of suspension based on liquid turbulence have been reported, both for stirred vessels [Baldi 1978, Molerus and Latzel 1987, Davies 1986, Buurman et al 1986, and Wichterle 1988] and for pipe-flow [Oroskar and Turian 1980 , Davies 1987 and Thomas 1979]. For example, for suspension in stirred vessels Baldi et al [1978] proposed that particle suspension is caused by a turbulent bursting process and suggested that a particle initially at rest at the base of the vessel is lifted into suspension when the turbulent-burst forces exerted on that particle become equal to the apparent weight of the particle. They assumed that fluid turbulence in the vessel was homogeneous and isotropic and that the size of the critical bursts (eddies) was of a scale of the order of the diameter of particles. The final expression for N_{js} was empirically corrected for the effect of solids concentration. While the final correlation that is proposed by Baldi et al [1978] is very similar to Eqn 15 proposed in this work, there are a number of experimental facts with which the turbulent-burst mechanism can not reconcile. Firstly, recent Experiments at University College London have demonstrated that particle motion and suspension occur in cases where liquid motion is well in the laminar region. Evidently, fluid turbulence is not a necessary requirement for particle suspension, an assumption which is at the centre of practically all turbulent-support mechanisms. Secondly, recent photographic observations of turbulent-bursts-particle interactions at the base of a duct have shown that the energy of these bursts decreases rapidly as they travel towards and through the boundary layer at the base of the duct. These observations have shown that very few turbulent bursts reach the base of the vessel. The clear implication is that turbulent-burst activity in the bulk of the liquid can not be responsible for suspension of particles that fall in the boundary layer at the base of the tank. Thirdly, detailed observations reported in this work regarding the various states of particle motion at the base of the vessel, from the initial state of rest to the subsequent rolling and sliding and the eventual suspension, are in total disagreement with any turbulent-burst mechanism: these observations indicate that the process of particle motion is orderly and the transition from on state of motion to the next is relatively smooth: it is difficult to see how any of these observations can be interpeted in terms of a turbulent-burst mechanism. Finally, Quraishi et al [1977] carried out some particularly interesting experiments in which they studied the influence of turbulent reducing polymeric additives on the minimum speed of impeller at complete suspension. Their results revealed that while these additives reduce the small scale of turbulence (those who advocate a turbulent-support mechanism often assume this scale of turbulence to be responsible for suspension) in the vessel, addition of these polymers had no observable effect on the just suspension speed. Recent studies at University College London [Ho 1990] confirm the observations of Quraishi et al [1977].

These experimental facts are incompatible with any suspension mechanism based on liquid turbulence regardless of whether or not the final equation that results from such an analyses correlates measured data. However, this does not mean turbulent flow can not cause particle suspension. On the contrary, particle suspension, like most other transport processes, is enhanced when fully turbulent flow is established in the vessel. The point that is emphasised here is that the existence of turbulence is not a necessary requirement for the initiation of suspension, that in most practical situations particle motion and suspension begin well before fully turbulent flow is established and that there is little experimental evidence to justify the application of turbulent theory to describe the processes by which particles are suspended in liquids in stirred vessels.

The basis of the mechanism of suspension proposed in this study is that particle suspension is caused by the bulk fluid flow/circulation in the vessel in general and the mean liquid velocity at the base of the tank in particular. Admittedly, in establishing the proposed model the complex hydrodynamic flow in the vessel has been simplified in order to arrive at an equation (Eqn 15) which can be compared with empirically developed, but well-estabilshed, correlations for just suspension speed, e.g. the Zwietering's correlation. In using Eqn 15, only the magnitude of the coefficient A needs to be determined experimentally. While some of the simplifying assumptions involved in setting-up Eqn 15 may be relaxed, thus leading to improved but somewhat more complex equations, the point that should be appreciated is that all the experimental facts discussed previously are in agreement with the hypothesis that bulk liquid flow/circulation, rather than liquid turbulence, is responsible for particle motion and suspension.

NOTATION

A constant in Eqn 15
A_p particle projected area
C_v solids concentration (volume fraction)
$C_{\%w}$ solids concentration (% wt. concentration)
C_D particle drag coefficient
C_L particle lift coefficient
D impeller diameter and pipe diameter
d_p particle diameter
F_D particle drag force
F_{Dr} particle drag force due to U_r
F_{Da} particle drag force due to U_a
F_L particle lift force
F_{Lr} particle lift force due to U_r
F_g gravitational force
F_E effective weight of particle
g acceleration due to gravity
K constant in Eqn 8
k constant in Eqn 17 (Table 2)
K_1 constant in Eqn 14
n_p number concentration of particles
N impeller speed
N_{is} critical impeller speed for incipient particle suspension
N_{js} critical impeller speed for suspension in tanks
P impeller power input
P_o impeller power number
Re impeller Reynolds number
T vessel diameter
U_r radial component of local velocity of liquid at the base of the vessel
U_{r1} U_r at the point of incipient particle motion
U_a axial component of local liquid velocity near the base of the direction
U_{a1} U_a at the point of incpient particle motion or particle suspension
U_m mean liquid velocity near the base of vessel
W width of baffle
w width of impeller blade

Greek letters
δ constant in Eqn 17 (Table 2)
ε_p energy dissipation for entrainment of single particle
ε_T total energy dissipation for complete suspension
ε_b mean liquid energy near the base of vessel
μ Newtonian viscosity
ν kinematic viscosity of liquid
ρ_f liquid density
ρ_p particle density
$\Delta\rho$ particle-liquid density difference

REFERENCES

Ayazi Shamlou P. and Edwards M.F., 1985, Chem.Eng.Sci., 40: 1773-1781.
Ayazi Shamlou P. and Koutsakos E., 1989, Chem.Eng.Sci., 44: 529-542.
Ayazi Shamlou P. and Zolfagharian A., 1987, Fluid Mixing III, Bradford, UK: 221-233.
Baldi G., Conti R. and Alaria E., 1978, Chem.Eng.Sci., 33: 21-25.
Buurman C., Resoort G. and Plaschkes A., 1986, Chem.Eng.Sci., 41:2865-2871.
Davies J.T., 1987, Chem.Eng.Sci., 42:1667-1670 and 1671-1676.
Harnby N., Edwards M.F., and Nienow A.W., (Eds), 1985, "Mixing in Process Industries",(Butterworths).
Ho S.H., 1990, "Suspension of Solid Particles in Liquids", Final year research project, University College London.
Molerus O., and Latzel, W., 1987, Chem.Eng.Sci., 42:1423-1430.
Nienow A. W., 1968, Chem.Eng.Sci.,14: 1453-1459.
O'Neill M.E., 1968, Chem.Eng.Sci., 23: 1293-1298.
Panesar B.S., 1988, "Suspension of Soild Particles in Liquids", Final year Research Project, University College London.
Quraishi A.Q., Mashelkar R.A. and Ulbrecht J.J., 1977, AIChE J., 23:487-492.
Saffman P.G., 1965, J. Fluid Mech.,22: 385-400.
Thomas D.G., 1962, AICHE, 8: 373-378.
Thomas, A.D., 1979, J. Multiphase Flow, 5:113-129.
Wichterle K., 1988, Chem.Eng.Sci., 43:467-471.
Yung B.P.K., Merry H., and Bott T.R., 1989, Chem.Eng.Sci., 44:873-882.
Zolfagharian A., 1990, PhD thesis, University of London (in preparation).
Zwietering T.N., 1958, Chem.Eng.Sci., 8: 244-253.

Acknowledgment
The author wishes to thank Mrs. Zolfagharian for carrying out the experiments reported in this work.

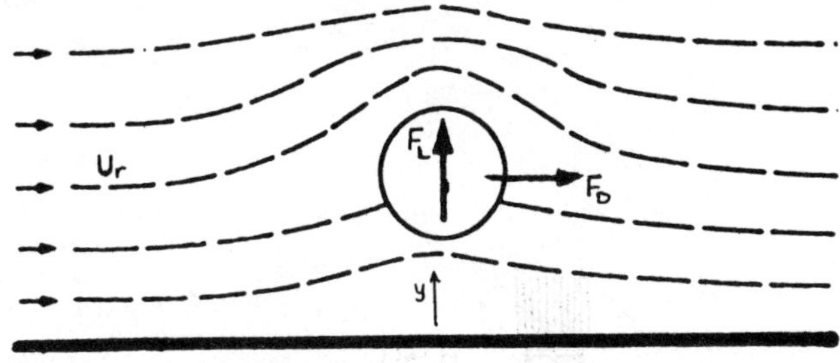

FIGURE 1a A fluid flowing over a particle close to a wall exerts two hydrodynamic forces on the particle: a lift force, F_L, perpendicular to the direction of flow and a drag force, F_D, parallel to the direction of flow.

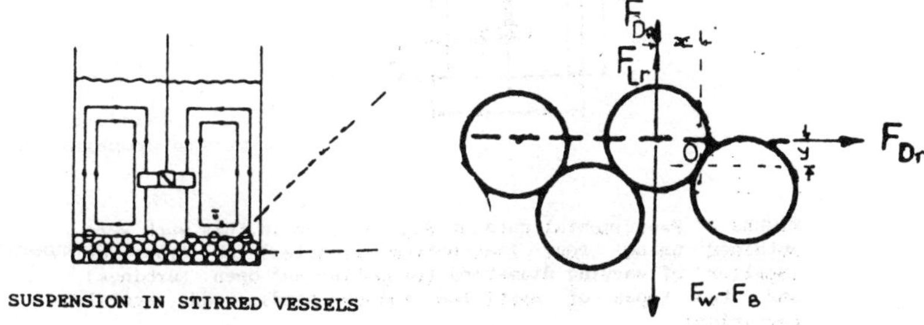

FIGURE 1b At the point of incipient particle motion the moment about point O of all forces acting on the particle is zero.

FIGURE 1c At the point of incipient particle suspension the balance of vertical component of all forces acting on the particle is zero.

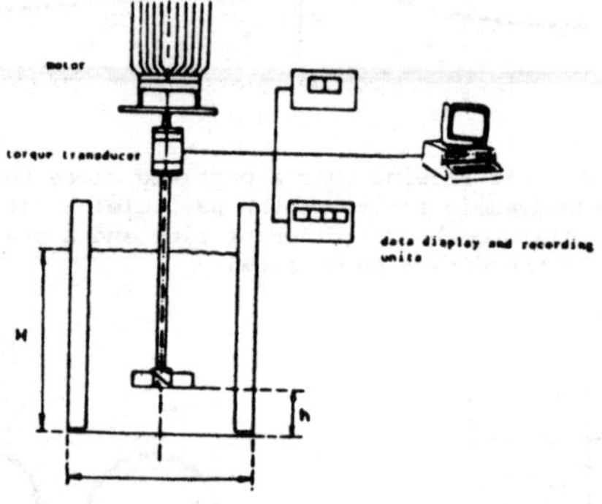

FIGURE 2 Experimental data on N_{js} reported in this work were obtained using four flat-bottom vessels, two types of impeller of varying diameters (propeller and open turbines) and two types of particles (glass beads and acetal particles).

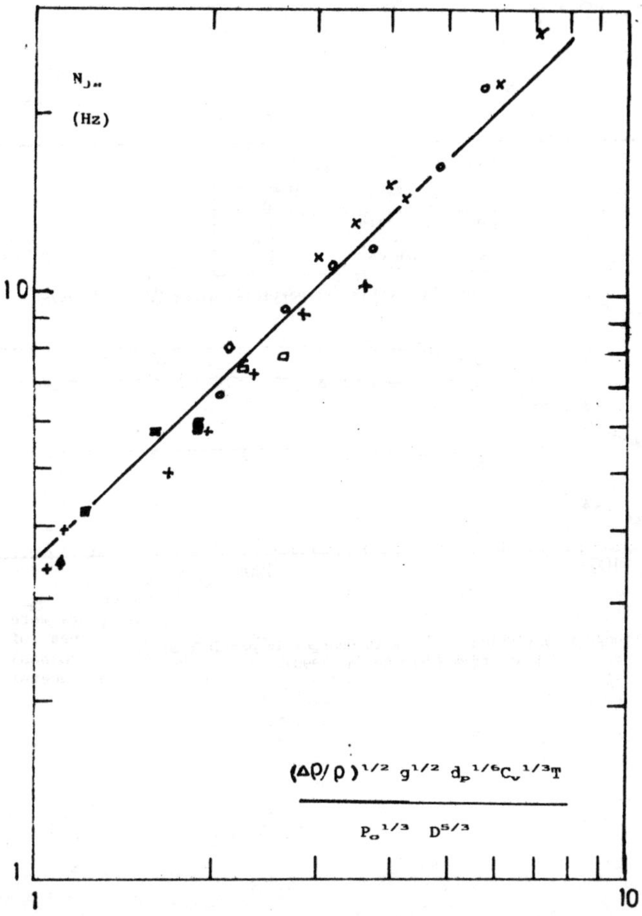

Key	impeller type	impeller dimaeter (cm)	tank diameter (cm)
×	propeller	6	30,24,19,15
○	(pumping downward)	7.6	30,24,19,15
□		12	30,24
◊		15	30
+	turbine	8	30,24,19,15
■	(pumping downward)	12	30,24

FIGURE 3 Variation of N_{js} with changes in system parameters is adequately described by model Eqn 15 (data from Zolfagharian 1990).

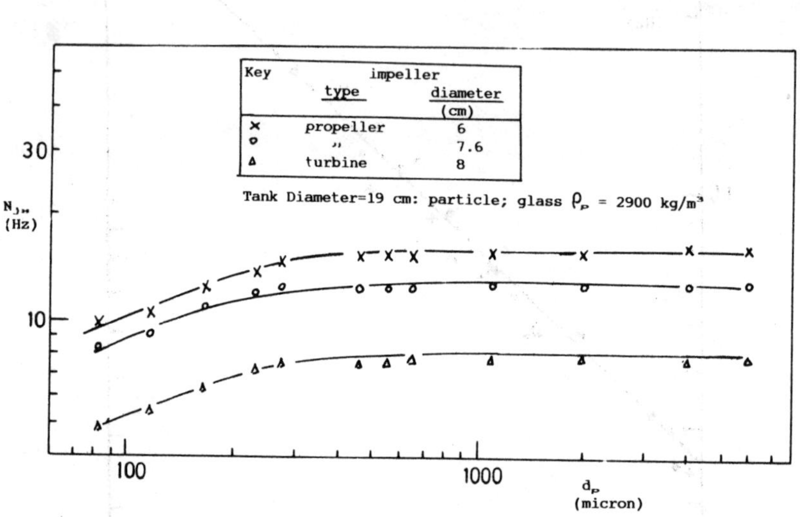

FIGURE 4 Variation of N_{js} with changes in particle diameter (data from Zolfagharian 1990).

CONTINUOUS MIXING OF VISCOUS FLUIDS

G.M. Gale*

Mixing techniques for viscous fluids are severely limited by the laminar shear flow and lack of turbulance, whilst in continuous mixing there are practical requirements of good mixing at low shear rates and short residence times to avoid excessive shear heating. The theoretical approach to overcoming these restraints by using regular cutting and turning actions to interrupt continuous shear can be achieved by a mechanism using overlapping rows of stationary and moving cavities. The viscous fluids considered are molten polymers but the theory and practice can be applied to the mixing of many viscous fluids.

INTRODUCTION

Although this paper is concerned with the mixing of molten polymers, the general principles apply to viscous fluids generally as they have in common the imposition of mixing by laminar shear; there being no turbulance. A frequent consequence is very poor mixing efficiency with most of the applied mechanical power being converted to heat. When this is combined with the usual poor heat transfer within the fluid, severe limitations are imposed on the mixing process.

Complications to any rational approach arise when considering mixing into a viscous fluid additive fluids of widely differing viscosities and also solids covering a wide range of particle sizes and surface energies. With polymer mixing, unlike clays, fats, etc., there is frequently the need to first convert solid pellets into a fluid by the combination of heat and shear, a process which can be severely disrupted (particularly when continuous) by the presence of additives.

In spite of all these difficulties millions of tonnes of polymer are produced every year with most of it containing one or more additives incorporated at some stage between polymerisation and product manufacture, both batch and

* Rapra Technology Limited, Shawbury, Shrewsbury, Shropshire, SY4 4NR.

continuous mixing being used. However with the introduction of "Just-in-time" concepts, the increasing need for manufacturing flexibility and the ever increasing prices of energy and raw materials, there is an accompanying need for better continuous mixing at the final product manufacturing stage. This leads to the problem of achieving a good mixing performance on manufacturing machinery where the main purpose is to convert polymer pellets into finished products. A consequence of this is illustrated by figure 1 in which transmitted light shows poor colour masterbatch distribution in a blow moulded bottle. Such examples of excess (and hence completely wasted) additive used to compensate for inadequate mixing is quite common, and can tip the balance from profit to loss.

The aim of this paper is to show that with a better understanding of the nature of viscous fluid mixing, novel processes are possible which provide many opportunities for reducing overall production and materials costs.

MIXING MECHANISMS

A convenient approach to viscous fluid mixing problems is to consider mixing to be divisible into two separate actions:-

(1) "Dispersive mixing: an operation which reduces the agglomerate size of the minor constituent to its ultimate particulate size".

(2) "Distributive mixing: an operation employed to increase the randomness of the spacial distribution of the minor constituent within the major base with no change in the size of the minor particle".

A good example is in the colouring of plastics. Machinery such as internal (batch) mixers and co-rotating (continuous) twin screw extruders provide the high shear forces required to disperse the pigment particles at a high concentration in the polymer to form a pigment masterbatch. Pellets of colour masterbatch are then blended with natural polymer and fed into the machines making the product. These machines, usually having a single screw to melt the pellets and pump the molten polymer into the forming stage are required to uniformly distribute the pigment throughout the polymer matrix. The possible outcome is shown diagramatically in Figure 2, the bottom left example representing the effect shown in Figure 1.

Some confusion arises over the term "mixing screw" which is usually applied to screws with improved melting performance and may contribute little to additive mixing. Further confusion is caused by reference to good dispersion by static mixers whereas their role is primarily distributive although they are unlikely to eliminate the patterns shown in Figure 1.

Dispersive mixing normally applies to the mixing of fine powders such as pigments, inorganic fillers, flame retardants, etc., and the problems and requirements are very comprehensively covered in a series of articles by M.J. Smith (1,2,3,4). The theoretical concepts and applications dealt with below are concerned entirely with distributive mixing although the division becomes questionable when considering immiscible blends as discussed below.

SOME THEORETICAL CONCEPTS

Although many authors have made contributions, the mechanism required by screw machines such as extruders, blow moulding and injection moulding machine has been most clearly demonstrated in experiments by Ng and Erwin(5) based upon concepts described by Spencer and Wiley(6). Reference to McKelvey(7) can also be usefully made, whilst Edwards and Shales(8) have shown that mixing in single screw extruders is even worse than that predicted by theory.

In the distributive mixing of two similar viscous lamina flow liquids, the degree of mixedness can be assessed by either the total interfacial area between them, or the striation thickness, the two being related.

$$\frac{\text{Interfacial area}}{2} \times \text{mean striation thickness} = \text{volume of liquid}$$

Referring to Spencer and Wiley(6); growth of interfacial area in a fluid subjected to shear follows the formula:-

$$\frac{A}{Ao} = 1 - 2S \cos\alpha \cos\beta + \cos^2\alpha \, S^2$$

where A is new area

Ao is original area

S is magnitude of shear stress

α and β are angles defining the orientation of the surface to the shear strain (7)

For large unidirectional simple shear, as found in the metering section of a single screw extruder (Figure 3):-

$$\frac{A}{Ao} = S \cos \alpha$$

The growth of interfacial area proceeds linearly with shear. Erwin's model(9) for a mixing section with flow interruption by pins, etc. (e.g. Figure 4) considers the extruder to subject the melt to large unidirectional shear and then distort it to randomise the orientation and then subject it to further shear.

In an example(9) a shearing section is assumed to impart a strain of 10^4 in a conventional extruder i.e. N = 1.

$$\frac{Af}{Ao} = f \left(\frac{S}{N}\right)^N = \frac{10^4}{2} = 5 \times 10^3$$

For an extruder with one mixing section there are N = 2 shearing sections separated by N-1 = 1 mixing section.

$$\frac{Af}{Ao} = \frac{10^4}{4} = 6.25 \times 10^6$$

i.e. 3 orders of magnitude improvement with little or no increase in total shearing of the fluid.

As large shear orients interfaces parallel to shearing planes, further improvements in mixing efficiency would be attained by each mixing section rotating the fluid so that the interfaces are more favourably oriented for the next shearing stage.

In the equation $\frac{A}{Ao} = S \cos\alpha$ this will occur when $\cos\alpha = 1$

If a shearing system has many mixing stages dividing the system into sections having large shear with equal magnitude, mixing will be:-

$$\frac{Af}{Ao} = \frac{S}{2N} \left(\frac{S}{N}\right)^{N-1} = \frac{1}{2} \left(\frac{S}{N}\right)^N$$

with an input of optimal orientation $\frac{Af}{Ao} = \left(\frac{S}{N}\right)^N$

This concept was demonstrated(5) in experiments using coloured polyethylene sheared in an annulus formed between two concentric cylinders. The procedure was to separately form rings of black and white polyethylene in the annulus, cut these into segments and replace them, alternating the black and white, with adjoining faces radial to the annulus (Figure 5).

The apparatus was heated in an oven at 175°C to melt the polyethylene, and the inner cylinder rotated to give an amount of shear strain:-

$$S = \frac{R\theta}{W} \qquad (\theta = \text{No. of radians turned})$$

After cooling to solidify, the sheared black and white annulus was cut into rectangular blocks and average striation thickness measured. The blocks were replaced, re-oriented such that the interfaces were perpendicular to the direction of shear. The melting, shearing, cooling, measurement of striation thickness and re-orienting were then repeated several times.

Plots of Af/Ao (derived from striation thickness ratio r_o/r_f) against total shear (known from the total amount the cylinder rotated) plotted on log scales gave close agreement to the line drawn for a slope of N = 1 and for the re-orientations of N = 2 and N = 3.

Although spaced circumferential rows of pins will divide and redirect striations at regular shearing intervals as required by the above theory, the mixing performance of such systems is only moderate(10). A better system is to alternate such rows of pins with similar rows of fixed pins (11), but although the mixing performance is good this type of configuration risks damage from engagement of fixed and moving parts in particular when influenced by differential thermal expansion.

An alternative approach is to have a rotor within a stator, each having cavities within their surfaces such that fluid flowing through has to follow an alternating path between cavities on opposite sides. The rotation of the rotor produces a chopping action whilst viscous drag between passing surfaces may provide a turning mechanism during transfer between cavities. The viscous fluid will be subjected to simple laminar shear mixing within the cavities between transfers. With this mechanical arrangement, relative axial movement will not normally produce mechanical damage.

Probably the first mixer of this type was one having Woodroffe key type cavities patented by Gerber(12) of Metal Box in 1961 (Figure 6), whilst a very similar geometry is achieved by Barmag(13). Streamlining of this configuration has been patented or illustrated, covering axial cavities in both surfaces(14), angled cavities in both surfaces(15) and angled cavities in the rotor combined with axial cavities in the stator(16).

The Cavity Transfer Mixer (or CTM)(17) differs from these others in two respects: the cavities are hemispherical in shape and they are arranged in a staggered configuration to minimise land areas (Figure 7). This is considered an advantage in minimising generation of heat.

MODEL EXPERIMENTS

Theoretical work has been carried out by Bromilow and Hulme(18) and by de Jong(19), who also made practical measurements as also have Edwards and Shales(8).

In our experiments with a transparent Perspex model the viscous fluid was a room temperature curing clear Silicone polymer. The CTM had 3 cavity rows, with 6 cavities in each row and was mounted vertically so that as it was filled, air was expelled. The catalysed Silicone (Silgard 184 - Dow Corning) was pumped through at a uniform rate of 103 ml/min using a pump which was essentially a motor driven syringe. Mixing behaviour was determined from observations of a coloured striation produced by injecting a stream of coloured silicone from a peristaltic pump. Rotational speeds were 2,4 and 5.5 rpm.

Observations were made by Video and still cameras, but most information was gained from observations of microtomed sections of samples removed after curing to an elastomer overnight.

The mixing mechanism was very complex but the overall effect was generally as follows. The injected striation travelled in a direction opposite to that of the moving rotor surface around the surface of the stator cavity and then crossed into a rotor cavity, turning sharply in the direction of rotor rotation. It then travelled across the cavity the following land folding it into a loop and cutting it. The loop was then deposited in the next cavity row in the stator where a build up of striations in parallel lines occurred. Laminar flow fields sheared these striations from the centre approximately at right angles. The mechanism is illustrated in Figures 8-10 and the photomicrograph (Figure 11). The overall pattern appeared to be one suitable for compliance with the theory although difficult to analyse in detail.

It was also concluded that a few large cavities would probably mix as well as many small cavities, such that in the former case the lower shear rates would be compensated by the longer residence time, and the reduced number of cutting actions per revolution would be compensated by the slower cavity transfer movements so that cut striation lengths would still be short. Mixers with a few large cavities will have advantages over ones with many small cavities as a result of lower pressure drops and reduced shear heating(11).

STRIATION THICKNESS MEASUREMENTS

Experiments were carried out in which clear flexible PVC was extruded through a 38 mm extruder fitted with a 7-row Cavity Transfer Mixer and a stream of black flexible PVC was injected at the mixer entrance from a 25 mm extruder.

Following stabilisation of conditions the extruders were stopped, the die removed and the mixer dismantled. The PVC was prised from the cavities and the average striation thickness measured under a microscope using 20 microns thick microtomed sections. (Similar experiments using HDPE with mixed colours and measurements by image analysis have been described by Edwards and Shales(8)).

From the theory described above,

$$\frac{r_o}{r_f} = \frac{S}{N}^{N}$$

where $\frac{r_o}{r_f}$ is striation thickness ratio.

S is the total amount of shear

N is the number of shear stages

In this case the shear in each mixer stage and the amount of cutting is fixed by the mixer design such that shear and cutting are related. Shear stages have not been equated with cavity stages of the mixer as material is flowing continuously through but they will be proportional to the number of cavities for a given flow rate and rotational speed.

It appeared reasonable to assume S = kN where k is a constant, and hence:

$$\frac{r_o}{f} = \frac{N}{kN}^{N} \quad \text{or} \quad \log \frac{r_o}{f} = N \log \frac{1}{k} = K_1 N = K_2 S$$

where K_1 and K_2 are constants.

Log striation thickness ratio will consequently be proportional to the amount of shear together with the associated amount of cutting which will be dependent on the distance travelled through the mixer. Using the microscopy measurements described above, graphs can be drawn of log striation thickness ratio ($\log \frac{r_o}{r_f}$) against mixer stage as shown in Figure 12, and these follow reasonably closely to a linear relationship.

From the shape of the graph we can obtain a measure of the mixer's performance. If the mixing were carried out using simple shear only, as in an extruder screw, the striation thickness ratio being proportional to S only would be very small by comparison.

THE DECOUPLED MIXER

It was initially assumed that commercial applications of the Cavity Transfer Mixer would be as extensions to extruders in which rotational speed and throughput are directly linked. In an evaluation of ten different dynamic mixers(11), two extruders fed natural and black polyolefins respectively into the entry of the mixer which had its own independent drive. This enabled the influence of rotational speed with a fixed throughput to be determined and isolated the mixer's performance from that of the extruder.

In Figure 13 comparisons are shown for polyethylene using

(1) A short extruder type flighted screw.

(2) A CTM with 6 cavities circumferentially and 7 rows axially.

(3) A CTM with 3 cavities circumferentially and 5 rows axially.

The trials showed that the Cavity Transfer Mixer achieved a "striation free" microtomed section at speeds 25% of that necessary for a pinned mixer (which was the next best) and with far less generation of heat. It also was evident that the decoupled arrangement made possible completely new processing systems as described below.

MIXING PROCESSES USING INDEPENDENTLY DRIVEN MIXERS

In the development of mixers having rotors attached to the extruder screw it was found that many applications were possible using blended additives introduced conventionally at the extruder feed and by injecting liquids, pastes and slurries directly into the mixer's entry through a non-return valve. However limits were reached whereby polymer blends with widely differing processing temperatures could not be accommodated by a single extruder, or levels of injected liquid were such that mixing was disrupted. These problems were overcome by using one of the combinations shown in Figure 14.

Particular interest has been shown for three types of applications:

(1) Polymer blending in which pellet blends are fed into one extruder.

(2) Polymer blending in which one extruder is fed with a thermoplastic and the other with a vulcanisable elastomer.

(3) Liquid incorporation in which up to 40% liquid is pumped directly into the mixer. The mixer would normally be fed by an extruder, but gear pumps can also be used, i.e. the unit is used as a pipeline mixer.

In view of the statement at the beginning of this paper that the theory applied to these mixers does not apply to dispersive mixing, the question arises of how the good mixing of immiscible blends has been achieved.

Polymer blends can be miscible, in which case final properties are usually intermediate between those of the constituents, or immiscible, in which case a number of structures can exist. Of particular interest are immiscible blends in which one polymer exists as tiny energy absorbing droplets within a comparatively brittle continuous phase. In the case of miscible blends the theories for distributive mixing will apply, but for immiscible blends the requirements are uncertain.

Several papers have described work based on the experimental technique of Taylor(20) using a simple shear field produced by movement of two parallel belts in opposite directions. Flumerfelt(21) using non-Newtonian fluids showed that in a simple shear field, a spherical droplet became elipsoid with the major axis inclined at about 45° from perpendicular to the shear field. Depending on relative viscosities, a critical shear rate was reached in which the droplet broke up into smaller droplets. A minimum size was eventually reached below which break up could not be achieved regardless of shear rate. Drop break up was less likely to exist where the droplet had either a relatively high or relatively low viscosity. However, in the latter case the inclined droplets had pointed ends where very small drops came off in a stream, unlike the relatively high viscosity droplets which showed an elipsoid shape nearly aligned with the flow direction (Figure 15). Under transient conditions, in which the shearing was stopped between speed increase settings, significantly lower shear rates for break up were required than for steady conditions.

Theodorou(22) used a similar type of arrangement in which the droplet was replaced by spherical agglomerates of acrylic powder bound together to a paste with silicone fluid (solids concentration about 70%). Three sets of cylinders covering different diameters were used. In the shear field, particles surrounding the agglomerate formed an inclined extended cloud, to some extent analogous to Flumerfelt's lower viscosity droplets, but the agglomerates themselves rotated.

In the extensional flow field, the particle cloud became aligned in the direction of maximum stress, whilst the agglomerates were subjected continuously to the maximum extensional force. This contrasted with the shear field in which the agglomerates rotated which consequently subjected them to alternating tension and compression. The extensional field was found to be 2 to 3 times more effective in dispersion than the shear field.

Polymer blending trials have been carried out using the configurations 1 and 3 in Figure 14. For EPDM rubber blended with polypropylene configuration 1 was used as both materials are available in pellet form with similar extrusion characteristics.

For natural rubber containing crosslinking additives blended with polypropylene configuration 3 was used. In this case the rubber was in strip form and extruded at only 100°C such that when blended in the Cavity Transfer Mixer with the polypropylene at 200°C "in-situ" crosslinking (vulcanisation) of the rubber phase occurred. Limits of viscosity difference were observed so that for good mixing the viscosity of the rubber was marginally reduced by adding up to 5% oil to the compound.

It has so far been assumed that the Cavity Transfer Mixer will achieve good polymer blending when the viscosities are not too widely similar as a result of the simple shear mechanisms of Flumerfelt being easily achieved by the mechanism of simple shear interrupted by a regular cutting and turning mechanism. In the case of the natural rubber- polypropylene blends, high shear as for dispersive mixing may be required, but an intermediate level could well be achieved in the Cavity Transfer Mixer as a result of extensional flow fields being present.

Incorporation of liquid additives, typically up to 6% by direct injection from a high pressure metering pump is possible in which considerable economic savings are possible. The injected liquids include colours, Polybutene-1 to make pallet wrap film tacky and silane for grafting such that crosslinked pipes and cables can be produced (23). The principle of the process is shown in the diagrammatic representation of a laboratory unit for producing XLPE hot water pipe (figure 16).

Mastics, adhesives and flexible materials can often require levels of liquid additive which will seriously interfere with the mixing process, in particular by causing internal slippage. Incorporation of up to 40% liquid in some cases is possible by using an independently driven mixer in which case the rotor's speed of rotation may be up to three times that of the extruder which feeds it.

CONCLUSIONS

It has been shown that the theory of mixing viscous fluids by Spencer and Wiley and demonstrated by Erwin can be applied to polymer processing, and that this makes possible many novel continuous mixing systems. However, in the case of blending immiscible polymers it is possible that extensional flowfields are also important but this is difficult to define.

Such systems enable mixing to be carried out on plastics and rubber processing equipment designed specifically for product manufacture rather than good mixing performance. As a result, costs savings can be achieved from lower materials costs and fewer rejects.

Although this paper has described mixing of molten polymers, the theories apply to most viscous fluids and process patents for the application of the mixer described exist for soap(24) and edible fats(25).

REFERENCES

1. Smith M.J. J. OCCA 56. 1973, 126-133
2. Smith M.J. J. OCCA 56. 1973, 155-164
3. Smith M.J. J. OCCA 56. 1973, 165-171
4. Smith M.J. J. OCCA 57. 1974, 36-43
5. Ng KY and Erwin L. 37th ANTEC of the SPE Vol 25 May 1979. 241.
6. Spencer R.S. and Wiley R.M. J. Colloid Sci. 6, 1951, 133.
7. McKelvey J.M. Polymer Processing (John Wiley) 1962.
8. Edwards M.F. and Shales R.W. "Mixing in single screw extruders" in Making the most of the Cavity Transfer Mixer, (Rapra Technology Ltd.) Oct. 1986 pp2.1-2.8.
9. Erwin L. Polym. Eng. Sci. 18 No.7 May 1978, 572.
10. Gale G.M. 41st ANTEC of the SPE 1983. 109.
11. GB 787764
12. GB 930339
13. GB 1475216
14. GB 2202785A
15. Kobe Steel (private communication)
16. Morrell SH, Plast and Rubb, Processing and Applications 2 No.4 Dec 1982, 319-329.
17. EP 0048590
18. Bromilow T.M. and Hulme A.T., "Towards a mathematical model of the Cavity Transfer Mixer" in Making the most of the Cavity Transfer Mixer (Rapra Technology Ltd.) Oct 1986 3.1-3.15.
19. de Jong E. "Mathematical Considerations of Emulsification in the Cavity Transfer Mixer", in Making More of the Cavity Transfer Mixer (Rapra Technology Ltd.) 15 Nov 1988 14-22.
20. Taylor G.I. Proc. Roy. Soc. Ser. A146, 1934, 501.
21. Flumerfelt R.W., Ind Eng Chem Fundam, 11 No.3, 1972, 312.
22. Theodorou T. MSc Thesis, Dept of Chem. Eng. and Chem. Technol. Imperial College, London, Sept. 1978.
23. EP 0198533
24. EP 0048590
25. GB 2118055, GB 2118056, GB 2118057, GB 2118058, GB 2118450, GB 2119666, GB 2118854.

THE DESIGN AND DEVELOPMENT OF AN ACTIVE TRANSPORT CATALYTIC REACTOR.

L. K. Doory, B. E. Said, U. Ullah, S. P. Waldram*

Promoting the formation of eddies close to a catalyst surface can enhance reaction rates under mass transfer limited conditions. Experiments are reported which show how the improved mixing which results leads to increased radial transport parameters for both monolithic and wall coated catalytic reactors. This is demonstrated for the oxidation of carbon monoxide over a platinum-rhodium catalyst operating at conditions typical of those found in automobile exhaust gas treatment: these are space velocities up to 100,000 h^{-1}, an inlet temperature of approximately 370°C and an inlet carbon monoxide concentration of 0.5 mol%.

INTRODUCTION

Heterogeneous catalytic reactors are designed in many different configurations depending on the process application for which they are intended: packed, trickle and fluidised beds are common as are slurry reactors and catalytic monoliths. In all situations where intrinsic reaction kinetics are very fast there is the possibility that actual rates of reaction are limited by mass transfer through the fluid adjacent to the catalyst surface. These situations are the subject of this paper. The ideas we develop are general in nature but are applied specifically to one catalytic reaction, namely the oxidation of carbon monoxide over a platinum-rhodium catalyst at conditions representative of those found in automobile exhaust gases. The catalytic treatment of car exhaust gases is of huge environmental, and commercial, importance and is the largest single application of catalysis worldwide when judged in terms of annual catalyst sales. Exhaust emission standards are becoming increasingly strict with proposals that in the USA, by 1994, CO emissions be reduced to 0.34 g/mile, NO_x to 0.4 g/mile and unburnt hydrocarbons to 0.25 g/mile. It is intended that catalytic treatment units should display a 100,000 mile lifetime. Proposed and actual European standards are more lax by a factor of about 2 to 10 on specific component concentrations.

* Author to whom correspondence should be addressed.
Chemical and Biochemical Engineering, University College London, Torrington Place, London WCIE 7JE

Catalytic monoliths for exhaust gas treatment

Most current exhaust gas treatment systems use an extruded monolith of cordierite with a cell density of the order of $62/cm^2$. Metallic supports of Fecralloy are becoming increasingly popular. The monolith substrate is treated with a washcoat consisting of alumina, stabilizers and promoters and then platinum group metals (platinum, rhodium and palladium) are dispersed on this by impregnation. The whole unit must operate at an air-fuel mass ratio of 14.7:1 $\pm$ 0.1 in order to perform simultaneously the required three way duty of oxidising CO and unburnt hydrocarbons and reducing NO_x. Space velocities within individual monolith channels range up to 100,000 h^{-1}, equivalent to an STP Reynolds number of approximately 300, or an actual Reynolds number of about 180. This means that at steady state conditions, fully developed laminar-flow will be present over as much as 95% of the monolith channel length, Langhaar (1). This in turn implies that radial transport of reactants to the catalytic walls of the monolith channel will largely be by molecular diffusion, a slow process which ensures that overall reaction rates are mass transfer limited on the timescale of the exhaust gas mean residence time. This is of the order of 0.02s. Great scope exists for promoting radial mixing so as to increase radial heat and mass transfer processes which will then lead to enhanced reaction rates. For a review of automobile exhaust gas treatment see Wei (2) and Cooper et al. (3).

Static, motionless or in-line mixers

Static mixers were originally developed for homogeneous blending of viscous materials. Subsequently they have been used to enhance heat transfer to, or from, a fluid flowing in a pipe in which they are inserted. Heat transfer enhancement factors of between 1.7 and 3.6 have been reported, Genetti (4). Houzelot and Villermaux, (5), have discussed the use of a co-axial cylinder to reduce the diffusional pathlength, and thereby increase reaction rate, in a laminar flow channel with catalytic walls. In this paper we report analogous enhancements in reaction rate. Flow disruption to promote radial transport rates has been achieved in two ways and results from each of these are described in the experimental section. The role of energy dissipation in mixing devices has been discussed by Villermaux, (6), but quoted mixing and drag coefficients for Sulzer and Kenics mixers are only applicable for Reynolds numbers two or more orders of magnitude smaller than in our experiments.

EXPERIMENTAL

A block diagram of the apparatus is shown in Figure 1. The equipment is designed to be used for a number of reaction engineering duties but in this case is supplied by cylinders of oxygen (99.97 vol %), nitrogen (99.99 vol%) and carbon monoxide (99.96 volume %.) Gas flow rates were set and monitored using Chell flow controllers and were mixed and preheated before being passed through the reactor. In these experiments no water vapour was added. Product and unreacted feed gases were monitored at the reactor exit using an infra-red gas analyser, (Analysis Automation Ltd, model 401D, N.D.I.R.) The reactor pressure drop was measured using a Chell differential pressure transducer. The 'IMPULSE' software package,

(Solartron, Schlumberger) was employed to monitor on an IBM PC the signals from all instruments, ie flow controllers, thermocouples, DP cell, IR, via a 20 channel input POD and PC input card (Schlumberger.)

The reactor consisted of either a monolith core or a reactor tube. The reactor tube could be operated over the same range of Reynolds numbers as in a single channel of the monolith. Both the monolith and the reactor tube were coated with a γ-alumina washcoat impregnated with platinum-rhodium catalyst. The one inch diameter core was supplied by Johnson Matthey and was hand finished so that a portion of it could be inserted into a tightly fitting stainless steel tube. The sample was 15 cm long, 1.25 cm diameter and contained 70 1.04 mm square channels. The reactor tube was of length 15 cm and diameter 1.5 cm and was made of Fecralloy. Insert devices of various configurations could be placed inside this tube: three of these are shown in Figure 2.

All experiments were carried out at inlet concentrations of CO and O_2 of 0.5 and 0.25 vol% respectively with the balance of feed being nitrogen. The inlet temperature was varied between 250°C and 400°C and inlet flow rates were up to 20 dm^3min^{-1} (STP). The temperature of the reactor could be maintained constant within ± 1°C.

RESULTS

CO oxidation in the monolith

The effect of deliberately produced eddies on conversion of CO oxidation in a monolithic reactor was studied by slicing the monolith into two, four, eight and sixteen axial segments successively. The sliced segments were arranged with an inter-segment spacing of 3mm with an angular rotation of 45° between segments. Figure 3 shows the conversion of CO at various flow rates and corresponding Reynolds numbers (actual and STP) as well as at STP space velocities. There is a loss of length of approximately 1mm each time a monolith segment is cut into two. Thus when the monolith is finally subdivided into sixteen pieces the total length of the catalyst is reduced by some 15mm i.e. 10%. Because of this, the lines of conversion versus flowrate in Figure 3 are not directly comparable. A more useful comparison is obtained when the results are re-expressed on the basis of the surface area of catalyst present in the original 15 cm monolith sample (normalised conversion Figure 4). These results show that at Re ~ 100, or an STP flow rate of 10.0 dm^3min^{-1}, conversion can be raised from approximately 50% to more than 90% when the monolith is cut into sixteen pieces.

Figure 5 shows the pressure drop across the monolith. The segmented monolith shows little change in pressure drop from the uncut data. The explanation of this surprising result is that the combined effects of length reduction and energy dissipation in induced eddies approximately compensate each other. The enhanced conversion is therefore achieved with essentially no increased pressure drop penalty.

CO oxidation in the tubular reactor

The enhancement of mass transfer to the catalyst on the walls of a tubular reactor was achieved by placing inserts of various configurations inside the tube.

The results of such experiments are shown in Figures 6 and 7. In Figure 6 curve 1 represents the conversion of CO at various Reynolds numbers in the empty tube. Inserts, as shown in Figure 2, are then in turn placed inside the tube. Conversion at a given Reynolds number is increased significantly but found to be totally insensitive to the insert configurations studied. Curve 2 in Figure 6 represents this fact and applies to all three static mixer designs. The IR analyser used for the measurement of CO was very sensitive and could measure changes in CO of 0.01%. The continuous lines of inserts were then successively replaced by three equally spaced elements within the same reactor tube. This was carried out for each insert design. Conversion was then lower than previously (compare curve 3 with curve 2 in Figure 6) but still quite insensitive to insert configuration.

Figure 7 clearly shows that the improvements in CO conversion shown in Figure 6 have been achieved at the cost of increased pressure drop. Although the performance of all static mixer inserts appears to be identical in enhancing reaction rates the Kenics static mixer achieves this at the lowest pressure drop penalty. The UCL insert results in the highest additional pressure drop.

DISCUSSION

The performance of any two chemical reactors can be compared when certain requirements are met. For all types of reactions (homogeneous or heterogeneous), temperatures, or the temperature profiles, must be analogous. Ideally, geometric, dynamic and chemical similarities should exist between the two reactors being compared. This can be achieved by matching the dimension ratios, Reynolds numbers and Damköhler numbers. However, it is very difficult to achieve both dynamic and chemical similarity. For example, if characteristic reaction times are preserved, the Reynolds numbers which also involve linear velocity cannot also be matched. Frequently, only some of the factors influencing the reaction are of major importance. For example, when external mass transfer is the rate controlling step in a heterogeneous reactor, geometric and dynamic similarity are sufficient for comparison.

The experiments for both types of reactors (monolithic and tubular) were conducted in the same range of channel Reynolds numbers, (50-400) and at an identical temperature (370 ± 1°C). The results show that enhanced conversion of CO (a factor of about 1.8 at Re ~ 100) is obtained by slicing the monolith into sixteen pieces without an accompanying increase in pressure drop. The axial slicing changes the ratio of developing flow length to overall channel length and from Hawthorn's correlation, (7), would be expected to result in a mass transfer coefficient enhancement of 1.5 between the uncut and 16 piece axially segmented experiments. The tubular reactor with inserts of any one of the configurations shown in Figure 2 gives a conversion of 100-94% in the range of Reynolds numbers studied. In the monolith with 16 segments the conversion varied from 99% to 88% in the same range of Reynolds numbers and identical temperature. The pressure drop for the tubular reactor and Kenics mixer ranged between 0.7 Torr to 0.9 Torr, whereas for the 16-piece monolith it was between 1.4 and 5.8. It appears that all the three inserts used in these experiments are overdesigned for the present function. These results also suggest that static mixers of simpler geometry than the ones used might give conversions of the same order of magnitude with lower pressure drop.

The design of insert devices is being studied and optimised using a commercial computational fluid dynamics package, 'FLUENT'. Both the mass transfer (obtained by analogy with heat transfer) and the pressure drop predictions from this package agree well with experimental observations made on prototype inserts.

ACKNOWLEDGEMENTS

This work was supported by a Science and Engineering Council Research grant as part of the Interfaces and Catalysis Initiative.

We are grateful to Johnson Matthey, Royston, for supplying the monolith core samples used in this work.

REFERENCES

1 Langhaar, H. L., 1942, ASME 64, A-55.

2 Wei, J., 1975, Advances in Catalysis 24, 57.

3 Cooper, B. J. et al., 1987, "Catalysis and automotive pollution control", Crucq A., Frennet A., eds., Elsevier,

4 Genetti, W. E., 1982, Chem. Eng. Commun. 14, 47.

5 Houzelot, J.L., Villermaux J., 1977, Chem. Engng Sci. 32, 1465.

6 Villermaux, J., 1988, Chem. Eng. Technol. 11, 276.

7 Hawthorn, R.D., 1974, AIChE Symp. Ser. 70, No. 137, 428.

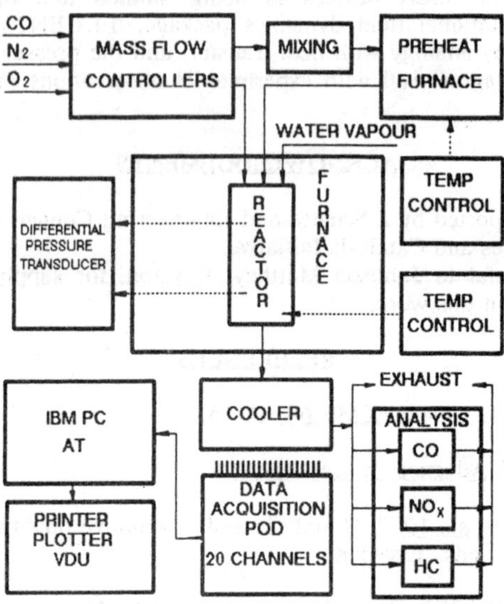

Figure 1 Apparatus block diagram

Figure 2 Static mixer insert devices, a Kenics, b Sulzer, c UCL design.

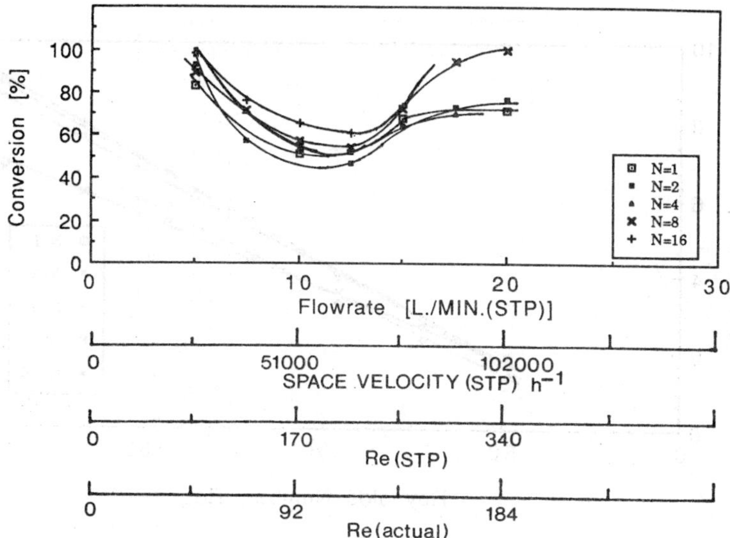

Figure 3 CO conversion as a function of STP flow rate, STP space velocity, STP Reynolds number or actual Reynolds number. Monolith core sample 15 cm length x 1.25 cm diameter. Unnormalised data. N indicates the number of axial segments into which the monolith has been cut. Temperature 371°C

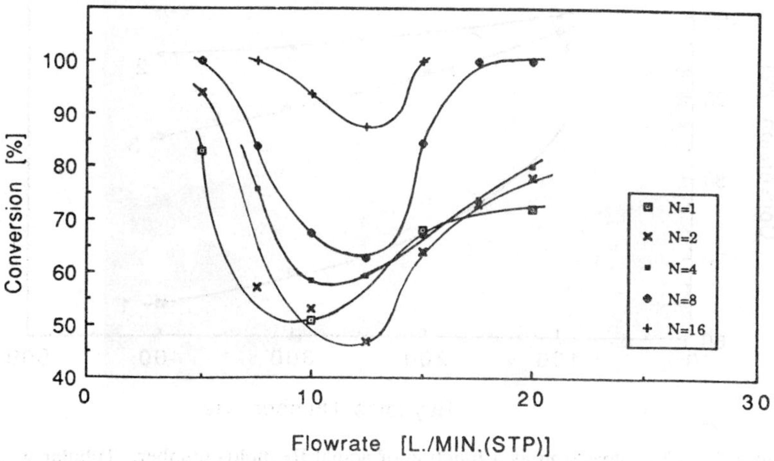

Figure 4 Data as for Figure 3 but normalised to account for changing total catalyst length.

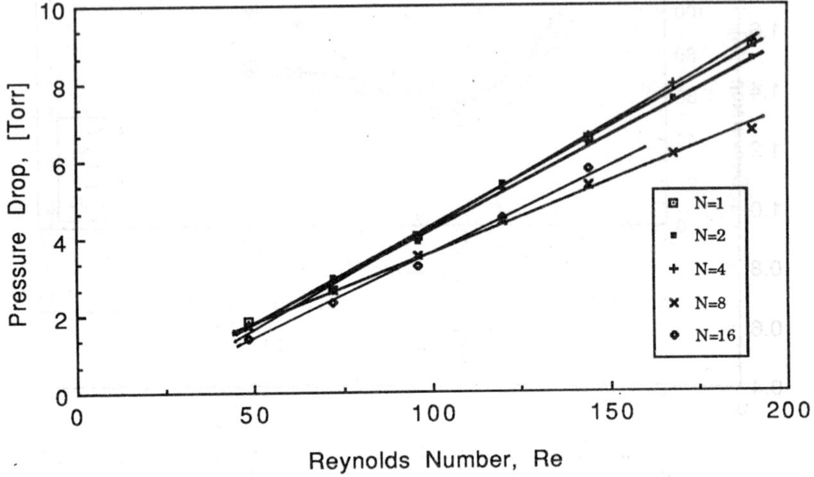

Figure 5 Pressure drop across the monolith reactor as a function of actual Reynolds number within a monolith channel. N indicates the number of axial segments into which the monolith has been cut. Data at 371°C.

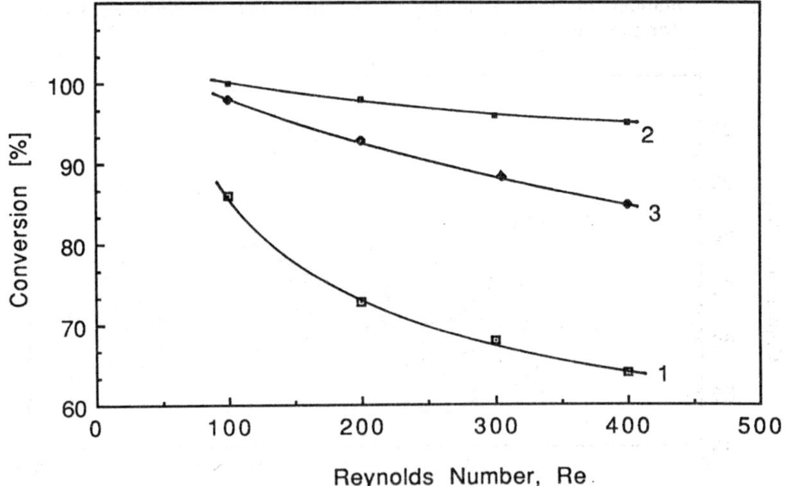

Figure 6 CO conversion as a function of actual Reynolds number. Tubular wall coated catalytic reactor empty, 1, with continuous lines of static mixers, Sulzer, Kenics or UCL, 2, and with three equally spaced static mixers, Sulzer, Kenics, UCL, 3. Data at 371°C

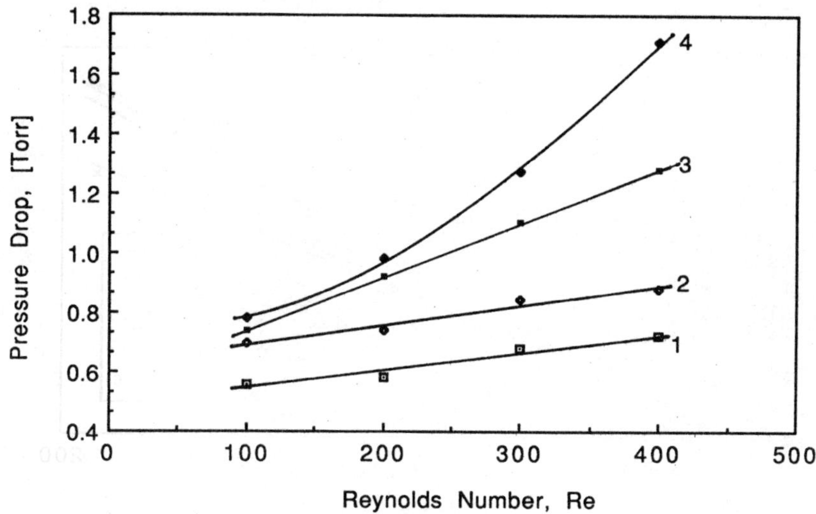

Figure 7 Pressure drop across the tubular wall coated reactor as a function of actual Reynolds number. Curve 1 tube empty, 2 with Kenics mixers, 3 with Sulzer mixers, 4 with UCL mixers. Data at 371°C.

DISSIPATIVE HEATING OF NEWTONIAN AND NON-NEWTONIAN VISCOUS FLUIDS IN MIXING VESSELS

D. Mewes* and M. Friederich

In viscous fluids temperature fields develop in the surrounding of a rotating stirrer because of dissipation. The temperatures are spacially distributed and time dependent. A tomographic method was used to reconstruct the local temperatures from holographic interferometric measurements taken instantaneously from various directions. The fluids used for the experiments were of high transparency and viscosity and of newtonian and non-newtonian behaviour. The dimensions of the mixing vessel used in the experiments were 0.1 m in diameter and 0.1 m in height. The types of stirrers used were flat plate turbines placed in one and two levels to the rotating shaft.
The time dependent spacial temperature fields are given by graphical representation. The experimental results are compared with theoretical ones.

INTRODUCTION

The production and processing of liquid products frequently involves mixing processes of very viscous liquids. During the mixing processes local gradients in concentration are equalized and homogenous chemical reactions may be accelerated. If vessels are used, according to the process the liquids are mixed either continuously or discontinuously. In spite of intensive research work, a complete theoretical prediction of the velocity-, temperature and concentration fields in a stirred vessel is not yet possible. This is partially due to the complex flow behavior, especially of highly viscous liquids, such as biosuspensions. The

* Institut für Verfahrenstechnik, Universität Hannover, Callinstr. 36, 3 Hannover 1, F.R.Germany

velocity field of these liquids depends on their complex rheological behaviour as well as on the type of stirrer and the mechanical energy input. The normal stress differences, which are present in viscoelastic liquids during the mixing process, affect the flow field significantly. The existence of critical shear stresses might leave unmixed regions in the stirred vessel.

Depending on the flow properties of the stirred liquids, the applied stirrer, its stirring speed and other equipment parameters, the energy input via the stirrer is dissipated in the fluid. Consequently, the temperature rises, especially in the vicinity of the stirrer. A temperature equalization takes place by convection and heat conduction. In liquids with shear-thinning effect, the limited remote effect of the stirrers leads to a reduced transfer of the dissipated energy flow to the vessel wall. Thus, the increase of the liquid temperature in the vicinity of the stirrer is much more pronounced than it would be in a Newtonian liquid.

MEASUREMENT TECHNIQUE

Three-dimensional temperature fields were recorded by interferometry from several directions. The interferograms were stored by holographic methods. The time dependent temperature fields were reconstructed by tomographic techniques.

Holographic Interferometry

With the use of the holographic recording method, it is possible to store the phase distribution of coherent light waves on a photographic plate and reproduce it at a later time. Coherent light waves are necessary to utilize the interference between two waves for recordings of the phase distribution. A laser is used as the light source. The light beam emitted by the laser is enlarged and divides into two partial beams with the beam splitter. The first beam, which is referred to as object wave, penetrates the stirred vessel. The second beam by-passes the vessel and is superimposed on the object wave behind it. This beam is called reference wave. The intensity distribution and phase position of the waves intefering with each other are forming a hologram which is on a photographic plate. The measurement technique is explained in Fig. 1. The interference pattern stored on the hologram acts as a diffraction grid with variable grid constant. If the hologram is irradiated only with the reference wave, part of it is weakened during penetration. The other part is deflected. This reconstructed wave is identical to the first object wave.

In Fig. 1 the real-time method is explained. It has the advantage that non-steady state processes can be recorded. In order to do so the instantaneous and the reconstructed wave interfere with each other behind the hologram. A change of the temperature in the measurement volume results in a change of the optical path and, consequently, a phase shift of the instantaneous object wave. A macroscopic interference pattern develops due to the interference between the instantaneous and the reconstructed wave. This can be recorded with photographic cameras.

A displacement of the reference wave by a small angle with respect to the recording of the hologram results in a regular interference pattern behind the holographic plate. The quantity and the orientation of the stripes in the interference pattern are determined by the size and the direction of the displacement angle of the reference wave. With this so-called finite-fringe-field method, the density of information of the interferograms can be predicted from the quantity of interference stripes.

Tomographical Reconstruction

The information contained in the interferograms represents the integral change of the refractive index along the path of a beam. The correlation between the local change of the refractive index and the phase shift is shown in Fig. 2 for two light beams. The light beams pass an inhomogenous density field which gives rise to variing refractive indices. The resulting phase shift is described mathematically by the equation

$$\varphi = \int_l \frac{\partial n}{\partial T}(T - T_{oo})dl \qquad (1)$$

l means the pass length of the light beam, n gives the refractive index that is changed by a variation of Temperature. The refractive index depends on all three coordinates and, in non-steday state processes, on the time. A recording of the process at discrete points of time and for certain cross-sectional planes leads to a reduction of one space coordinate. The refractive index then depends only on the x and y coordinates. There is only a one-dimensional equation O(x) for the two-dimensional dependence of the refractive index n(x,y). Therefore, a mathematical calculation of the field of refractive indices is possible only with the use of information from several irradiation directions. Such methods are known as tomographical methods. In non-steady state processes, the information must be recorded simultaneously from all projecting directions. This is not possible with the measurement methods commonly used in medicine and material testing, such as x-ray computer tomography /1/.

The algorithms used here require irradiations of the measurement volume from several hundred directions. This requires long experimental time intervals and is not suitable for experimental investigations of non-steady state processes. The method which was used here requires a considerably smaller number of irradiations and is called the sample method which was described by Sweeney /2/.

The calculation effort for the reconstruction of the field function can be reduced with the use of parallel beams. This requires a special design of the experimental set-up. The light beams penetrating the measurement volumes must be made parallel by using lenses. The largest commercial lenses have a diameter of approximately 150 mm. Consequently, the diameter of the light bundles and the measurement volumes is limited. The volume of cylindrical vessels is therefore limited to approximately 1 liter.

EXPERIMENTS

The experimental set-up used for the investigations of non-steady state temperature fields caused by dissipation can be divides into an optical section and a stirring section.

Optical set-up

Fig. 3 gives a simplified representation of the optical set-up, which is mounted on a vibration-free table. The argon-ion laser, which is used as a light source, is installed on a second vibration-free table. The coherent light beam emitted by the laser has a wave length of 514.5 ***nm***. It is divided into an object wave and a reference wave with beam splitters.

The enlarged object waves which penetrate the measurement chamber and the stirred vessel located in its center are staggered by 45°. The object waves leaving the vessel are recollected by lenses, before they fall on a holographic plate in groups of two. The object waves are bundled in order to increase the intensity of the light in the hologram planes. Thus, short exposure times can be achieved for the holographic recordings as well as short shutter times for the cameras installed behind the holograms. This is very important for the observation of the non-steady state processes in the measurement chamber. The reference wave is divided into two reference waves in a beam splitter. Each of these reference waves on a holographic plate, where it interferes with the object waves. The cameras with synchronic shutter releases record the interferograms that originate behind the holograms.

Experimental set-up

Fig. 4 gives a perspective representation of the octagonal measurement chamber. It consists of an octagonal outer vessel with glass walls. The cylindrical stirred vessel is installed in the center of the outer measurement chamber. The stirred vessel consists of a cylindrical glass cup and a two-piece cover. The interior diameter of the cylinder is D = 94 mm, its height is H = 105 mm. A glass plate is inserted at a height of about 10 mm above the bottom of the vessel. It divides the vessel into a measurement volume above the glass plate and a comparison volume below. This separation is necessary for the evaluation of the interferograms. The top and bottom of the measurement chamber as well as the two-piece cover of the stirred vessel are made of polyethylene. Due to the low heat conductivity of this material, momentary temperature variations outside of the measurement chamber cannot affect the measurements of the small temperature differences in the stirred vessel. The two-piece cover of the stirred vessel consists of an outer spherical ring, which holds the glass cup. The stirred vessel is mounted to the top of the measurement chamber and centered with this ring. The second, interior cover has an opening for the impeller. With this cover, the impeller can be quickly exchanged, without having to remove the stirred vessel from the measurement chamber. Impellers with a diameter of up to d_r = 79 mm can be inserted into the stirred vessel. The liquids filled into the cylindrical stirred vessel and the octagonal measurement chamber have the same refractive index as the glass used for the stirred vessel. Thus, the four object waves staggered by 45° fall on the plane glass surfaces of the measurement chambers at 90° angels without being deflected. The requirement of parallel light paths in the measurement volume is met.

The experimental set-up given in Fig. 5 consists of the cylindrical stirred vessel, the impeller motor and a meter for the impeller speed. It is installed in a tempered chamber, together with the optical set-up. The stirred vessel is mounted on a vibration-free table. The impeller motor and the speed meter are installed on a bridge structure to avoid a transfer of vibrations between motor and stirred vessel. The glass impeller is connected to the motor shaft via a rubber hose to reduce the transfer of vibrations from the motor. The motor speed is measured with a forked light barrier at a perforated disk mounted to the motor shaft and indicated by a frequency meter.

When the temperature in the measurement chamber is constant, the holograms are recorded and repositioned. The reference wave is displaced by a small angle to create a reference pattern. They are recorded with two cameras. The stirrer motor and the cameras are activated simultaneously with one switch. The duration between two exposures is controlled with a timer. When the test is completed, the exposed films are removed from the cameras and developed.

Processing of experimental results

The information about the timely and spatial changes of the temperature field in the measurement chamber contained in the films must be transformed from analogous values into digital information for the tomographical reconstruction. The procedure for this transformation is described in Fig. 6. It contains an example for the reconstruction of the temperature field in one cross-sectional plane. It can be shown as either a pseudo three-dimensional representation or as isotherms. A grid with 2500 grid points is used for the reconstruction of the temperature field in one cross-sectional plate. The applied tomographical method does not provide an exact solution, but an approximation for the ill-conditioned system of equations. The mean error of the reconstruction is between $1.3\% \leq \bar{\varepsilon} \leq 6.0\%$ for the selected test functions and with the use of imprecise entry data. The maximum error is approximately $\varepsilon_{max} = 26\%$. In general, the portions of higher frequency are dampened in the iteration process and, therefore, maxima of too low values and local minima of too high values are reconstructed. Good reconstruction results are obtained with 15 iteration steps.

EXPERIMENTAL RESULTS

Rheological behavior of the examined liquids

Polymer solutions with a mass portion of 1.5 percent methylhydroxyethylcellulose are used for the experimental investigations. Fig. 7 gives the shear stresses of a polymer solution with four mass percent water as a function of the velocity gradient.

The measured values are valid for the temperature ϑ = 24°C. They represent the steady state final values. A rotation viscosimeter with coaxial cylinders is used for the measurements. The velocity gradient is varied in the range 0.62 1/s ≤ $\dot{\gamma}$ ≤ 592 1/s. An initial shear stress is necessary in the polymer solution to induce a velocity gradient. Rheological behaviour is correlated with the equation of Herschel-Bulkley

$$\tau = \tau_0 + (\eta_h \dot{\gamma})^m \tag{2}$$

For the polymer solution with four mass percent water, the constants have the following values:

τ_0 = 4,52 Pa
η_h = 3,66 $Pa^{1/m}s$
m = 0,57

The Reynolds number is defined as

$$Re \equiv \frac{n d_r^2 \rho}{\eta}. \tag{3}$$

In eq. (3) n gives the rotation of the impeller, d_r its diameter, ρ the density of the liquid and η the viscosity. For Newtonian liquids, the viscosity is independent of the velocity gradient, for non-Newtonian liquids, it depends on the velocity gradient. The correlation by Metzner and Otto /3/ for the velocity gradient

$$\dot{\gamma} = C \cdot n. \tag{4}$$

is used to calculate the velocity gradient in the vicinity of the impeller as well as the viscosity. It consists of the impeller speed and a constant value that is specific for the applied impeller type. For disk impellers and inclined blade disk impellers, such as the ones used here, this impeller constant has the value C ≈ 11.5. The calculated velocity gradient is entered into the law of flow by Herschel and Bulkley

$$\eta = \frac{\tau_0 + (\eta_h \dot{\gamma})^m}{\dot{\gamma}}. \tag{5}$$

to calculate the viscosity. The viscosity is obtained as a function of the velocity gradient $\dot{\gamma}$, the flow index n, a dimensional constant and the initial shear stress .

<u>Calculated velocity fields in non-Newtonian liquids</u>

Liquids with initial shear stresses are used in the experimental investigations. Due to the shear stress, unmixed regions are present in the stirred vessel. It is therefore important to determine how the initial shear stress affects the velocity field and the size of the mixed region. The description of the velocity field in the mixed region is based on the assumption of a spherical stirrer given in Fig. 8. The mixed region has the shape of a sphere. Therefore, spherical coordinates are selected to calculate the velocity field. The radius of the spherical stirrer is r_r, the radius of the spherical mixed region is r_c. According to Solomon et al. /4/, the required momentum of the spherical stirrer is as follows:

$$M = \tau \pi^2 r^3 \tag{6}$$

It is proportional to the shear stress and the third power of the radius. At the point $r = r_c$, the shear stress is identical to the initial shear stress:

$$M = \tau_0 \pi^2 r_c^3. \tag{7}$$

The two equations can be combined and solved with respect to the shear stress:

$$\tau = \tau_0 \left(\frac{r_c}{r}\right)^3. \tag{8}$$

The statement by Herschel and Bulkley is used to describe the flow behavior of the applied polymer solution:

$$\tau_0 + (\eta_h \dot{\gamma})^m = \tau_0 \left(\frac{r_c}{r}\right)^3. \tag{9}$$

An equation for the calculation of the velocity gradient is obtained by further transformation:

$$\dot{\gamma} = \frac{1}{\eta_h} \{\tau_0 \left[(\frac{r_c}{r})^3 - 1 \right] \}^{\frac{1}{m}}. \tag{10}$$

The velocity gradient is calculated in spherical coordinates with the equation

$$\dot{\gamma} = -r \frac{d}{dr} \left(\frac{w_\varphi}{r}\right) \tag{11}$$

An analytical solution of the integral of the velocity gradient is possible for the flow index m = 0.5 which is close to m = 0.57, which was determined for the applied polymer solution. The equation for $w_\varphi(r)$ is obtained by integration over the distance r_r to r_c, with the boundary condition $w_\varphi(r_c) = 0$.

$$w_\varphi(r) = \frac{r \tau_0^2}{\eta_h} \left[\frac{1}{6}\left(\frac{r_c}{r}\right)^6 - \frac{2}{3}\left(\frac{r_c}{r}\right)^3 - ln\left(\frac{r_c}{r}\right) + \frac{1}{2} \right]. \tag{12}$$

The tangential velocity $w_\varphi(r)$ is related to the maximum tangential velocity $w_\varphi(r_r)$ and given in Fig. 9 in dependence of the dimensionless distance

$$y^* = \frac{r - r_r}{r_c - r_r} \tag{13}$$

r_r is the radius of the stirrer, r_c the maximum radius of the mixed region. The velocities are divided by the circumferential speeds of the stirrer $w_\varphi = 2\pi n r_r$ and valid for different ratios (r_c/r_r).

As soon as the mixed region reaches the edge of the stirred vessel, eq. (12) is no longer valid. The momentum transferred to the liquid cannot be calculated beyond the edge of the mixed region.

Calculated and measured velocity fields

Fig. 10 gives the calculated tangential velocities as well as the measured velocities. For small Reynolds numbers, the values coincide well. However, for Reynolds numbers Re = 24 and Re = 68, there is a considerable difference between the calculated and measured values. The measured velocities are always higher than the calculated ones. This is due to the presence of turbulent flow in some areas of the vessel. The model is based on purely laminar and tangential flow, which means than the velocity field is not described correctly.

In Fig. 11 calculated and measured cavern sizes are compared. The measured dimensionless cavern diameter d_c/d_r are plotted against the Reynolds number. The cavern size increases with increasing Reynolds numbers and with increasing stirring time. After a stirring time of approximately 3 hours, the cavern diameters are approximately 20% larger than the ones measured after a stirring time of approximately 30 minutes. This increase is caused by the increase of the liquid temperature due to dissipation and the thixotropic flow behavior. The initial shear stress decreases and the mixed region becomes larger.

To make a comparison possible, the dimensionless cavern diameters d_c/d_r calculated with equation (12) are entered as well. The calculated cavern diameters are less dependent of the Reynolds number than the measured diameters. The cavern diameters obtained from theoretical observations do not represent the flow conditions correctly, since these theoretical observations are based on laminar, purely tangential flow. This assumption does not hold for Reynolds numbers Re > 10. For these values, the shape of the mixed region changes more and more from a sphere to a rotational ellipsoid with decreasing ratio cavern height to cavern diameter h_c/d_c. The model, however, provides surprisingly exact values for the cavern sizes measured for long stirring times.

The equation by Solomon et al. /4/ provides better values of the cavern sizes measured for short stirring times. The autors developed a model for the calculation of the cavern size which contains no constants that need to be determined empirically. The shape of the cavern is idealized as a sphere. Since the energy input from the stirrer is completely dissipated, the cavern diameter can be calculated for a given stirring speed, a known cavern geometry and a known initial shear stress. The cavern diameter is obtained in dimensionless form:

$$\left(\frac{d_c}{d_r}\right)^3 = \frac{4\rho Ne\, n^2 d_r^2}{\pi^3 \tau_0}. \tag{14}$$

In Eq. (14) ϱ means the density, and the initial shear stress is given by τ_0, the stirrer diameter by d_r, its stirring speed by n and the Newton number is given by Ne. The graph also contains the cavern diameters calculated with the equation by Elson. The values of the cavern diameter calculated with this equation and our own values coincide very well. With this model, the cavern diamter can be calculated in the correct magnitude, if the energy input into the liquid is calculated with the use of the statement Ne = 80/Re in the laminar region of the flow and the initial shear stress according to Casson.

Temperature fields

With the knowledge of the velocity fields, a qualitative estimation of the mixing processes in the stirred vessel is possible. There are only few theoretical methods to calculate mixing processes in non-Newtonian liquids. Especially the appearance of normal stress differences of flow boundaries leads to considerable difficulties for the description of mixing processes. Optical tomography is a suitable method for the analysis of the time-dependent temperature fields.

In order to compare the different measured temperature fields, it is necessary to indicate dimensionless parameters for the stirring time and the energy input into the liquid. In general, the Fourier number and the Brinkmann number are used as parameters to describe the heat transfer. Brauer /5/ did theoretical investigations on the heat transfer in stirred vessels for Newtonian liquids. The Brinkmann number defined by Brauer indicates the ratio of frictional heat flux and the heat flux through the vessel wale. If the temperature gradient close to the wall is zero, the Brinkmann number becomes infinitely large.

A different Brinkmann number is defined, which allows a comparison of the temperature fields caused by dissipation

$$Br \equiv \frac{P}{d_c/2\lambda(T_{max} - T_0)}. \tag{15}$$

It describes the ratio of the energy flow P dissipated and a fictive heat flux per area discharged by heat conduction from the mixed region. In liquids with a critical shear stress, the diameter of the mixed region is identical to the diameter of the stirrer for small Reynolds numbers. The vessel diameter is the maximum possible value. For liquids without critical shear stress, the diameter of the mixed region is always identical to the vessel diameter.

The energy P dissipated per time unit is equal to the energy input at the steady state condition of the liquid. If the dependence of the Newton number on the Reynolds number is known for the applied stirrer, this energy can be calculated with the equation

$$P = \eta n^2 d_r^3 Ne Re \qquad (16)$$

The Fourier number is obtained with the diameter of the mixed region:

$$Fo \equiv \frac{4\lambda t}{c_v \rho d_c^2}. \qquad (17)$$

For the comparison of the measured local temperatures, a dimensionless local temperature is required as a third dimensionless parameter. In dimensionless form

$$T^* \equiv \frac{T - T_0}{T_{max} - T_0} \qquad (18)$$

it describes the ratio local temperature difference to maximum temperature difference in the liquid. T_0 is the temperature at the beginning of the stirring process. It can also be considered as the continuously constant wall temperature, since the temperature of the vessel wall is identical to the temperature of the liquid at the beginning of the stirring process. With the applied method, the temperature fields can be measured as long as the temperatures close to the vessel wall are constant. Consequently, the temperature T_0 does not change for the entire duration of the test.

The Prandtl number is given by

$$Pr \equiv \frac{\eta c_v}{\lambda} \qquad (19)$$

For the heat transfer in stirred, non-Newtonian liquids it is most effective to use the representative viscosity according to its definition by Metzner and Otto /3/. The Prandtl number depends not only on parameters such as the specific heat capacity c_v and the heat conductivity , but also on the flow conditions in the region surrounding the stirrer. With this Prandtl number it is possible to describe the heat transfer in stirred vessels for non-Newtonian liquids with the correlations found for Newtonian liquids /6,7/.

Instationary temperature fields caused by dissipation

The timely change of the temperature fields caused by the dissipation of the kinetic energy of the stirrer is indicated in Fig. 12 by a shift of the lines of interference given by photographic taken of one angle of view for different moments during one test. The stirrer consists of a disk impeller, which is installed at about one third of the distance between vessel bottom and vessel top. The Reynolds number is Re = 8.

In the beginning at t = 0, the disk impeller is motionless. The interferogram shows the initial and reference condition in the liquid. The interference pattern consists of almost vertical stripes. At the beginning of the stirring process, the pattern does not change. The disturbed interference pattern at the vessel wall indicates that the experimental set-up is installed vibration-free. After a stirring time t = 18 s, a deflection of the stripes becomes visible in the vicinity of the impeller. The deflection of the interference pattern increases with increasing stirring times. However, even after a stirring time t = 60 s, it remains restricted to the vicinity of the impeller.

The temperature fields shown in Fig. 13 were reconstructed with the use of optical tomography. At the time t = 18 s, two isotherms are entered in the vicinity of the disk impeller, 19 s later the temperature has increased further in this area. A minor expansion in radial and axial direction of the temperature field caused by dissipation can be observed at t = 60 s. The heated region is more pronounced above the impeller than below. The shape of the heated region is more cylindrical than spherical.

Variation of the stirrer type

Fig. 14 gives the temperature fields caused by three different stirrers in dimensionless form. A comparison is made between the temperature fields caused by the disk impeller, the inclined blade disk impeller stirring in downward direction, and the stirrer system. The applied liquid is a polymer solution with four mass percent water.

The Fourier numbers indicated for the three stirrers result from different stirring times. The energy input to the liquid is similar for the indicated Fourier numbers. During the entire observation time, there is no heat transfer through the vessel walls. The entire dissipated energy is stored in the liquid. Therefore, the temperature can be compared to each other. Statements can be made about the effect of the stirrer type on the dissipation of the energy input and the heat transfer in the liquid.

In all three cases, the maximum temperature differences are located in the region between and above the impeller blades. The shape of the heated regions is elliptical. The heated region expands more in axial direction, if an inclined blade disk impeller is used. The expansion is more in radial direction, if a disk impeller or a stirrer system is used. This coincides with the shapes of the mixed regions of the individual stirrers.

A comparison of the temperature profiles in radial direction for the three stirrers is given in Fig. 15. The temperature profiles located in the cross sections of the stirrer levels are shown. The different temperature profiles are very similar. The maximum temperature differences occur in the region of the impeller blades. The temperatures drop towards the vessel wall. The boundary of the mixed region is marked in all three graphs. It is calculated with the equation given by Elson et al. /8/. The temperature gradients change inside the cavern. In the mixed volume, heat is transferred by convection and by conduction to the outside of the cavern. Outside of the mixed region, the liquid is motionless and only heat conduction takes place.

Variation of the Reynolds number

The temperature profiles recorded in the region of laminar flow are about identical for the disk impeller and the inclined blade disk impeller as well as the stirrer system consisting of two disk impellers installed one above the other. Since the variation of the Reynolds number in the range $6 \leq Re \leq 8$ is very small, differences are hardly visible.

The effect of the Reynolds number on the temperature field caused in the liquid is discussed for the inclined blade disk impeller. The temperature fields developing in the cross section of the impeller are recorded. A comparison of the temperature fields is only effective for long stirring times and high Fourier numbers, since the dissipation of the kinetic emergy takes place mainly in the vicinity of the stirrer and since the heat transfer towards the boundary of the cavern must be established. If these conditions are present, the effect of conductive and convective transfer for different Reynolds numbers can be investigated on the basis of the temperature fields. In Fig. 16 gives the temperature profiles in the cross section of the impeller level for three different Reynolds numbers. For the Reynolds number $Re = 1$ and the Fourier number $Fo = 5.9 \cdot 10^{-2}$, the shape of the temperature field is identical to the shape at the beginning of the test. This indicates that the heat transfer takes place by heat conduction only. If the Reynolds number is increased to $Re = 6$, the shape of the temperature field is identical to the one found for the Reynolds number $Re = 1$. A clear difference between the temperature fields becomes obvious, if the Reynolds number is increased to $Re = 12$. The temperature is almost constant in the mixed region; it decreases considerably at the cavern edge. This means that there is a pronounced convective transfer in radial direction within the mixed region. On the basis of the temperature field, the velocity in radial direction can be estimated at approximately $w_r = 10^{-4}$ m/s. The rheological properties of the liquid lead to a strong decrease of the radial and axial transfer in the mixed region. This may be caused by the normal stress differences present in the liquid.

COMPARISON WITH THE RESULTS OF NEWTONIAN LIQUIDS

The results by Ostendorf /9/ are used for the comparisons with the temperature fields caused in Newtonian liquids. Ostendorf also used optical tomography for measurements of the temperature fields caused by dissipation. In Fig. 17 a comparison between temperature fields caused by a disk impeller in a Newtonian and a non-Newtonian liquid is given. The different stirring times result from the impeller speeds adjusted for the individual test parameters and the parameters of the applied liquids. The energy values are almost identical for the indicated times. For almost identical energy inputs, a strong radial expansion of the temperature field is obvious for glycerin. Especially in the impeller level, the temperature field expands almost to the vessel wall.

In the polymer solution, no pronounced deformation of the temperature field is present in the cross section of the impeller. In spite of the Reynolds number Re = 8, which is twice as high in the polymer solution compared to the Reynolds number in the Newtonian liquid, the flow in radial direction and, consequently, a convective transfer towards the edge of the mixed region in the polymer solution is too small to be measured. The diameter of the mixed region is about half the size of the diameter of the vessel. Within the mixed region of the polymer solution, the molecular mass transfer dominates over the convective mass transfer. If the Reynolds numbers are increased further, the convective transfer dominates over the molecular transfer. In Newtonian liquids, the convective transfer dominates.

For the described experiment a polymer solution with $Re^2 Pr = 413000$ was used. In glycerin, this value is approximately one third, i.e. $Re^2 Pr = 140000$. According to Eisermann /10/, the convective transfer should no longer be negligible over the molecular transfer for values $R^{e2} Pr \geq 1000$ in Newtonian liquids. In the examined polymer solutions this is the case for much higher values of the parameter product $Re^2 Pr$. For values $Re^2 Pr \geq 800000$, a convective transfer can be observed in the liquids.

SUMMARY

Experimental and theoretical investigations of temperature fields in stirred vessels caused by dissipation are presented. In the experiments polymer solutions, which have a non-Newtonian behavior were tested. Reconstructions of the instationary and spacial temperature fields in the stirred vessel are obtained with the use of holographic interferometry and the application of a tomographical analysis method. This allows observations of the heat production and the heat transfer in the stirred vessel. A cylindrical, enclosed stirred vessel without flow boundary is used. The investigations are conducted for laminar flow.

Polymer solutions have a shear thinning flow behavior. The steady state flow is described with the equation by Herschel und Bulkley. The presence of a flow boundary leads to incomplete mixing of the liquid in the stirred vessel for the examined flow region. Good mixing is present in the vicinity of the stirrer, however, the liquid is motionless at large distances from the stirrer.

With the model of a spherical stirrer, the velocity field can be predicted well for the region of laminar flow, in consideration of the flow properties. Compared to liquids without flow boundary, large velocity gradients are present in the vicinity of the stirrer for liquids with flow boundaries. This leads to an increased energy input in the region of laminar flow.

For Reynolds numbers $Re \leq 9$, the dissipated energy is transferred mainly by heat conduction. For higher Reynolds numbers, a pronounced convective transport is present in the mixed region. Outside of the mixed region, heat flux is present only due to heat conduction. Compared to the temperature fields in Newtonian liquids measured by Ostendorf /10/, the convective transfer is considerably smaller and the heated region is much closer to the stirrer.

LITERATURE

/1/ R. Gordon, R. Bender, G.T. Herman: Algebraic rekonstruction technique (ART) for three-dimensional electron microscopy and x-ray photography; J. of Theor. Biol. 29(1970), 471/481 VDI-Verlag, 1988

/2/ D.W. Sweeney: Interferometric measurement of three-dimensional temperature fields; Ph. D. Thesis, University of Michigan, 1972

/3/ A.B. Metzner, R.E. Otto: Agitation of non-Newtonian fluids; AICHE-Journal 3(1957)1, 3/10

/4/ J. Solomon, T.P. Elson, A.W. Nienow, G.W. Pace: Cavern sizes in agitated fluids with a yield stress; Chem. Eng. Commun. 11(1981), 143/164

/5/ H. Brauer: Wärmetransport in Rührgefäßen; Wärme- und Stoffübertragung 13(1980), 105/113

/6/ P. Ayazi Shamlou, M.F. Edwards: Heat transfer to viscous Newtonian and non-Newtonian fluids for helical ribbon mixers; Chemical Engineering Science 41(1986)8, 1957/1967

/7/ W. Kai, Y. Shengyao: Heat transfer and power consumption of non-Newtonian fluids in agitated vessels; Chemical Engineering Science 44(1989)1, 33/40

/8/ T.P. Elson, D.J. Cheesman, A.W. Nienow: X-ray studies of cavern sizes and perfomance with fluids possessing a yield stress; Chemical Engineering Science 41(1986)10, 2555/2562

/9/ W. Ostendorf, D. Mewes: Measurement of three-dimensional unsteady temperature profiles in mixing vessels by optical tomography; Chem. Eng. Technology 11(1988)3, 148/155

SYMBOLS

c	$J\ kg^{-1}K^{-1}$	specific heat of liquid
d_c	m	diameter of the cavern of mixed liquid in the surrounding of the stirrer
d_r	m	diameter of the stirrer
D	m	cylinder diameter
l	m	pass length of the light beam
m	-	flow index according to eq. (2)
M	Nm	momentum
n	s^{-1}	speed of rotation
n	-	refractive index
r	m	radius
r_c	m	radius of the cavern of mixed liquid in the surrounding of the stirrer
r_r	m	radius of the stirrer
T	K	temperature
T_o	K	temperature at the beginning of experiments
w_φ	$m\ s^{-1}$	velocity in circumsperical direction
x	m	coordinate
y	m	coordinate
$\dot{\gamma}$	s^{-1}	velocity gradient, shear velocity
η_h	$(Nm^{-2})^{-m}s$	viscosity according to eq. (2)
η	$Nm^{-2}s^{+1}$	viscosity
ϕ	m	phase shift of coherent light
λ	m	wave length
λ	$J\ m^{-1}s^{-1}K^{-1}$	heat conductivity
P	$J\ s^{-1}$	energy dissipated per time
ϱ	$kg\ m^{-3}$	density of liquid
τ_o	$N\ m^{-2}$	critical shear stress
τ	$N\ m^{-2}$	shear stress

$Br \equiv \dfrac{P}{d_c \lambda (T_{max} - T_o)}$	*Brinkmann number*
$Fo \equiv \dfrac{4 \lambda t}{c_v \varrho d_c^2}$	*Fourier number*
$Ne \equiv \dfrac{P}{n^3 d_r^5 \varrho}$	*Newton number*
$Pr \equiv \dfrac{\eta c_v}{\lambda}$	*Prandtl number*
$Re \equiv \dfrac{n d_r^2 \varrho}{\eta}$	*Reynolds number*
$T^* \equiv \dfrac{T - T_o}{T_{max} - T_o}$	*dimensionless temperature*
$y^* \equiv \dfrac{r - r_r}{r_c - r_r}$	*dimensionless radial coordinate*

FIGURES

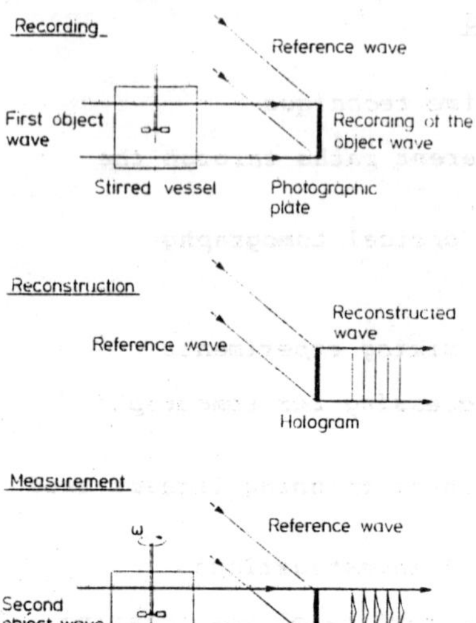

Fig. 1: *Explanation of the real time technique*

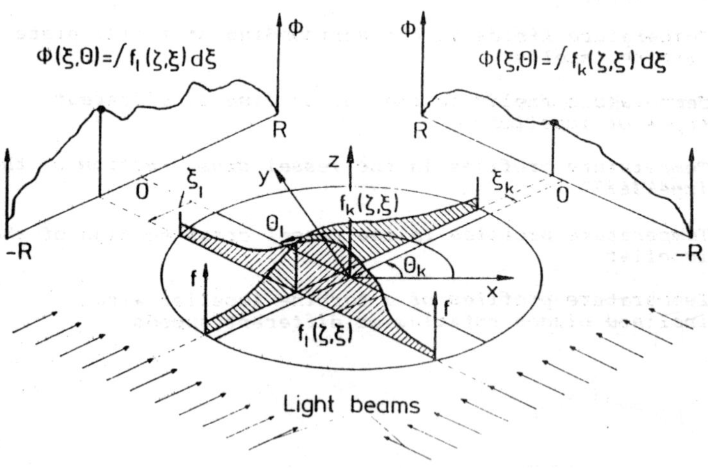

Fig. 2: *Line integrals along different paths through the measurement volume*

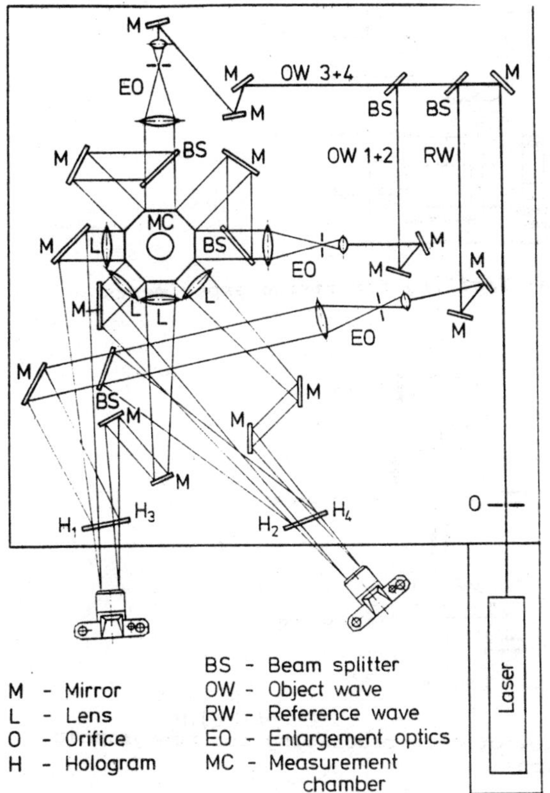

Fig. 3: *Experimental facility for optical tomography*

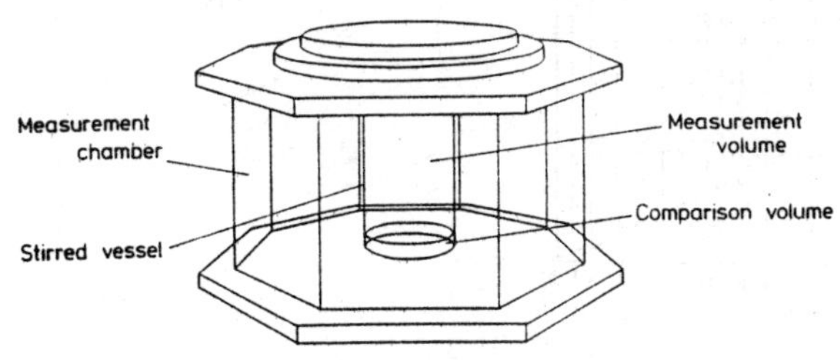

Fig. 4: *Mixing vessel*

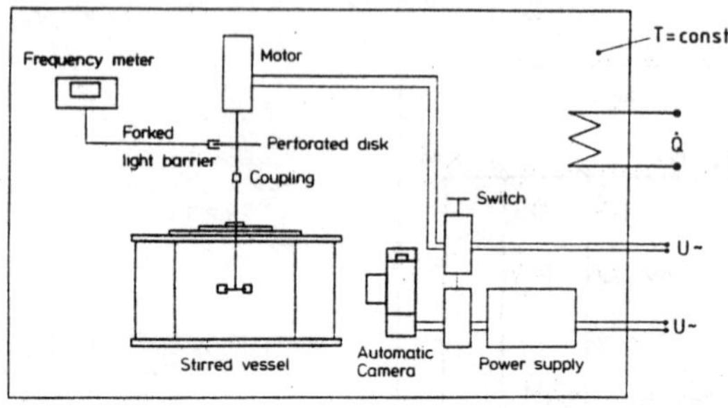

Fig. 5: Experimental facility for mixing experiments

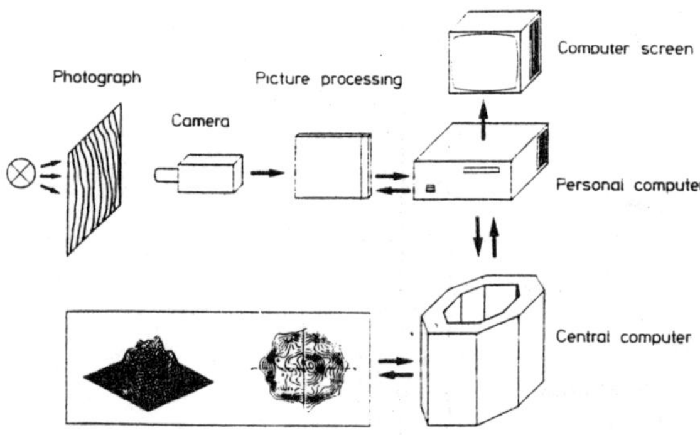

Fig. 6.: Schematic view of data processing for tomographic reconstruction

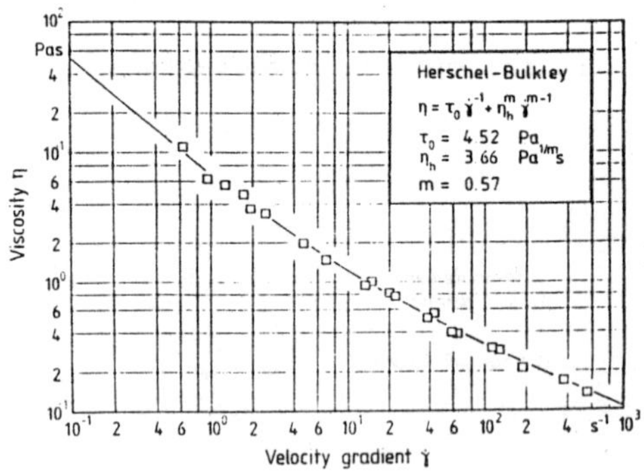

Fig. 7: Rheological behaviour of shear thinning liquids used for experiments

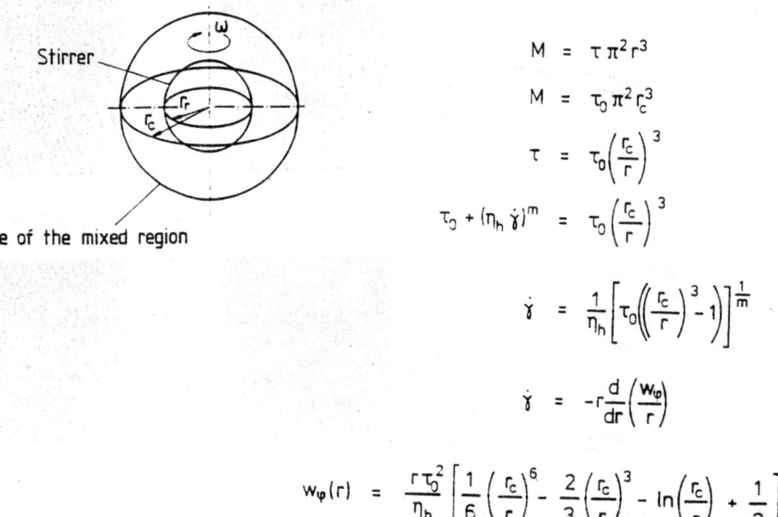

Fig. 8: Modell used for theoretical investigations

Fig. 9: Velocity profiles near the surface of the rotating spherical model stirrer

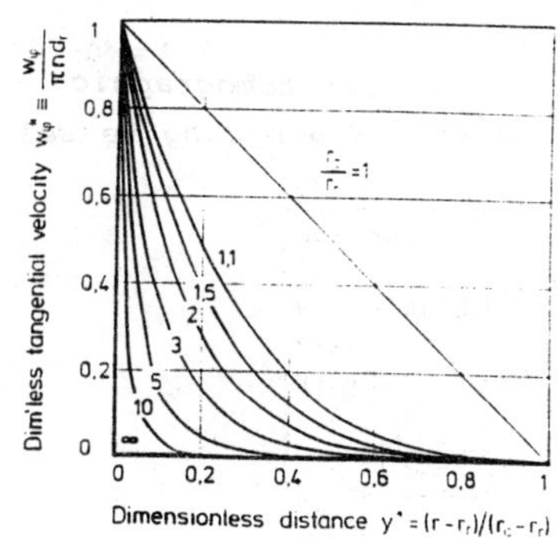

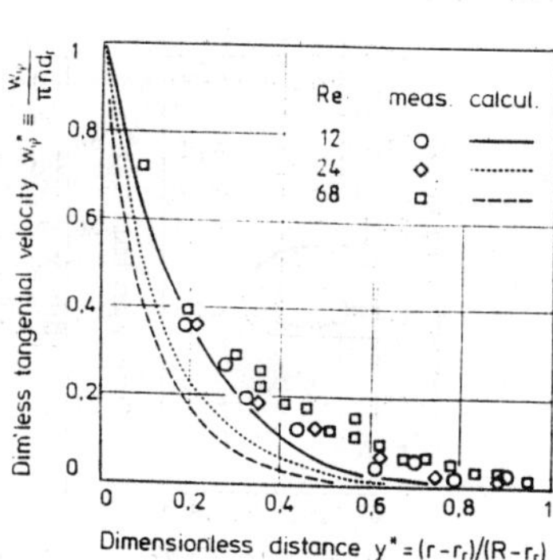

Fig. 10: Comparison of measured and calculated tangential velocity profiles

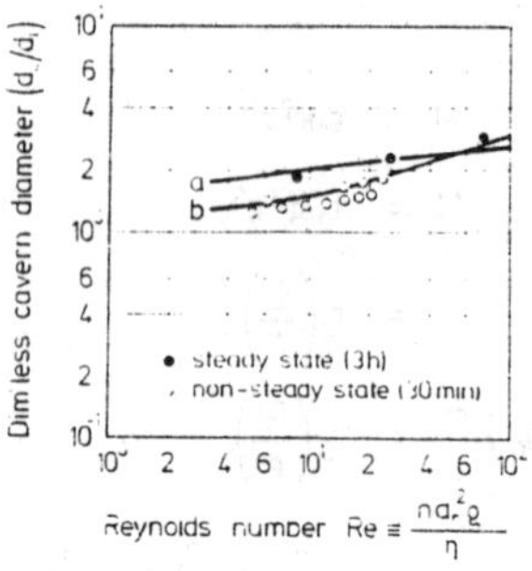

Dimless cavern diameter (d_c/d_r)

• steady state (3h)

, non-steady state (30 min)

Reynolds number $Re \equiv \frac{n d_r^2 \varrho}{\eta}$

a) Own model

b) Solomon, Elson, Nienow

$$\left(\frac{d_c}{d_r}\right)^3 = \frac{4 \varrho Ne\, n^2 d_r^2}{\pi^3 \tau_0}$$

Fig. 11: *Diameter of the cavern in the near vacinity of the stirrer*

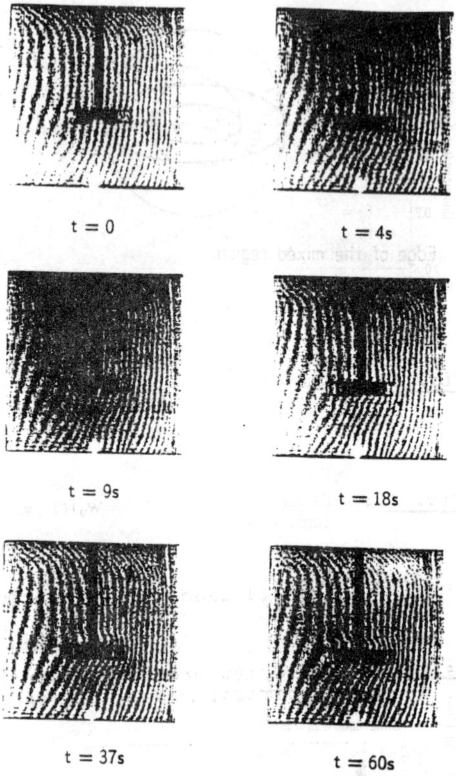

t = 0 t = 4s

t = 9s t = 18s

t = 37s t = 60s

Fig. 12: *Interferograms at different times for a mixing vessel*

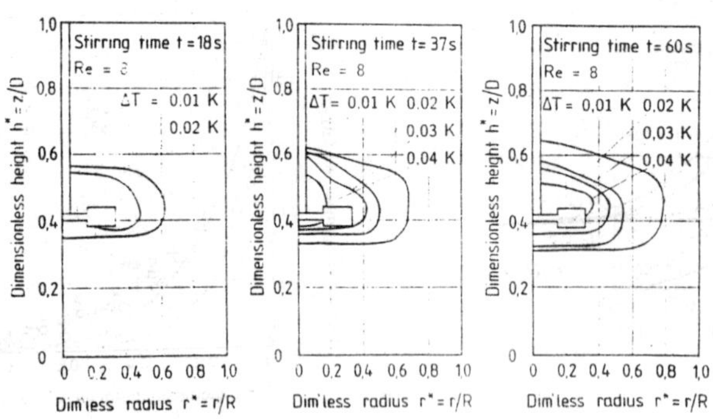

Fig. 13: *Temperature fields in the surrounding of a flat-plate turbine impeller*

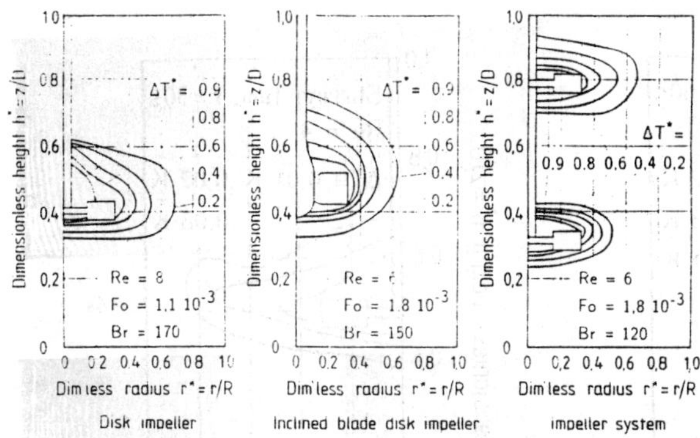

Fig. 14: *Temperature fields in the surrounding of different types of impeller*

Fig. 15: *Temperature profiles in the vessel cross section of the impeller*

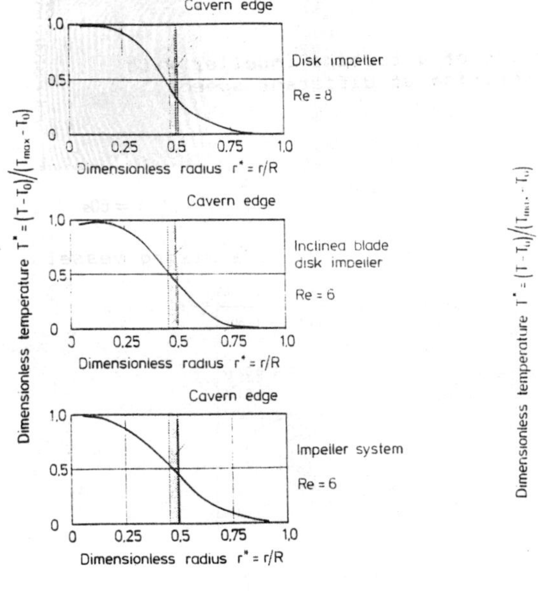

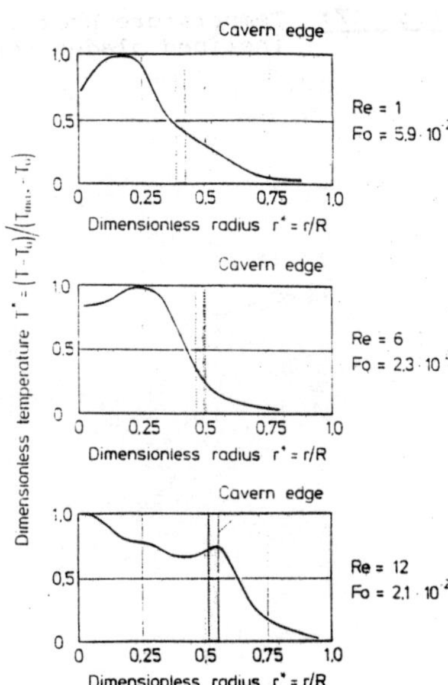

Fig. 16: *Temperature profiles in the vessel cross section of the impeller*

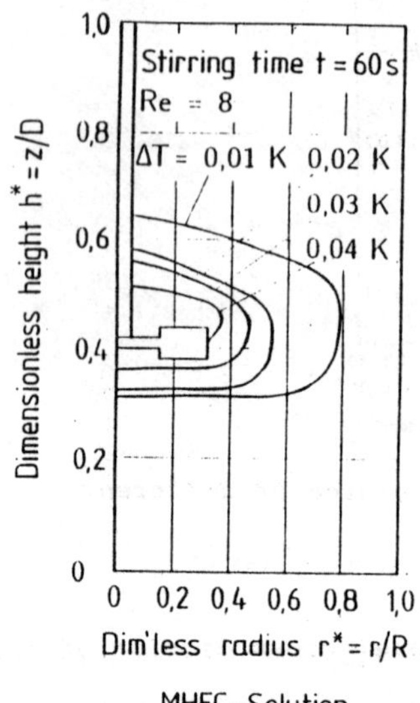

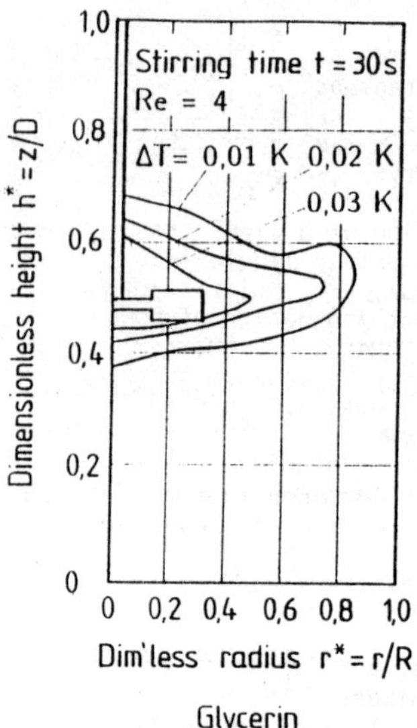

Fig. 17: *Temperature profiles of a turbine impeller with inclined blades rotating at different speeds*

AUTHOR/TITLE/KEYWORD

S

T

U